Chickpea: Crop Wild Relatives for Enhancing Genetic Gains

AF760712

Chickpea: Crop Wild Relatives for Enhancing Genetic Gains

Edited by

Mohar Singh

Principal Scientist (Plant Breeding)
ICAR National Bureau of Plant Genetic Resources
Shimla, Himachal Pradesh
India

ELSEVIER

ACADEMIC PRESS
An imprint of Elsevier

Academic Press is an imprint of Elsevier
125 London Wall, London EC2Y 5AS, United Kingdom
525 B Street, Suite 1650, San Diego, CA 92101, United States
50 Hampshire Street, 5th Floor, Cambridge, MA 02139, United States
The Boulevard, Langford Lane, Kidlington, Oxford OX5 1GB, United Kingdom

Copyright © 2020 Elsevier Inc. All rights reserved.

No part of this publication may be reproduced or transmitted in any form or by any means, electronic or mechanical, including photocopying, recording, or any information storage and retrieval system, without permission in writing from the publisher. Details on how to seek permission, further information about the Publisher's permissions policies and our arrangements with organizations such as the Copyright Clearance Center and the Copyright Licensing Agency, can be found at our website: www.elsevier.com/permissions.

This book and the individual contributions contained in it are protected under copyright by the Publisher (other than as may be noted herein).

Notices
Knowledge and best practice in this field are constantly changing. As new research and experience broaden our understanding, changes in research methods, professional practices, or medical treatment may become necessary.

Practitioners and researchers must always rely on their own experience and knowledge in evaluating and using any information, methods, compounds, or experiments described herein. In using such information or methods they should be mindful of their own safety and the safety of others, including parties for whom they have a professional responsibility.

To the fullest extent of the law, neither the Publisher nor the authors, contributors, or editors, assume any liability for any injury and/or damage to persons or property as a matter of products liability, negligence or otherwise, or from any use or operation of any methods, products, instructions, or ideas contained in the material herein.

Library of Congress Cataloging-in-Publication Data
A catalog record for this book is available from the Library of Congress

British Library Cataloguing-in-Publication Data
A catalogue record for this book is available from the British Library

ISBN: 978-0-12-818299-4

For information on all Academic Press publications visit our website at
https://www.elsevier.com/books-and-journals

Publisher: Charlotte Cockle
Acquisitions Editor: Nancy Maragioglio
Editorial Project Manager: Andrea Dulberger
Project Manager: Poulouse Joseph
Cover Designer: Alan Studholme

Typeset by TNQ Technologies

Working together to grow libraries in developing countries
www.elsevier.com • www.bookaid.org

Contents

Contributors.. xi
About the Editor.. xv
Preface xvii

CHAPTER 1 Introduction.. 1
Mohar Singh
1.1 Introduction... 1
1.2 Germplasm characterization and evaluation......................... 2
1.3 Production-related problems.. 6
1.3.1 Fusarium wilt 6
1.3.2 Ascochyta blight 7
1.3.3 Botrytis gray mold.. 7
1.3.4 Pod borer... 7
1.3.5 Bruchids.. 8
1.3.6 Cold tolerance 8
1.3.7 Drought tolerance... 8
1.3.8 Heat tolerance... 9
1.4 Origin, distribution, gene pools 10
1.5 Chickpea genetic resources: collection, conservation, characterization, and maintenance 10
1.6 Conventional cytogenetic manipulations 11
1.7 Embryo rescue and chromosomal manipulations 11
1.8 Gene pyramiding and multiple character breeding.............. 11
1.9 Molecular markers and marker-trait associations................ 12
1.10 Genetic transformation 13
1.11 Chickpea economy in India... 13
References.. 14

CHAPTER 2 Origin, distribution, and gene pools 19
Aqeel Hasan Rizvi and Ashutosh Sarker
2.1 Introduction.. 19
2.2 Origin and distribution 21
2.3 Taxonomy.. 23
2.4 Gene pools... 23
2.4.1 Primary gene pool 27
2.4.2 Secondary gene pool... 28
2.4.3 Tertiary gene pool 31

2.5 Barriers in interspecific hybridization and crossability groups 31
References 32

CHAPTER 3 Chickpea genetic resources: collection, conservation, characterization, and maintenance 37
Rahul Chandora, Gayacharan, Neelam Shekhawat and Nikhil Malhotra
3.1 Introduction 37
3.2 Species distribution and gene pools 39
3.3 Germplasm conservation 40
3.3.1 Ex situ conservation 41
3.3.2 In situ conservation 42
3.3.3 Duplicates 45
3.3.4 Safety duplication 45
3.4 Germplasm characterization and evaluation 45
3.4.1 Genomic resources and their utilization, characterization and evaluation 49
3.4.2 Development of core and mini-core collection 51
3.5 Future challenges and prospects 53
References 53

CHAPTER 4 Conventional cytogenetic manipulations 63
Gopal Katna, S.D. Nitesh and Kamal Dev Sharma
4.1 Introduction 63
4.2 Production-related problems 64
4.2.1 Ascochyta blight 65
4.2.2 Fusarium wilt 65
4.2.3 Botrytis gray mold 66
4.2.4 Dry root rot 66
4.2.5 Powdery mildew 66
4.2.6 Pod borer 66
4.2.7 Bruchids 66
4.2.8 Abiotic stresses 67
4.3 Narrow genetic bases: a bottleneck for enhancing productivity 67
4.4 Conventional chickpea breeding strategies 68
4.5 Cytogenetics and chickpea improvement 69
4.6 Exploitation of marker-assisted selection 70
4.7 Generation of additional variability through hybridization 70

4.8 Generation of variability through mutagenesis....................71
4.9 Wild *Cicer* species..72
4.10 Utilization of wild species for crop improvement................74
4.10.1 Direct hybridization 74
4.10.2 Chromosome doubling prior to hybridization............74
4.10.3 Other techniques..74
4.11 New germplasm/varieties developed through wide hybridization 75
4.12 Wide hybridization for transfer of resistance to biotic stresses 77
4.13 Future strategy and conclusions 79
References...81

CHAPTER 5 Embryo rescue and chromosomal manipulations 95
Biswajit Mondal, S.K. Chaturvedi, Alok Das, Yogesh Kumar, Aneeta Yadav, Shiv Sewak and N.P. Singh
5.1 Introduction..95
5.2 Origin and taxonomy 96
5.3 Genus *cicer*...97
5.4 Gene pools and wide hybridization....................................98
5.4.1 Embryo rescue..102
5.4.2 Chickpea embryology 106
5.4.3 Somatic embryogenesis in chickpea..........................111
5.4.4 Considerations regarding somatic embryogenesis.........113
5.4.5 Considerations in the seedlings derived from the hybrid embryos..115
5.4.6 Embryo rescue protocols in chickpea........................116
5.5 Chromosomal manipulation..121
5.6 Cytogenetics and distant hybridization 122
5.6.1 Chromosome morphology and DNA content..............122
5.6.2 Chromosomal manipulations, visualization, and other techniques..125
References..126

CHAPTER 6 Gene pyramiding and multiple character breeding.. 131
Ashutosh Kushwah, Soma Gupta, Shayla Bindra, Norah Johal, Inderjit Singh, C. Bharadwaj, G.P. Dixit, P.M. Gaur, Harsh Nayyar and Sarvjeet Singh
6.1 Introduction... 131
6.2 Genetic resources in chickpea... 133

6.3 Genomic resources 134
6.3.1 Transcriptomics resources 136
6.3.2 Genetic and physical maps 138
6.3.3 Functional or transcript maps 138
6.3.4 Genome sequencing and resequencing 143
6.3.5 Trait mapping 143
6.3.6 Genomic-assisted breeding 145
6.3.7 Marker-assisted backcrossing 149
6.3.8 Marker-assisted recurrent selection 150
6.3.9 Genomic-assisted breeding through MAGIC lines 151
6.3.10 Advanced backcross QTL analysis—based breeding 151
6.3.11 Genomic selection 153
6.4 Speed breeding 154
6.5 Conclusion 155
References 156

CHAPTER 7 Molecular markers and marker trait associations 167
Kamal Dev Sharma, Surinder Singh Chandel and Rajeev Rathour
7.1 Introduction 167
7.2 DNA-based markers for chickpea 169
7.2.1 Simple sequence repeat markers 169
7.2.2 Expressed sequence tag—simple sequence repeat markers 171
7.2.3 Transcription factor gene—derived microsatellite markers 172
7.3 DNA-based markers other than SSRs 173
7.3.1 Genic markers 173
7.3.2 Nongenic markers 173
7.4 Genotyping-by-sequencing 175
7.4.1 Transferability of chickpea SSRs to other taxa and vice versa 176
7.5 Development of linkage maps 177
7.6 Genetic mapping of disease resistance and agro-morphological traits in chickpea 181
7.6.1 Disease resistance 181
7.7 Genetic mapping of abiotic stress tolerance 188
7.8 Marker-assisted breeding 190
7.8.1 Introgression of wilt and blight resistance 190
7.8.2 Introgression of double podding character 191

7.9 Genomic selection for prediction of agronomic traits 191
7.10 Future perspectives 192
References 194

CHAPTER 8 Genetic transformation 205
Alok Das, Shallu Thakur, Alok Shukla, Prateek Singh, Jamal Ansari and N.P. Singh
8.1 Introduction 205
8.2 Genetic base of chickpea 205
8.3 Constraints to chickpea productivity 206
8.4 Genetic engineering as complement to plant breeding 206
8.4.1 *Agrobacterium tumefaciens*—mediated genetic transformation 207
8.4.2 Gene gun—mediated genetic transformation 208
8.5 Selectable/scorable markers 218
8.6 Root transformation 219
8.7 Conclusion 220
References 221

CHAPTER 9 Chickpea economy in India 225
Subhash Sharma, Samriti and Ravinder Sharma
9.1 Introduction 225
9.2 Global pulse economy: an overview 228
9.3 Global scenario of chickpea 229
9.4 Indian pulse scenario 229
9.5 Per capita availability of pulses in India 232
9.5.1 Indian scenario of chickpea 232
9.6 Growth in chickpea area, production, and productivity 232
9.7 Contribution of chickpea area to the total pulses area in India 233
9.8 Availability status of pulses 233
9.9 Availability status of chickpea in India 236
9.9.1 Different state's contribution in area and production of total pulses in India 236
9.9.2 Different states' contribution in area and production of chickpea 237
9.10 Export and import of chickpea 239
9.11 Trade destination of chickpea 240
9.12 Import duty on important pulses 243
9.13 Cost of cultivation of chickpea 244
9.14 MSP for major pulses 244

9.15 Prices of chickpea in major domestic markets 245
9.16 Market segmentation in chickpea market 245
9.16.1 The direct food use market 246
9.16.2 The split (dal) chickpea market 246
9.16.3 The chickpea flour market 247
9.16.4 The roasted and puffed chickpea market 247
9.17 Marketing channel of chickpea in India 247
9.18 Government intervention for chickpea in India 248
9.19 Conclusion 249
References 249

Index 251

Contributors

Jamal Ansari
Division of Plant Biotechnology, ICAR-Indian Institute of Pulses Research, Kanpur, Uttar Pradesh, India

C. Bharadwaj
ICAR-Indian Agricultural Research Institute, New Delhi, Delhi, India

Shayla Bindra
Punjab Agricultural University, Ludhiana, Punjab, India

Surinder Singh Chandel
Department of Agricultural Biotechnology, CSK Himachal Pradesh Agricultural University, Palampur, Himachal Pradesh, India

Rahul Chandora
ICAR-National Bureau of Plant Genetic Resources, Shimla, Himachal Pradesh, India

S.K. Chaturvedi
College of Agriculture, RLBCAU, Jhansi, Uttar Pradesh, India

Alok Das
Division of Plant Biotechnology, ICAR-Indian Institute of Pulses Research, Kanpur, Uttar Pradesh, India

G.P. Dixit
ICAR-Indian Institute of Pulses Research, Kanpur, Uttar Pradesh, India

P.M. Gaur
International Crops Research Institute for the Semi-Arid Tropics, Patancheru, Telangana, India

Gayacharan
ICAR-National Bureau of Plant Genetic Resources, New Delhi, Delhi, India

Soma Gupta
ICAR-Indian Institute of Seed Science, Mau, Uttar Pradesh, India

Norah Johal
Punjab Agricultural University, Ludhiana, Punjab, India

Gopal Katna
Department of Organic Agriculture and Natural Farming, CSK Himachal Pradesh Agricultural University, Palampur, Himachal Pradesh, India

Yogesh Kumar
Division of Crop Improvement, ICAR-Indian Institute of Pulses Research, Kanpur, Uttar Pradesh, India

Ashutosh Kushwah
Punjab Agricultural University, Ludhiana, Punjab, India

Nikhil Malhotra
ICAR-National Bureau of Plant Genetic Resources, Shimla, Himachal Pradesh, India

Biswajit Mondal
Division of Crop Improvement, ICAR-Indian Institute of Pulses Research, Kanpur, Uttar Pradesh, India

Harsh Nayyar
Panjab University, Chandigarh, Punjab, India

S.D. Nitesh
Department of Crop Improvement, CSK Himachal Pradesh Agricultural University, Palampur, Himachal Pradesh, India

Rajeev Rathour
Department of Agricultural Biotechnology, CSK Himachal Pradesh Agricultural University, Palampur, Himachal Pradesh, India

Aqeel Hasan Rizvi
Food Legume Breeder, South Asia and China Regional Program, International Center for Agricultural Research in the Dry Areas, New Delhi, India

Samriti
Department of Social Sciences, Dr YS Parmar University of Horticulture and Forestry, Solan, Himachal Pradesh, India

Ashutosh Sarker
Regional Coordinator, South Asia and China Regional Program, International Center for Agricultural Research in the Dry Areas, New Delhi, India

Shiv Sewak
Division of Crop Improvement, ICAR-Indian Institute of Pulses Research, Kanpur, Uttar Pradesh, India

Kamal Dev Sharma
Department of Agricultural Biotechnology, CSK Himachal Pradesh Agricultural University, Palampur, Himachal Pradesh, India

Ravinder Sharma
Department of Social Sciences, Dr YS Parmar University of Horticulture and Forestry, Solan, Himachal Pradesh, India

Subhash Sharma
Department of Social Sciences, Dr YS Parmar University of Horticulture and Forestry, Solan, Himachal Pradesh, India

Neelam Shekhawat
ICAR-National Bureau of Plant Genetic Resources, Jodhpur, Rajasthan, India

Alok Shukla
Division of Plant Biotechnology, ICAR-Indian Institute of Pulses Research, Kanpur, Uttar Pradesh, India

Inderjit Singh
Punjab Agricultural University, Ludhiana, Punjab, India

Mohar Singh
ICAR-National Bureau of Plant Genetic Resources, Shimla, Himachal Pradesh, India

N.P. Singh
ICAR-Indian Institute of Pulses Research, Kanpur, Uttar Pradesh, India

Prateek Singh
Division of Plant Biotechnology, ICAR-Indian Institute of Pulses Research, Kanpur, Uttar Pradesh, India

Sarvjeet Singh
Punjab Agricultural University, Ludhiana, Punjab, India

Shallu Thakur
Division of Plant Biotechnology, ICAR-Indian Institute of Pulses Research, Kanpur, Uttar Pradesh, India

Aneeta Yadav
Faculty of Agriculture and Allied Industries, RAMA University, Kanpur, Uttar Pradesh, India

About the Editor

Dr. Mohar Singh is currently serving as Principal Scientist to the National Bureau of Plant Genetic Resources Regional Station, Shimla, India. He obtained his Doctoral degree in Plant Breeding from Himachal Pradesh Agricultural University, Palampur, India. He has made an outstanding contribution to the management of plant genetic resources for food and agriculture in the country. His research interest reflects a continuum of high-quality basic and strategic research in pulses. He has developed 3 core sets, 2 reference sets, registered 5 genetic stocks, 25 gene sequences, 6 farmer varieties, and 2 lentil varieties developed through distant hybridization for rainfed areas of north-western Indian Himalaya. He has conducted 10 explorations on crop wild relatives (CWRs) and explored >900 wild germplasm of cereals, oilseeds, and pulses. He is instrumental to initiate prebreeding in chickpea and lentil for securing national nutritional demand in India. His pioneer research work on understanding the population structure and diversity assessment of global wild species of lentil and chickpea is very well known. This has led to the identification of most target gene sources in the secondary and tertiary gene pools of chickpea and lentil for biofortification of cultivated varieties including several yield and major biotic and abiotic stress—related traits that were successfully incorporated in cultivated backgrounds of these two important pulse crops. Successful deployment of marker-assisted breeding for introgression of two most promising superior haplotypes with high seed weight and pod number from cultivated and wild species into high yielding varieties of chickpea for improving their overall yield and productivity. Dr. Singh has published more than 60 research articles in international journals of repute viz., *DNA Research*, *Scientific Reports*, *Plant Breeding*, *Crop Science*, *Euphytica*, *PLoS ONE*, *Frontiers in Plant Science*, *Genetic Resources and Crop Evolution*, *Journal of Genetics*, *Journal of Genetics and Breeding*, *Plant Genetic Resources*: *Cambridge*, and *Advances in Horticulture Science*. He also holds two text books and five edited books to his credit.

Preface

The majority of cultivated grain legume crop species including chickpea have been exploited up to their maximum level of productivity, and there is a projection that by 2050, the world population will be more than 9 plus billion, which require an astonishing increase in food production. To attain further breakthroughs in increasing grain yield and improving stability in future cultivars, new sources of genes and alleles need to be identified in the plant genetic resources (PGR) including crop wild relatives (CWRs) and introgressed into the elite genetic backgrounds so that new plant types for different conditions could be tailored. Furthermore, in the face of climate change, this great resource may prove to be instrumental for future food and nutritional security. Therefore, it is imperative to motivate crop researchers to look into new sources of variation in untapped germplasm including CWRs and identify target traits of interest using appropriate tools and techniques in order to make the selection more efficient and reliable. In view of this context, an effort has been made to bring together the rather scattered research work done in this useful area in the form of an edited collection, a compilation that should be of great value to the chickpea researchers across the globe. The book comprises a total of nine chapters contributed by eminent researchers from various reputed organizations of the world. An introductory chapter describes some key issues linked to bottlenecks and potential of species utilization on current trends of wide hybridization. The subsequent chapters deal with different aspects related to potential resources and techniques for enhancing genetic gains of cultivated gene pool. The editor is extremely thankful to all authors for their significant contributions to this volume. The entire process of preparing the manuscript was marked by cordial collegiality. I am also indebted to the staff of Academic Press for their excellent professional support in the completion of this project. Despite one round of galley proof-reading, the book may still have some scientific, technical, and printing errors. I will appreciate if these omissions are brought to my notice, so that they may be addressed in future editions.

Mohar Singh

CHAPTER

Introduction

1

Mohar Singh

ICAR-National Bureau of Plant Genetic Resources, Shimla, Himachal Pradesh, India

1.1 Introduction

Chickpea (*Cicer arietinum* L.) is the third most important grain legume species cultivated in 59 nations contributing to world nutritional security owing to 23% protein content (FAO, 2016). It is a self-pollinating true diploid ($2n = 2x = 16$) belonging to family Fabaceae. The species is grouped into desi and kabuli types: desi chickpeas generally have small, colored seeds, whereas kabulis produce large, creamish colored seeds (Archak et al., 2016). Kabuli chickpea is usually utilized as whole grains, while desi is processed into flour. It is a cool-season annual crop performing suitably in 20–25°C daytime and 15–20°C night temperatures. The crop produces good grain yields in drier conditions, because of having deep tap root system. Chickpea has a moderately sized genome of around 750 Mbp (Varshney et al., 2013) evolved from its immediate wild progenitor *C. reticulatum* through natural selection. The genus *Cicer* comprises 44 species and is divided into two subgenera. The cultivated species, *C. arietinum* is found only in cultivation and cannot colonize successfully without human intervention. The wild species occur in weedy habitats, mountain slopes among rubble, and broad-leaf or pine forests, which can be accommodated into 9 annual and 35 perennial species.

Despite considerable international investment in traditional breeding, productivity of the crop has not yet been significantly enhanced. The reasons underlying marginal improvements are series of biotic and abiotic stresses, which reduce crop yield and its adaptation (Varshney et al., 2012). Nonavailability of genetically improved varieties remains a major bottleneck in achieving higher productivity levels. Furthermore, constraints in breeding of domesticated chickpea include narrow genetic base, because of its single domestication event and a very high self-pollination rate (Abbo et al., 2003). Adequate sources of resistance to prevailing biotic and abiotic stresses including productivity-related traits are often not available within cultivated gene pool, and this has forced to take the interest in crop wild relatives (CWRs) for the genetic improvement of cultivated crop varieties (Mallikarjuna et al., 2007). There are practical evidences available for the use of wild progenitors as donors of productivity alleles in some crops such as rice, tomato, and chickpea. Therefore, to attain further breakthrough for enhanced yield and stability in future crop cultivars, new traits of interest are needed to be identified and incorporated into

Chickpea: Crop Wild Relatives for Enhancing Genetic Gains. https://doi.org/10.1016/B978-0-12-818299-4.00001-4
Copyright © 2020 Elsevier Inc. All rights reserved.

cultivated varieties for developing novel germplasm. Hence, to meet the dietary requirements of growing world population, the consolidated efforts are needed to increase the nutrition and yield potential of chickpea. So, an immediate thrust is needed to widen the genetic base of chickpea cultivars by introgression of diverse set of genes and alleles available in distantly related wild *Cicer* taxa. To broaden the genetic base and maximize gains from the selection, it is therefore necessary to accumulate other complementary alleles from CWRs.

Substantial improvement in any crop species depends upon the whole spectrum of genetic diversity available in its germplasm repository. Systematic work on germplasm conservation is prerequisite for any genetic intervention. A lot of interest has recently been developed in CWRs with the notion that they will gain significance as changing climate put both traditional and advanced cultivars under increasing stress, leading to a prerequisite for plant breeding to produce new crop varieties able to grow under the new climatic system. Traditionally, the approach to the conservation of CWRs has been ex situ, while the need to maintain CWRs in their natural habitats as in situ has been supported. Large-scale collection and conservation efforts have been made to protect chickpea biodiversity, and ex situ gene banks have been established for the collection and conservation of useful plant genetic resources. There are a large number of gene banks conserving over 98,000 germplasm accessions of chickpea comprising of landraces, modern cultivars, genetic stocks, mutants, and wild *Cicer* species. Globally, International Crops Research Institute for the Semi-Arid Tropics (ICRISAT), Patancheru, Hyderabad, India, has the largest collection comprising 19,959 accessions of cultivated chickpea and 308 accessions of 18 wild *Cicer* species from about 60 countries. It is followed by the National Bureau of Plant Genetic Resources (NBPGR), New Delhi, India, having a total of 17,053 accessions. Other major gene banks holding chickpea germplasm include International Centre for Agricultural Research in Dry Areas holding 13,818 accessions; Australian Temperate Field Crops Collection Victoria, 8660 accessions; and Washington State University Pullman, 6789 accessions. The important regions for in situ conservation of wild chickpea include Turkey and Syria for *C. reticulatum*, *C. bijugum*, *C. echinospermum*, and *C. pinnatifidum*; Israel, Jordan, and Lebanon for *C. judaicum*; and Afghanistan for *C. yamashitae*. Furthermore, Lahaul and Spiti in North-West Indian Trans-Himalayas is the main region for in situ conservation of *C. microphyllum*, a perennial wild *Cicer* species.

1.2 Germplasm characterization and evaluation

It is widely recognized that characterization is the basic step to classify the crop gene pool into distinct phenotypic categories using crop descriptor states, while evaluation is prerequisite to understand the value of genetic resources for its successful utilization against the target traits. A large number of chickpea germplasm accessions have been characterized and evaluated over the years in various institutions across globe. The CWRs provide the broadest range of genetic diversity in grain legumes, including chickpea, and have the ability to provide the wide range of germplasm resources for the incorporation of elite traits of agro-economic importance (Singh et al., 2014a,b). The wild *Cicer* species not only consists of useful variation for morphological characteristics and protein content but are also rich sources of resistance to various biotic and abiotic stresses (Croser et al., 2003; Sandhu et al., 2006; Singh et al., 2014a,b), biochemical traits (Kaur et al., 2010), and yield components (Singh and Ocampo, 1997; Singh et al., 2014a,b). The studies pertaining to wild *Cicer* species vis-à-vis agro-morphological traits, biotic and abiotic stresses along with nutritional aspects at ICRISAT, ICARDA, NBPGR, and other institutions across the world have been successful in initiating chickpea improvement programs. It was reported that linkage drag of undesirable traits hampered the efforts of transferring elite traits from wild species to cultivated chickpea. Therefore, Aryamanesh et al. (2010) developed a genetic linkage map of chickpea using an interspecific F_2 population between cultivated and *C. reticulatum*. They identified a closely related marker for plant growth habit and quantitative trait loci (QTLs) for ascochyta blight resistance and flowering time. Thereafter, Singh et al. (2014a,b) have made detailed characterization and evaluation of global wild annual *Cicer* accessions for useful traits of interest. They identified multiple gene resistant sources for various agronomic traits including prevailing biotic stresses. It was revealed that the frequency distribution of wild *Cicer* species exhibited a wide range of intraspecific variation for some of the important morphological plant characteristics. Plant pigmentation showed variation in *C. reticulatum*, *C. judaicum*, and *C. pinnatifidum* along with light pubescent leaves, except *C. yamashitae*, where it was densely pubescent. In most of the *Cicer* species, seed shape was angular, with the exception of *C. bijugum* where it was irregular, rounded, and pea shaped. Testa texture was rough in *C. reticulatum*, *C. judaicum*, *C. pinnatifidum*, and *C. yamashitae*, but appeared tuberculated in *C. bijugum* and *C. echinospermum*. Similarly, substantial variation in seed color was observed in *C. reticulatum*, *C. judaicum*, and *C. pinnatifidum*. Although QTLs controlling agro-morphological traits in chickpea were also identified, a genome-wide scanning of wild *Cicer* accessions was revealed by the studies of Saxena et al. (2014) and Das et al. (2015). Moreover, a comprehensive comparative transcriptome profiling and high-resolution QTL mapping of *C. microphyllum* revealed molecular machinery regulating agronomic traits in chickpea (Srivastava et al., 2016a,b), which was further followed by detailed analysis of differentially expressed transcripts related to traits ranging from seed growth and metabolic processes to elite traits of interest in chickpea improvement programs (Srivastava et al., 2016a,b; Sagi et al., 2017). The summary of important traits identified in different wild *Cicer* species has been presented in Table 1.1.

Chickpea is susceptible to various abiotic and biotic stresses; therefore, CWRs can provide the wide range of germplasm resources for their incorporation into the cultivated gene pool, because the narrow genetic base of cultivated chickpea is one of the major limitations in improving chickpea production and

Table 1.1 Gene sources identified in wild annual *Cicer* species.

Agro-morphological		
Species	**Trait of interest**	**Gene pool**
Cicer reticulatum	Resistant to ascochyta	Primary
Cicer echinospermum	Resistant to ascochyta	Primary
Cicer judaicum	Ascochyta blight and botrytis gray mold	Secondary
Cicer pinnatifidum	Ascochyta blight, botrytis gray mold, and pod numbers	Secondary
Cicer bijugum	Ascochyta blight, botrytis gray mold, and pod numbers	Secondary

Reproduced from Singh, M., Bisht, I.S., Dutta, M., Kumar, K., Basandrai, A.K., Kaur, L., Sirari, A., Khan, Z., et al., 2014a. Characterization and evaluation of wild annual Cicer *species for agro-morphological traits and major biotic stresses under Northwestern Indian conditions. Crop Sci. 54, 229–239. Singh, S., Singh, I., Kapoor, K., Gaur, P.M., Chaturvedi, S.K., Singh, N.P., Sandhu, J.S., 2014b. Chickpea. In: Singh, M., et al. (Eds.), Broadening the Genetic Base of Grain Legumes. Springer, pp. 51–73.*

productivity worldwide. CWRs are species closely related to crops, including crop progenitors, identified as critical resources which are vital for wealth creation, food security, and environmental stability (Maxted et al., 2008). Systematic screening of wild *Cicer* collections by ICRISAT, ICARDA, and NBPGR have promoted genetic base broadening flagship activities in chickpea and identified several useful target traits of interest. These institutions have characterized over 400 wild annual *Cicer* accessions and landraces for various agro-morphological traits and biotic and abiotic stresses. Wild *Cicer* species have been extensively screened, and several of them have been reported to have very high levels of tolerance to many biotic and abiotic stresses, which includes resistance to ascochyta blight (Pande et al., 2010), botrytis gray mold (BGM; Basandrai et al., 2008), fusarium wilt (Nguyen et al., 2004), multiple disease resistance against ascochyta blight, BGM, root knot nematode (Singh et al., 2014a,b), cold (Toker, 2005), drought, and heat (Canci and Toker, 2009). The transfer of specific genes is frequently associated with the transfer of large alien chromosome segments having undesirable traits. Owing to linkage drag, the genes for primitive or wild traits are often introduced along with desirable traits. Breaking linkages with unwanted type and restoring the genotype associated with accepted agronomic background may take a long time. The first report on interspecific hybridization involving *C. arietinum* with *C. reticulatum* and *C. cuneatum* was published by Ladizinsky and Adler (1976). The cross between *C. arietinum* and *C. reticulatum* was achieved successfully. Subsequently, wide hybridization was attempted between *C. arietinum* and *C. echinospermum* (Singh and Ocampo, 1993). Due to the use of in vitro technique, success has been made in achieving crosses between *C. arietinum* and *C. bijugum* and *C. arietinum* and *C. judaicum*. Badami et al. (1997) reported successful hybridization between *C. arietinum* and *C. pinnatifidum* using embryo rescue technique. However, Singh et al. (2015, 2018) have also attempted interspecific crosses between

C. arietinum × *reticulatum* and *echinospermum* and transferred important yield component traits. Further, using various techniques, interspecific hybrids have been produced between *C. arietinum* × *C. cuneatum*, *C. arietinum* × *C. judaicum*, *C. arietinum* × *C. bijugum*, and *C. arietinum* × *C. pinnatifidum* (Singh et al., 1999; Mallikarjuna and Jadhav, 2008) to exploit an introgression of desirable alien genes from wild *Cicer* species into the cultivated gene pool. An attempt has also been made successfully to hybridize two wild annual *Cicer* species with three cultivated chickpea cultivars which resulted in high level of heterosis for number of pods/plant and seed yield/plant. Cross-combinations of cultivated chickpea cultivars with *C. reticulatum* and *C. echinospermum* exhibited higher variability for important yield component traits. Using wild *Cicer* accessions, promising donors, and popular chickpea cultivars as recipients, introgression of traits in elite material is being pursued by traditional and/or molecular breeding approaches, and the prebreeding populations are being developed for useful traits following simple (*C. arietinum* × *C. reticulatum* or *C. echinospermum*) and/or complex three-way (*C. arietinum* × [*C. reticulatum* × *C. echinospermum*]) crosses as reported by Sharma (2017). These interspecific hybrids have contributed significantly toward the development of genomic resources for chickpea improvement. Chickpea breeding programs are aimed at developing early-maturing cultivars especially to increase the crop adaptation by avoiding abiotic stresses affecting the crop production globally. Wild relatives of chickpea viz. *C. reticulatum*, *C. echinospermum*, *C. bijugum*, and *C. pinnatifidum* have been shown to possess early flowering and maturity accessions (Singh et al., 2014a,b). The wild *Cicer* species also harbor beneficial alleles/genes for high seed protein and improvement of agronomic traits in cultivated chickpea. The productivity genes/alleles have been introgressed from *C. reticulatum*, *C. echinospermum*, and *C. pinnatifidum* species (Singh and Ocampo, 1997; Singh et al., 2012). Of the eight annual wild *Cicer* species, only *C. reticulatum* is readily crossable with cultivated chickpea resulting in a fertile hybrid, whereas for exploitation of the remaining seven annual wild *Cicer* species for chickpea improvement, specialized techniques such as application of growth hormones, embryo rescue, ovule culture, and other tissue culture techniques have been suggested by various researchers (Mallikarjuna and Jadhav, 2008). Promising high-yielding lines with good agronomic base and seed traits such as early flowering and 100-seed weight have also been obtained from crosses involving *C. reticulatum* and *C. echinospermum* with cultivated chickpea (Singh et al., 2005; Upadhyaya, 2008).

Furthermore, biotechnological interventions can help to increase chickpea productivity by using the marker-assisted selection (MAS) approach in improvement programs. Recent developments in genomics technology have helped to explain the mechanism of complex traits controlling chickpea productivity and the genetic architecture of traits of economic importance to accelerate breeding programs. The extensive use of molecular markers in chickpea genetics and breeding started only after the development of simple sequence repeat (SSR) markers. Over 2000 SSR markers are available for chickpea molecular analysis. The draft genome sequence of chickpea identified over 48,000 SSRs suitable for PCR primer design for use as genetic markers (Varshney et al., 2013) while a draft sequence of *C. reticulatum* (PI 489777) measuring 327.07 Mb was assembled to the eight linkage groups with 25,680 protein coding genes (Gupta et al., 2017). A number of marker-trait associations have been identified in chickpea along with the dense genetic maps which have allowed MAS to become a routine in breeding programs. Molecular markers have been identified for the genes/QTLs linked to resistance to several abiotic stresses such as drought and salinity resilience; biotic stresses, viz., fusarium wilt, ascochyta blight, BGM, and yield contributing characteristics in chickpea. Recent advances in the development of markers and identification of molecular markers linked to genes/QTLs controlling elite traits of interest have encouraged applications of marker-assisted backcrossing in chickpea improvement. Moreover, genome-wide selection has been proposed as a potential approach for improving complex traits governed by many genes/QTLs (Bajaj et al., 2015). In this approach, both phenotyping and genotyping data are used to predict genomic estimated breeding values of progenies. Additionally, emerging novel approach of genome editing technologies namely CRISPR/Cas9, transcription activator-like effector nucleases (TALENs), zinc finger nucleases could be potentially used in genomics-assisted selection in chickpea for rapid genetic gain (Bortesi and Fischer, 2015).

1.3 Production-related problems

1.3.1 Fusarium wilt

Chickpea wilt occurs in 32 countries across six continents in the world (Nene et al., 1991; Singh and Sharma, 2002). It is caused by *Fusarium oxysporum* f. sp. *ciceri*. The yield losses caused by it vary from 10% to 90% (Jimenez-Diaz et al., 1989; Singh and Reddy, 1991). At present there are eight distinct physiological races of *F. oxysporum*, viz., 0, 1A, 1B/C, 2, 3, 4, 55, and 6. Out of them four races have been identified as 1, 2, 3, and 4 which are prevalent in India (Haware and Nene, 1982) and race 0, 5, and 6 are reported from Spain (Jimenez-Diaz et al., 1989). Breeding for fusarium wilt is of prime importance because of nature of the pathogen as it can persist in soil year after year even in the absence of the host (Haware et al., 1996). Due to the presence of such a number of races, it is very tedious to breed a variety, which shows stability to the disease across different agro-ecological regions. Most of the resistance against fusarium wilt is of vertical nature (Sharma et al., 2005) as it is mostly governed by a single major gene. Chickpea genotypes differ in the development of initial symptoms of wilt indicating different degrees of resistance which is controlled by segregation of a single gene (Upadhyaya et al., 1983). Such individual genes which are part of oligogenic resistance mechanism delay the onset of disease symptoms and the phenomenon is called "late wilting." The yield losses are significantly less in case of genotypes which show the phenomenon of late wilting due to combination of two recessive genes for resistance. However, resistance to race 0 is due to two independent genes (Rubio et al., 2003), while resistance is digenic or trigenic for race 1A, 2, and 4. Resistance to race 3 and 55 is controlled by single gene (Sharma et al., 2005). Singh et al. (2012) have also reported resistance against fusarium wilt in the indigenous chickpea genetic resources.

1.3.2 Ascochyta blight

Ascochyta blight is the most important and major devastating foliar disease of chickpea in many parts of the world. It is caused by *Ascochyta rabiei* and leads to yield losses ranging from 10% to 100% (Nene and Reddy 1987; Singh, 1990). Among the most affected segments are Indian subcontinent and Mediterranean region as the prevailing climatic conditions are quiet conducive for pathogens. Ascochyta blight disease epidemics are common and had occurred in India, Pakistan, the United States, Pacific northwest, Australia, and Syria in the past (Malhotra et al., 2003). *A. rabiei* isolates have been classified into either a two or three-pathotype system (I, II, and III) according to their levels of virulence (Udupa et al., 1998; Chen et al., 2004; Jayakumar et al., 2005). The prime reason for epidemics is the nature of pathogen, which is unstable and continuously evolves new races which break down the host resistance, and thus shorten the whole life of varieties under chickpea production system. The development of more virulent pathotype has been reported in Syria (Reddy and Kabbabeh, 1985), Italy (Stamigna et al., 2000), and Pakistan (Jamil et al., 2000). The genetics of resistance to ascochyta blight is digenic in nature; both recessive and dominant genes are involved in its genetic control (Bhardwaj et al., 2010). This has stimulated breeders to devise a suitable approach for developing varieties with durable genetic resistance.

1.3.3 Botrytis gray mold

Botrytis gray mold (BGM) is another major foliar disease of chickpea and is prevalent in about 15 countries including India, Bangladesh, Nepal, Pakistan, Australia, Argentina, Myanmar, Canada, Columbia, Hungary, Mexico, Spain, Turkey, the United States, and Vietnam. Long back there was no reliable source of resistance to BGM in India (Singh and Reddy, 1991), but in recent years interspecific derivative lines from the crosses of *C. arietinum* and *C. pinnatifidum* had shown high level of genetic resistance against BGM (Kaur et al., 2013). This resistance can be transferred into elite lines to develop high-yielding chickpea cultivars with durable genetic resistance. It has also been suggested that the resistance introgressed from wild *C. echinospermum* to cultivated chickpea was found to be monogenic in nature revealing that resistance to BGM can be easily introgressed into elite background of chickpea.

1.3.4 Pod borer

Pod borer caused by *Helicoverpa armigera* is the most damaging insect worldwide, which is quiet prevalent in Asia, Africa, Australia, and some other chickpea growing areas. As pod borer is a polyphagous insect that attacks more than 182 plant species. Development of resistant cultivars to *H. armigera* could be integrated in the pest management strategy particularly in the developing countries (Fitt, 1989; Sharma and Ortiz, 2002). At ICRISAT more than 14,000 chickpea germplasm accessions were screened under real-field conditions, which have resulted into identification of some degree of tolerance toward *H. armigera* (Lateef and Sachan, 1990). But still complete resistance is lacking against the insect in cultivated germplasm, for which wild species need to be screened and exploited into cultivated background for diversification of domesticated gene pool.

1.3.5 Bruchids

Apart from field diseases, chickpea is also damaged in the storage and one such insect is bruchids (*Callosobruchus chinensis*). It causes huge loss of grains in the Mediterranean region, where infestation levels approaching 13% (Dias and Yadav, 1988) to total loss (Weigand and Tahhan, 1990). Resistance in the cultivated chickpea germplasm is lacking against bruchid, and wild chickpea accessions have shown some degree of resistance to storage pests (Singh et al., 1998; Padmavati et al., 2015). Owing to crossing barrier, it has not been possible to transfer this trait to the cultivated background, for which contemporary breeding approaches are required to exploit the useful traits for chickpea improvement.

1.3.6 Cold tolerance

Temperature is another important abiotic stress factor, which affect the crop when it is exposed to freezing stress during vegetative growth in most of the cold climatic situations in the world. Chickpea crop grown in winter season is more productive than the traditionally sown in spring season in the Mediterranean region. This is due to long growing season and better moisture availability. The problem encountered by winter season crop is flower drop and pod abortion, which leads to major yield loss, when mean temperature of day fall below 15°C. Chaturvedi et al. (2009) reviewed the research work on cold tolerance in chickpea and highlighted the significance of cold tolerance at reproductive stage. Deployment of genes for cold tolerance will protect crop from getting damaged and reduce yield losses. As far as genetic control of cold tolerance is concerned, it was found to be controlled by at least five genes with both additive and nonadditive gene effects. Screening of wild *Cicer* species showed promising traits for cold tolerance, but until now there are no reports of introgression from wild to the cultivated chickpea. *C. microphyllum*, a perennial wild *Cicer* species, has also been reported tolerant to cold conditions, since the species is growing in cold desert Himalaya.

1.3.7 Drought tolerance

Drought is the second most important abiotic stress which contributes to the losses in chickpea production globally. To counter drought stress, development of early-maturing varieties will make judicious use of the available soil moisture and produce relatively higher grain yield. Drought tolerance may be under polygenic or oligogenic genetic control with additive and nonadditive gene effects. Root-related traits have been given due importance in recent years, because lines with longer root systems have shown better drought tolerance/avoidance. *C. microphyllum* had also shown drought-responsive genes, and secondly it has deep tap root system, which contributes toward drought tolerance. As chickpea is grown under receding soil moisture, the long root trait plays an important role in countering drought. Apart from this, early flowering genotypes are considered to be very important for drought escape. There is a single recessive locus involved in controlling early flowering (Kumar and van Rheenen, 2000). Any accession with recessive allele in homozygous state may escape drought and give a desirable yield, and such alleles could be easily transferred to the drought-susceptible varieties using suitable breeding approach for developing drought-tolerant varieties. Apart from this, wild *Cicer* species have been screened and few accessions of *C. pinnatifidum* and *C. reticulatum* were found to be resistant against drought (Toker et al., 2007). ICRISAT has identified one genotype ICC 4958 which is considered as potential donor for drought tolerance. Some chickpea cultivars with improved drought tolerance have been released using ICC 4958 as one of the donor in south India and Kenya.

1.3.8 Heat tolerance

Most of the period heat stress occurs in combination with drought stress (Toker and Canci, 2009), and it is very difficult to make distinction whether the crop is under heat or drought stress. Furthermore, chickpea is grown in post rainy period in south Asia, which results in the exposure of crop to drought along with high temperature regime, and it is predicted that climate change would result in temperature rise by 3–4°C over current levels by 2050 (Basu et al., 2009). Chickpea is usually grown in winter season in northern India, it experiences a high temperature during the reproductive phase and most sensitive organs of plant to heat are flowers (Toker and Canci, 2006). During the reproductive period, if temperature rises than the threshold level, it would adversely affect the pod formation and seed set, and thus, results in reduced grain yield (Wang et al., 2006; Basu et al., 2009). Furthermore, adverse effects of high temperature occurs in seed germination, respiration, membrane stability, photosynthesis, hormone level, nutrient absorption, protoplasmic movement, quality of seeds, fruit maturation, fertilization, materials transport, withering, burning of lower leaves, desiccation of poorly developed plants, stunting flower and pod abortion, reduced root nodulation, nitrogen fixation, and seed yield. In chickpea germplasm, large genetic variations have been observed for heat tolerance when reference set was screened against heat stress at several locations in India including ICRISAT evaluation (Krishnamurthy et al., 2010). A field screening technique for heat tolerance has been developed, and several useful sources of heat tolerance were identified by the ICRISAT. To accomplish abovementioned constraints, the present book contains useful chapters for enhancing genetic gains of cultivated gene pool of chickpea and the summary of all chapters is given below.

1.4 Origin, distribution, gene pools

The genus *Cicer* has 44 species, out of which 43 are wild and 1 is cultivated. These genetic resources are classified into three gene pools, i.e., primary, secondary, and tertiary, based on their taxonomical identity for improving the cultigen. The primary gene pool consists of cultivated species and landraces. The secondary gene pool consists of the progenitor species *C. reticulatum* and *C. echinospermum*, the species that are crossable with *C. arietinum*, but with reduced fertility of the resulting hybrids and enhanced progenies. The tertiary gene pool consists of all the annual and perennial *Cicer* species, which are not crossable with cultivated species. No perennial *Cicer* species are known to cross readily with the cultivated species and produce mature viable seeds.

1.5 Chickpea genetic resources: collection, conservation, characterization, and maintenance

In most of the chickpea growing regions, the cultivated varieties which are being grown by the farmers have narrow genetic base, and to attain further breakthrough for increasing yield and improving stability in future cultivars, new gene sources need to be identified in wild species and introgressed into cultivated varieties for diversification of crop gene pool, because success of any crop improvement program depends on the level of genetic diversity and the availability of useful target traits. In this direction, wild *Cicer* species are playing a critical role in the broadening of cultivated genetic base and tailoring gene germplasm using these species. However, as a result of the loss of natural habitats due to drastic climatic changes, wild *Cicer* species are rapidly becoming endangered species. Therefore, sustaining chickpea yield has become a challenging task for breeders. In such a scenario, ex situ germplasm collections are being utilized increasingly and they are playing a vital role in developing climatic resilience in chickpea varieties. Major constraints in chickpea improvement are its plant type, which is less amenable for mechanization, devastating pests and diseases, and its highly sensitive nature to heat and salinity stress. Recently, climatic changes have worsened the situation. Sudden temperature rises, short spells of winter, and erratic rains have become a common phenomenon which may reduce chickpea production in coming future. Therefore, focused breeding programs on traits like heat, drought, and salinity tolerance are highly warranted. To some extent, niche-specific breeding programs combined with molecular marker technology, genomics, and advanced phonemics tools may sustain chickpea yield.

1.6 Conventional cytogenetic manipulations

Wild *Cicer* species harbor several important target traits of interest, which are not present adequately in the cultigen, e.g., resistance to cold, drought, pod borer, and ascochyta blight and hence these wild species could be a wonderful resource for their introgression purposes. The cytogenetic studies have been used to karyotype the chromosomes and characterize hybrids cytologically, and introgression of traits also can be monitored using marker-assisted selection and fluorescence in situ hybridization approaches. Marker-assisted selection also offers added advantages of introgression of multiple traits or traits that are difficult to assess through phenotyping.

1.7 Embryo rescue and chromosomal manipulations

Considerable progress has been made in chickpea through embryo rescue in wide hybridization in terms of recovery of hybrid plantlets, but development of introgressed donor lines and varietal release is yet to be accomplished. The information generated pertaining to embryo culture of chickpea facilitated researches in somatic embryogenesis, which had immense role in development of plantlets from genetically transformed tissues. Biology of the developing somatic embryo in chickpea has been investigated and developmental stages have been identified. Embryo culture conditions has been standardized for culturing the mature and immature embryos, and liquid media with filter paper bridge set up have been identified to be congenial to support the growth of the immature embryo. Karyotypes of different species of *Cicer* have been developed and corresponding DNA contents have been reported. C-banding patters of the chromosomes have been developed and their fine structure during pachytene has also been studied. Linkage groups have been identified on the basis of the location of the centromere, secondary constriction, and arms ratio in different *Cicer* species.

1.8 Gene pyramiding and multiple character breeding

Several genomic resources and next-generation sequencing platforms like skim sequencing, genotyping by sequencing, and whole genome resequencing approaches deliver the genome-wide marker information, which facilitates high-resolution trait mapping and prolifically execution of QTL analysis. This genome information is exploited for mapping studies like linkage mapping and association mapping and finally organized for development of promising cultivars using genomics-assisted breeding approaches such as marker-assisted backcrossing, marker-assisted recurrent selection, genomic-assisted breeding through multiparent advanced generation intercrossing lines, advanced backcross QTL analysis—based breeding and genomic selection. Advancement in these genomics technologies combined with the accessibility of high-throughput genotyping as well as sequencing platforms provides stimulation for developing climate-resilient crops. In addition, speed breeding will accelerate the chickpea breeding by improving stability and increasing chickpea yield to address the need of rising population. Breeding for multiple traits including yield and its component traits, quality, and resistance/tolerance to various abiotic and biotic stresses is of paramount importance in order to develop cultivars that meet consumer's demands and thrive well under variable climatic and farming conditions. Integrating multiple traits into single genetic background through conventional breeding approaches is feasible but challenging. Recent advancement and forthcoming molecular breeding approaches present immense opportunities in chickpea. Genomic selection, a future molecular breeding tool, has great potential in chickpea breeding to increase the efficiency of selection and improve multiple traits simultaneously. In summary, chickpea improvement is heading toward integration of modern genomics approach with existing breeding programs, leading to resultant enhanced yield.

1.9 Molecular markers and marker-trait associations

Since inception, molecular markers have become important genomic resources for plants in ascertaining hybridity, diversity/relatedness, developing linkage maps, mapping genes/QTLs, and monitoring introgression of genes into progeny plants including gene pyramiding, because the chickpea has narrow genetic base and consequently many DNA-based markers were unsuccessful in revealing desired polymorphism. The problem was overcome by the development of SSR and single nucleotide polymorphism markers that revealed adequate polymorphism and these became markers of choice in chickpea. In addition, transcription factor gene–derived microsatellite markers, intron-targeted primers, expressed sequence tag polymorphisms, diversity arrays technology, and insertion–deletion markers have also been developed for chickpea. Genotyping by sequencing is also gaining popularity. Saturated linkage maps of chickpea with adequate marker density on each linkage group have also been constructed; however, several of the recently developed markers are still to be placed on linkage maps. The markers have subsequently been used to map genes or identify QTLs for biotic stress (fusarium wilt, ascochyta blight, phytophthora root rot, and BGM) resistance, abiotic stress (salinity, cold, and drought) tolerance, and agro-morphological traits. Using reference genomes of chickpea and QTL intervals, candidate genes in QTL regions for several of these characters have been identified. Notably, genes involved in flowering in chickpea have been identified. Same QTL was also found to govern more than one trait, e.g., drought tolerance and plant vigor, salinity tolerance and higher plant vigor, salinity tolerance and higher reproductive success. The markers linked closely to genes/QTLs of interest have also been exploited to introgress wilt and blight resistance, and double podding character into agronomically superior cultivars.

Genomic selection (GS) can potentially be used for prediction of phenotype in chickpea; however, the technique at present is in state of infancy. Markers developed so far in chickpea have been useful for gene mapping as well as gene introgression; additional markers are, however, needed to develop denser maps possessing markers more closely associated to the traits.

1.10 Genetic transformation

A well-defined regeneration protocol is prerequisite for the genetic transformation of chickpea, although *in planta* methods were also successfully reported and demonstrated. Important parameters for successful transformation of chickpea include inter alia media composition, compatibility of the *Agrobacterium* strains, vectors, co-cultivation period, and type of explants. Chickpeas were reported amenable to both direct and indirect methods, though indirect method employing *Agrobacterium tumefaciens* strains is common. Reports of successful genetic transformation chickpea, mostly for insect-resistant (IR) trait (gram pod borer, bruchids, and aphids), are plenty; however, lower transformation efficiency, genotype dependency, and long and laborious methods are still the major challenges for researchers. Notably, root transformation using *Agrobacterium rhizogenes* has emerged as an alternative to traditional transformation methodologies for reverse genetics studies in chickpea. Here, we highlight advancement and recent findings of genetic transformation efforts of the important grain legume, chickpea. Genetic engineering should preferably be employed for traits which are not identified in crossable gene pool and subsequently employ transgenic lines in breeding program. Significant progress has been made toward genetic transformation of chickpea, once considered recalcitrant to transformation. Genetic engineering of chickpea is very essential considering the narrow genetic base and absence of useful traits in chickpea gene pool. Until now, confined field trials of transgenic chickpea for insect-resistant trait are reported; however, the efforts require constant support to bring forth viable transgenes, addressing all regulatory issues. Efforts should be directed toward development of large number of transgenic lines, evaluating them in open field conditions, and identifying elite event(s) with high trait efficacy. Efforts should also be initiated for targeted integration of transgene in genome using genome editing techniques (CRISPR/Cas9 systems).

1.11 Chickpea economy in India

Among pulses, share of chickpea to the total pulse production has increased up to 35.22% in 2017–18, but yield of chickpea is comparatively low. India's pulse consumption is steadily rising, creating a supply–demand gap which the government has attempted to fill with imports through 2017. The cost of production of chickpea has increased over time. The Indian chickpea market is highly segmented and the demand depends on end use. Importers who are licensed to import chickpeas into India sell directly to a broker, commission agent, middlemen, or directly to a secondary wholesaler, miller, or processor. The minimum support prices (MSP) has increased over the years but remained below the domestic prices. The issue of pulses is high on the minds of governments and farmers, both inside and outside India right now, with numerous policy actions geared toward increasing Indian domestic pulse prices. Thus to enhance domestic chickpea production and productivity, improved production technologies and favorable policy measures are two requisites.

References

Abbo, S., Jens Berger, J., Turner, N.C., 2003. Evolution of cultivated chickpea: four bottlenecks limit diversity and constrain adaptation. Funct. Plant Biol. 30, 1081–1087.

Archak, S., Tyagi, R.K., Harer, P.N., Mahase, L.B., Singh, N., Dahiya, O.P., Abdul Nizar, M., Singh, M., et al., 2016. Characterization of chickpea germplasm conserved in the Indian National Gene Bank and development of core set using qualitative and quantitative data. Crop J. 4, 417–424.

Aryamanesh, N., Nelson, M.N., Yan, G., Clarke, H.J., Siddique, K.H.M., 2010. Mapping a major gene for growth habit and QTLs for ascochyta blight resistance and flowering time in a population between chickpea and *Cicer reticulatum*. Euphytica 173, 307–331.

Badami, P.S., Mallikarjuna, N., Moss, J.P., 1997. Interspecific hybridization between *Cicer arietinum* and *C. pinnatifidum*. Plant Breed. 116, 393–395.

Bajaj, D., Saxena, M.S., Kujur, A., Das, S., Badoni, S., Tripathi, S., et al., 2015. Genome-wide conserved non-coding microsatellite (CNMS) marker-based integrative genetical genomics for quantitative dissection of seed weight in chickpea. J. Exp. Bot. 66, 1271–1290.

Basandrai, A.K., Pande, S., Basandrai, D., 2008. Recent advances in host-parasite interaction, chickpea *Ascochyta rabaie* and *Botrytis cinerea* systems. In: Setia, R.C., Nayyar, H., Setia, N. (Eds.), Crop Improvement, Strategies and Applications. IK International, New Delhi, India, pp. 265–301.

Basu, P.S., Ali, M., Chaturvedi, S.K., 2009. Terminal heat stress adversely affects chickpea productivity in Northern India – strategies to improve thermo-tolerance in the crop under climate change. In: Panigrahy, S., Shankar, S.R., Parihar, J.S. (Eds.), ISPRS Archives XXXVIII-8/W3 Workshop Proceedings: Impact of Climate Change on Agriculture, 23–25 February, New Delhi, India, pp. 189–193.

Bhardwaj, R., Sandhu, J.S., Kaur, L., Gupta, S.K., Gaur, P.M., Varshney, R., 2010. Genetics of resistance to ascochyta blight in chickpea. Euphytica 171, 337–343.

Bortesi, L., Fischer, R.B., 2015. The CRISPR/Cas9 system for plant genome editing and beyond. Biotechnol. Adv. 33, 41–52.

Canci, H., Toker, C., 2009. Evaluation of annual wild Cicer species for drought and heat resistance under field conditions. Genet. Resour. Crop Evol. 56, 1–6.

Chaturvedi, S.K., Mishra, D.K., Vyas, P., Mishra, N., 2009. Breeding for cold tolerance in chickpea. Trends Biosci. 2 (2), 1–6.

Chen, W., Coyne, C.J., Peever, T.L., Muehlbauer, F.J., 2004. Characterization of chickpea differentials for pathogenicity assay of ascochyta blight and identification of chickpea accessions resistant to *Didymella rabiei*. Plant Pathol. 40, 401–407.

Croser, J.S., Ahmad, C.F., Clarke, H.J., Siddique, K.H.M., 2003. Utilisation of wild *Cicer* in chickpea improvement – progress, constraints, and prospects. Aust. J. Agric. Res. 54, 429–444.

Das, S., Upadhyaya, H.D., Bajaj, D., Kujur, A., Badoni, S., Kumar, V., Tripathi, S., Gowda, C.L., Sharma, S., Singh, S., Tyagi, A.K., Parida, S.K., 2015. Deploying QTL-seq for rapid delineation of a potential candidate gene underlying major trait-associated QTL in chickpea. DNA Res. 22, 193–203.

Dias, C.A.R., Yadav, T.D., 1988. Incidence of pulse beetles in different legume seeds. Indian J. Entomol. 50, 457–461.

FAO, 2016. FAOSTAT. http://www.faostat.fao.org.

Fitt, G.P., 1989. The ecology of heliothis species in relation to agro-ecosystems. Annu Rev Entomol 34, 17–52.

Gupta, S., Nawaz, K., Parween, S., Roy, R., Sahu, K., Kumar Pole, A., Khandal, H., Srivastava, R., Kumar Parida, S., Chattopadhyay, D., 2017. Draft genome sequence of *Cicer reticulatum* L., the wild progenitor of chickpea provides a resource for agronomic trait improvement. DNA Res. 24, 1–10.

Haware, M.P., Nene, Y.L., Pundir, R.P.S., Narayana Rao, J., 1996. Screening of world chickpea germplasm for resistance to fusarium wilt. Field Crops Res 30, 147–154.

Haware, M.P., Nene, Y.L., 1982. Races of *Fusarium oxysporum* f. sp. *ciceri*. Plant Dis. 66, 809–810.

Jamil, F.F., Sarwar, N., Sarwar, M., Khan, J.A., Geistlinger, J., Kahl, G., 2000. Genetic and pathogenic diversity within *Ascochyta rabiei* (Pass.) Lab. populations in Pakistan causing blight of chickpea (*Cicer arietinum* L.). Physiol. Mol. Plant Pathol. 57, 243–254.

Jayakumar, P., Gossen, B.D., Gan, Y.T., Warkentin, T.D., Banniza, S., 2005. Ascochyta blight of chickpea: infection and host resistance mechanisms. Can. J. Plant Pathol. 27, 499–509.

Jimenez-Diaz, R.M., Trapero-Casas, A., Cabrera de la Colina, J., 1989. Races of *Fusarium oxysporum* f. sp. *Ciceris* infecting chickpea in southern Spain. In: Tjamos, E.C., Beckman, C.H. (Eds.), Vascular Wilt Disease of Plants, vol. H28. NATO ASI Series Springer, Berlin, pp. 515–520.

Kaur, L., Sirari, A., Kumar, D., Sandhu, J.S., Singh, S., Singh, I., Kapoor, K., Gowda, C.L.L., Pande, S., Gaur, P., Sharma, M., Imtiaz, M., Siddique, K.H.M., 2013. Harnessing ascochyta blight and botrytis grey mould resistance in chickpea through interspecific hybridization. Phytopathol. Mediterr. 52 (1), 157–165.

Kaur, A., Sandhu, J.S., Gupta, S.K., Bhardwaj, R., Bansal, U.K., Saini, R.G., 2010. Genetic relationships among annual wild *Cicer* species using RAPD analysis. Indian J. Agric. Sci. 80 (4), 309–311.

Krishnamurthy, L., Gaur, P.M., Basu, P.S., Chaturvedi, S.K., Tripathi, S., Vadez, V., Rathore, A., Varshney, R.K., Gowda, C.L.L., 2010. Large genetic variation for heat tolerance in the reference collection of chickpea (*Cicer arietinum* L.) germplasm. Plant Genet. Resour. Charact. Util. 9 (1), 59–69.

Kumar, J., van Rheenen, H.A., 2000. A major gene for time of flowering in chickpea. J. Hered. 91, 67–68.

Ladizinsky, G., Adler, A., 1976. The origin of chickpea *Cicer arietinum* L. Euphytica 25, 211–217.

Lateef, S.S., Sachan, J.N., 1990. Host plant resistance to *Helicoverpa armigera* (Hub.) in different agro-economical conditions. In: Chickpea in Nineties: Proceedings of the Second International Workshop on Chickpea, 4–8 December 1989. ICRISAT, Hyderabad, India, pp. 181–189.

Malhotra, R.S., Baum, M., Udupa, S.M., Bayaa, B., Kababbeh, S., Khalaf, G., 2003. Ascochyta blight research in chickpea-present status and future prospects. In: Sharma, R.N., Srivastava, G.K., Rahore, A.L., Sharma, M.L., Khan, M.A. (Eds.), Proceedings of the International Chickpea Conference Chickpea Research for the Millennium, 20–22 January 2003, Raipur, Chhattisgarh, India, pp. 108–117.

Mallikarjuna, N., Jadhav, D.R., 2008. Techniques to produce hybrids between *Cicer arietinum* L. × *Cicer pinnatifidum* Jaub. Indian J. Genet. 68, 398–405.

Mallikarjuna, N., Sharma, H.C., Upadhyaya, H.D., 2007. Exploitation of wild relatives of pigeonpea and chickpea for resistance to *Helicoverpa armigera*. EJ SAT Agric. Res. Crop Improv. 3 (1), 4–7.

Maxted, N., Kell, S.P., Ford-Lloyd, B.V., 2008. Crop wild relatives conservation and use: establishing the context. In: Maxted, N., Ford-Lloyd, B.V., Kell, S.P., Iriondo, J.M., Dulloo, E., Turok, J. (Eds.), Crop Wild Relatives, Conservation and Use. CAB International, Wallingford, pp. 3–30.

Nene, Y.L., Reddy, M.V., 1987. Chickpea diseases and their control. In: Saxena, M.C., Singh, K.B. (Eds.), The chickpea. CAB International, Wallingford, pp. 233–270.

Nene, Y.L., Reddy, M.V., Haware, M.P., Ghaneka, A.M., Amin, K.S., 1991. Field Diagnosis of Chickpea Diseases and Their Control. Info Bull No. 28. ICRISAT, Hyderabad, India.

Nguyen, T.T., Taylor, P.W.J., Redden, R.J., Ford, R., 2004. Genetic diversity in *Cicer* using AFLP analysis. Plant Breed. 123, 173–179.

Padmavati, G., Kuldeep, T., Singh, M., Bisht, I.S., Bhalla, S., 2015. Searching for resistance in wild *Lens* species against pulse beetle (*Callosobruchis chinensis* (L.). Legume Res. https://doi.org/10.5958/0976-0571.2015.00103.4.

Pande, S., Sharma, M., Gaur, P.M., Gowda, C.L.L., 2010. Host Plant Resistance to *Ascochyta* Blight of Chickpea. Information Bulletin No. 82, Project Report. ICRISAT.

Reddy, M.V., Kabbabeh, S., 1985. Pathogenic variability in *Ascochyta rabiei* (Pass.) Lab. in Syria and Lebanon. Phytopathol. Mediterr. 24, 265–266.

Rubio, J., Hajj Moussa, E., Kharrat, M., Moreno, M.T., Millan, T., Gill, J., 2003. Two genes and linked RAPD markers involved in resistance to *Fusarium oxysporum* f.sp. *ciceris* race 0 in chickpea. Plant Breed. 122, 188–191.

Sagi, M.S., Deokar, A.A., Tar'an, B., 2017. Genetic analysis of NBS-LRR gene family in chickpea and their expression profiles in response to ascochyta blight infection. Front. Plant Sci. 8, 838.

Sandhu, J.S., Gupta, S.K., Singh, G., Sharma, Y.R., Bains, T.S., Kaur, L., Kaur, A., 2006. Interspecific hybridization between *Cicer arietinum* L. and *Cicer pinnatifidum* Jaub. et. Spach for improvement of yield and other traits. In: 4th International Food Legumes Research Conference, October 18–22, 2005. Indian Society of Genetics and Plant Breeding, New Delhi, p. 192.

Saxena, M.S., Bajaj, D., Das, S., Kujur, A., Kumar, V., Singh, M., Bansal, K.C., Tyagi, A.K., Parida, S.K., 2014. An integrated genomic approach for rapid delineation of candidate genes regulating agro-morphological traits in chickpea. DNA Res. 21, 695–710.

Sharma, H.C., Ortiz, R., 2002. Host plant resistance to insects: an eco-friendly approach for pest management and environment conservation. J. Environ. Biol. 23, 11–35.

Sharma, K.D., Chen, W., Muehlbauer, F.J., 2005. Genetics of chickpea resistance to five races of Fusarium wilt and a concise set of race differentials for *Fusarium oxysporum* f. sp. *ciceris*. Plant Dis. 89 (4), 385–390.

Sharma, S., 2017. Prebreeding using wild species for genetic enhancement of grain legumes at ICRISAT. Crop Sci. 57, 1132–1144.

Singh, G., Sharma, Y.R., 2002. Fungal diseases of pulses. In: Gupta, V.K., Paul, Y.S. (Eds.), Diseases of Field Crops. Indus Publishing, New Delhi, India, pp. 155–192.

Singh, K.B., 1990. Pattern of resistance and susceptibility to races of Ascochyta rabiei among germplasm accessions and breeding lines of chickpea. Plant Dis 74, 127–129.

Singh, K.B., Ocampo, B., Robertson, L.D., 1998. Diversity for abiotic and biotic stress resistance in the wild annual *Cicer* species. Genet. Resour. Crop Evol. 45, 9–17.

Singh, I., Singh, R.P., Singh, S., Sandhu, J.S., 2012. Introgression of productivity genes from wild to cultivated *Cicer*. In: Sandhu, S.K., Sidhu, N., Rang, A. (Eds.), International Conference on Sustainable Agriculture for Food and Livelihood Security. Crop Improvement Society of India, Ludhiana, India, pp. 155–156.

Singh, K.B., Ocampo, B., 1993. Interspecific hybridization in annual *Cicer* species. J. Genet. Breed. 47, 199–204.

Singh, K.B., Ocampo, B., 1997. Exploitation of wild *Cicer* species for yield improvement in chickpea. Theor. Appl. Genet. 95, 418–423.

Singh, K.B., Reddy, M.V., 1991. Advances in disease resistance breeding in chickpea. Adv. Agron. 45, 191–222.

Singh, M., Bisht, I.S., Dutta, M., Kumar, K., Basandrai, A.K., Kaur, L., Sirari, A., Khan, Z., et al., 2014a. Characterization and evaluation of wild annual *Cicer* species for agromorphological traits and major biotic stresses under Northwestern Indian conditions. Crop Sci. 54, 229–239.

Singh, M., Kumar, K., Bisht, I.S., Dutta, M., Rana, M.K., Rana, J.C., Bansal, K.C., Sarker, A., 2015. Exploitation of wild annual *Cicer* species for widening the gene pool of chickpea cultivars. Plant Breed. 134, 186–192.

Singh, M., Rani, S., Malhotra, N., Katna, G., Sarker, A., 2018. Transgressive segregations for agronomic improvement using interspecific crosses between *C. arietinum* L. × *C. reticulatum* Ladiz. and *C. arietinum* L. × *C. echinospermum* Davis species. PLoS One. https://doi.org/10.1371/journal.pone.0203082.

Singh, N.P., Singh, A., Asthana, A.N., Singh, A., 1999. Studies on inter-specific crossability barriers in chickpea. Indian J. Pulses Res. 12, 13–19.

Singh, S., Gumber, R.K., Joshi, N., Singh, K., 2005. Introgression from wild *Cicer reticulatum* to cultivated chickpea for productivity and disease resistance. Plant Breed. 124, 477–480.

Singh, S., Singh, I., Kapoor, K., Gaur, P.M., Chaturvedi, S.K., Singh, N.P., Sandhu, J.S., 2014b. Chickpea. In: Singh, M., et al. (Eds.), Broadening the Genetic Base of Grain Legumes. Springer, pp. 51–73.

Srivastava, R., Bajaj, D., Malik, A., Singh, M., Parida, S.K., 2016a. Transcriptome landscape of perennial wild *Cicer microphyllum* uncovers functionally relevant molecular tags regulating agronomic traits in chickpea. Sci. Rep. 6, 33616.

Srivastava, R., Singh, M., Bajaj, D., Parida, S.K., 2016b. A high-resolution InDel (insertion–deletion) markers-anchored consensus genetic map identifies major QTLs governing pod number and seed yield in chickpea. Front. Plant Sci. 7, 1362.

Stamigna, C., Crino, P., Saccardo, F., 2000. Wild relatives of chickpea: multiple disease resistance and problems to introgression in the cultigen. J. Genet. Breed. 54, 213–219.

Toker, C., 2005. Preliminary screening and selection for cold tolerance in annual wild *Cicer* species. Genet. Resour. Crop Evol. 52, 1–5.

Toker, C., Canci, H., 2006. Selection for drought and heat resistance in chickpea under terminal drought conditions. In: Kharkwal, M.C. (Ed.), Food Legumes for Nutritional Security and Sustainable Agriculture, 4th International Food Legumes Research Conference, October, 2006. Indian Agricultural Research Institute, New Delhi, pp. 18–22.

Toker, C., Canci, H., 2009. Evaluation of yield criteria for drought and heat resistance in chickpea (*Cicer arietinum* L.). J. Agron. Crop Sci. 195, 47–54.

Toker, C., Canci, H., Yildirim, T., 2007. Evaluation of perennial wild Cicer species for drought resistance. Genet. Resour. Crop Evol. 54, 1781–1786.

Udupa, S.M., Weigand, F., Saxena, M.C., Kahl, G., 1998. Genotyping with RAPD and microsatellite markers resolves pathotype diversity in the Ascochyta blight pathogen of chickpea. Theor. Appl. Genet. 97, 299–307.

Upadhyaya, H.D., 2008. Crop germplasm and wild relatives: a source of novel variation for crop improvement. Korean J. Crop Sci. 53, 12–17.

Upadhyaya, H.D., Smithson, J.B., Haware, M.P., Kumar, J., 1983. Resistance to wilt in chickpea. II. Further evidence for two genes for resistance to race 1. Euphytica 32 (3), 749–755.

Varshney, R.K., Kudapa, H., Roorkiwal, M., Thudi, M., Pandey, M.K., Saxena, R.K., et al., 2012. Advances in genetics and molecular breeding of three legume crops of semi-arid tropics using next-generation sequencing and high-throughput genotyping technologies. J. Biosci. 37, 811–820.

Varshney, R.K., Song, C., Saxena, R.K., Azam, S., Yu, S., Sharpe, A.G., et al., 2013. Draft genome sequence of chickpea (*Cicer arietinum*) provides a resource for trait improvement. Nat. Biotechnol. 31, 240–246.

Wang, J., Gan, Y.T., Clarke, F., McDonald, C.L., 2006. Response of chickpea yield to high temperature stress during reproductive development. Crop Sci. 46, 2171–2178.

Weigand, S., Tahhan, O., 1990. Chickpea insect pests in the Mediterranean zones and new approaches to their management. In: Chickpea in the Nineties: Proceedings of the Second International Workshop on Chickpea Improvement, 4–8 December 1989. ICRISAT, Hyderabad, India, pp. 169–175.

CHAPTER

Origin, distribution, and gene pools

2

Aqeel Hasan Rizvi[1], Ashutosh Sarker[2]

[1]*Food Legume Breeder, South Asia and China Regional Program, International Center for Agricultural Research in the Dry Areas, New Delhi, India;* [2]*Regional Coordinator, South Asia and China Regional Program, International Center for Agricultural Research in the Dry Areas, New Delhi, India*

2.1 Introduction

Chickpea (*Cicer arietinum* L.) is the world's second most cultivated food legume species, grown in over 14.5 million hectare in 2017, from the Mediterranean basin and West Asia to the Indian subcontinent, Eastern Africa, Australia, and North and South America. The present global production of chickpea is more than 14.7 million tons, and the productivity is about 1014.6 kg/ha (FAO, 2019). The major chickpea producing countries are India, Australia, Myanmar, Ethiopia, Turkey, Russian Federation, Pakistan, the United States, Iran, and Mexico (Figs. 2.1 and 2.2), which contributed collectively more than 90% of the global chickpea production.

Chickpea has earned the reputation of being the earliest cultivated grain by man owing to its presence in the Middle Eastern archaeological sites dating 7500–6800 BC (Zohary and Hopf, 2000). Vernacular names associated with chickpea are *Hamas*, *Hommos*, *Humz*, *Nakhi*, and *Melanch* (Arabian countries), *Keker* (Netherlands), *Kichererbse* (Germany and Belgium), *Ceseror* and *Cicerolle* (France), *Ceci* (Vatican City and Switzerland), *Simbra* (Ethiopia), *Lablabi* (Turkey), *Garbanzo* or *Garbanzobean* (Spain), *Gravanço* (Portugal), and *Ovetichie* (Russia).

India's lingual, regional, and cultural variability has contributed to the various appellations that chickpea is known by *Chana* or *Gram* or *Bengal gram* or *Chani* (Haryana, Rajasthan, Uttarakhand, Uttar Pradesh, Madhya Pradesh, Chhattisgarh, Bihar, Jharkhand), *Chhole* (Punjab, Jammu and Kashmir, and Delhi), *Chola* (West Bengal), *Harbhara* (Maharashtra), *Boot* (Orissa), *Sanagulu* (Andhra Pradesh), *Kadale* (Karnataka), *Kadalai* (Tamil Nadu), and *Kadala* (Kerala), indicating its widespread cultivation and knowledge of utilization (Singh et al., 2014).

According to Cubero (1975), seed shape, size, and color have attributed to two distinct cultivated forms. These are desi-type and the kabuli-type chickpea. The desi type has pink flowers, angular, brown, small seeds with high percentage of fiber, grown mostly in South Asia and Africa, whereas, kabuli type has white flowers, owl head–shaped, beige, large seeds with low percentage of fiber, grown in Mediterranean countries. Kabuli types are grown in about two-thirds of the chickpea-growing

Chickpea: Crop Wild Relatives for Enhancing Genetic Gains. https://doi.org/10.1016/B978-0-12-818299-4.00002-6
Copyright © 2020 Elsevier Inc. All rights reserved.

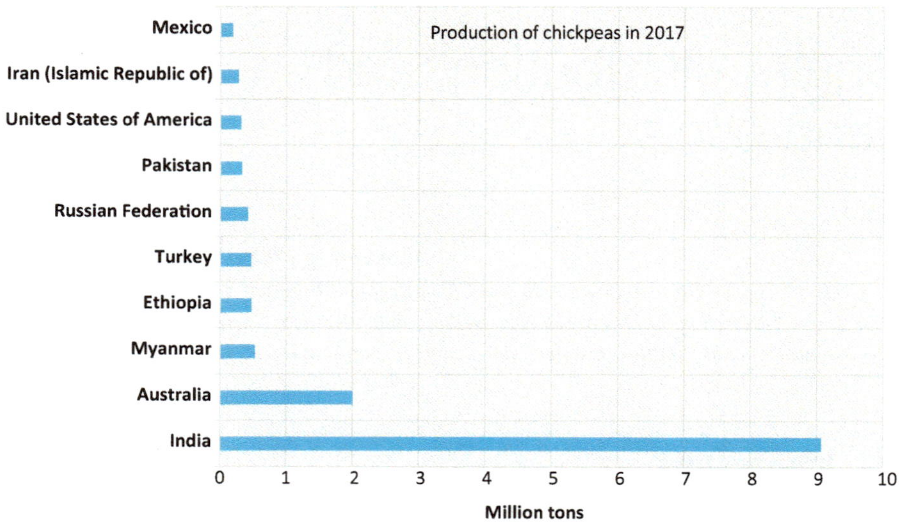

FIGURE 2.1

Chickpea production of top 10 countries during 2017.

Source: FAOSTAT Assessed May 29, 2019.

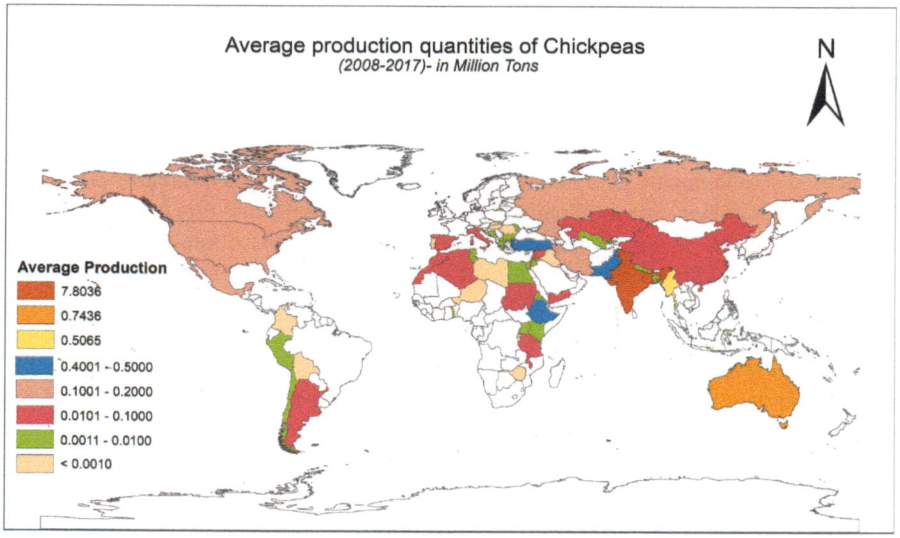

FIGURE 2.2

Chickpea production of the world during 2008–17.

Source: FAOSTAT Assessed May 29, 2019.

countries, but desi type predominates in chickpea production and accounts for about 85% share, while kabuli accounts for about 15% of the world chickpea production (Sharma et al., 2013). It is a member of Fabaceae family of the plant kingdom and is self-pollinated true-diploid ($2n = 2x = 16$) with genome size of 738 Mbp (Varshney et al., 2013). Singh (1997) has comprehensively listed the various traits of chickpea that have rendered it to become the most important food legume plant in sustainable agriculture production system. His list includes low production cost, wider adaptation, ability to fix atmospheric nitrogen, fit in various crop rotations, and deep tap root system which helps in opening up the soil to the deeper strata, ensuring better texture and aeration of soil for the next season crop.

Chickpea plant in a symbiotic association with *Rhizobium* can fix atmospheric nitrogen up to 140 kg/ha, thus meeting up to 80% of its nitrogen requirement (Saraf et al., 1998). This plays an important role in crop diversification, sustainable farming systems, and enhancing the soil quality for subsequent cereal crop cultivation by adding organic matter for the maintenance of soil health and ecosystem. The plant has high nutritional value for human consumption. It is a rich source of protein (20%–22%), fiber, minerals (calcium, potassium, phosphorus, magnesium, iron, and zinc), unsaturated fatty acids, and β-carotene (Jukanti et al., 2012). Its β-carotene content is higher than genetically engineered "Golden rice" (Abbo et al., 2005). The high fiber content helps to regulate cholesterol and blood sugar after meals, thus making it a healthy food for diabetic patients (McIntosh and Miller, 2001; Pittaway et al., 2006). Chickpea does not contain any antinutritional factors (Mallikarjuna et al., 2011). Every plant part of chickpea is beneficial for consumption by and large. The acid exudates from the leaves have been medicinally used in India for aphrodisiac, bronchitis, cholera, constipation, diarrhea, dysentery, snakebite, sunstroke, and warts. Sprouted seeds are eaten as a vegetable or salad. Young leaves, stems, and green pods are eaten like vegetables. The dried seeds may be used in soups and ground into flour. Grain husks, stems, and leaves may be used in livestock feed. In the United States and Europe, chickpea is marketed dried, canned, or in various vegetable mixtures. Mashed chickpea mixed with oils and spices (hummus) is a popular hors d'oeuvre in the Mediterranean Middle East (Singh et al., 2014).

2.2 Origin and distribution

Chickpea has been a subject of interest to research studies, and unearthing of seeds during excavations has added to plethora of assumptions and conclusions drawn by these workers. The diversity in terms of the centers of its origin, closely related wild species, dates of its earliest domestication, and use pattern of different plant parts has remained a subject of comprehensive discussions and advocacy of differing opinions by eminent scientists.

Chickpea was domesticated as part of the evolution of agriculture in the Fertile Crescent 12,000–10,000 years ago (Zohary and Hopf, 1973; Bar-Yosef, 1998; Abbo et al., 2003). This region extends from western Iran through Iraq, Jordan, and Israel to southeast Turkey. These researchers have reported that *C. bijugum*,

C. echinospermum, and *C. reticulatum* are most closely related wild species and would have attracted early food collectors. Furthermore, observations have revealed that the seed size of *C. bijugum* and *C. reticulatum* is quite reasonable, and unlike other species, they do not shatter their seeds immediately upon ripening.

Ladizinsky (1975) opined that the region of Turkey and the ancient city of Jericho domesticated chickpea around 7500 BC. It is therefore, considered as the center of origin of this legume. However, van der Maesen (1987) believed that the southeastern part of Turkey adjoining Syria was the possible center based on the presence of the closely related annual species, i.e., *C. reticulatum* Ladizinsky and *C. echinospermum* P.H. Davis. Ladizinsky (1975) was the first to describe *C. reticulatum* Ladizinsky; it could also be classified as a wild variety or subspecies of *C. arietinum*. He further endorsed that *C. reticulatum* can be considered as a progenitor, or may have had a common ancestor with chickpea. Since early botanists did not recognize these close relatives, various diffuse origins have been postulated for chickpea crop.

Earliest records for chickpea domestication in the Old World at Tell el-Kerkh (10th millennium BC) in Syria, Cayönü (7250–6750 BC) and Hacilar (c.6700 BC) in Turkey, and Jericho (8350–7370 BC) in the West Bank have been reported by Sharma et al. (2013). Tell el-Kerkh lies at a significant distance from the native lands of *C. reticulatum*. This is also the site where both *C. arietinum* and its immediate progenitor *C. reticulatum* were clearly identified. Thus, Tanno and Willcox (2006) concluded that domestication of chickpea in the southeastern part of Turkey began much before.

Excavations at Hac near Burdur have unearthed seeds dating back to 5450 BC in Turkey (Helbaek, 1970). Chickpea converged into two directions from Turkey, i.e., the western province (grown in spring and summer) and eastern and southern part of region (grown as cool dry season). From 3300 BC onward, cultivation of chickpea has been well documented in Egypt and the Middle East (van der Maesen, 1972).

According to van der Maesen (1972), Redden and Berger (2007), the primary center of diversity is in the Fertile Crescent, where the crop was domesticated. Secondary centers of diversity emerged with the geographic spread of chickpea, some of them being as old as 2000 years spread across Mediterranean Europe, the Indian subcontinent, and Northeast Africa, and more recently in Mexico and Chile.

Vavilov (1951) has advocated four primary centers of diversity apparently from the Mesopotamian records and archaeological evidences, viz., Mediterranean, Central Asia, Near East, and India; and Ethiopia as a secondary center of origin. He observed that, like other grain legumes, large-seeded cultivars flourished around the Mediterranean basin, whereas small-seeded cultivars predominated eastward.

From the present day distribution of *C. reticulatum* Ladiz., the wild progenitor (Ladizinsky and Adler, 1976), it is evident that chickpea is likely to have been domesticated in Southeast Anatolia. It spread westward to modern Greece in the late Neolithic era. Chickpea had been disseminated widely by the Bronze Age to Crete in west, upper Egypt in the south, eastward through contemporary Iraq to the Indian subcontinent, where remains have been found in Harappan settlements in Pakistan

(Vishnu-Mittre and Savithri, 1982) and a variety of locations in India (Uttar Pradesh and Maharashtra). By consolidating its distribution in South and West Asia, it seems to have appeared in Ethiopia for the first time by the Iron Age (Table 2.1).

The human migration toward the west and south has led to the spread of chickpea via the Silk Route (Singh et al., 1997). Spanish and Portuguese brought chickpea to the New World in the 16th century, while kabuli types were brought to India through Central Asia in the 16th century (Redden and Berger, 2007). During the Bronze Age, chickpea spread to Western Europe especially Italy and Greece. In the New World, it is important in Mexico, Argentina, Chile, Peru, the United States, and Australia. Wild species have been found to be most abundant in Turkey, Iran, Afghanistan, and Central Asia (Duke, 1981).

Currently, *Cicer* species occur from sea level to over 5000 m near glaciers in the Himalayas. Chickpea cultivation has also spread to near-equatorial Uganda and Ethiopia, and northern latitudes within Canada and Russia (Fig. 2.3) (Croser et al., 2003).

2.3 Taxonomy

Chickpea is the only domesticated species under genus *Cicer*, family Fabaceae or Leguminosae, subfamily Papilionoideae, and tribe *Cicereae* Alef. Previously, genus *Cicer* was classified in tribe Vicieae Alef.; later it was reported to belong to its own monogeneric tribe *Cicereae* Alef (Kupicha, 1981). The genus includes 9 annuals (8 wild and 1 cultivated) and 35 perennials (van der Maesen, 1972; Muehlbauer, 1993) (Table 2.2). The genus is further divided into two subgenera, Pseudononis and Viciastrum. Characteristic features of the Pseudononis are small flowers (normally 5–10 mm), subregular calyx, with hardly gibbous base and with sublinear nearly equal teeth. The subgenus, Viciastrum includes perennial species and is characterized by medium to large flowers (medium 12–15 mm or large 17–27 mm), calyx strongly gibbous at the base, with unequal teeth. These subgenera are further divided into four sections, viz., Monocicer, Chamaecicer, Polycicer, and Acanthocicer (van der Maesen, 1987). The section Monocicer is the most important section for chickpea improvement and includes eight annual species, viz., *C. arietinum*, *C. reticulatum*, *C. echinospermum*, *C. judaicum*, *C. bijugum*, *C. pinnatifidum*, *C. cuneatum*, and *C. yamashitae*, whereas second section Chamaecicer includes one annual species *C. chorassanicum* and one perennial species *C. incisum* (Kazan and Muehlbauer, 1991; Muehlbauer et al., 1994). Sections Polycicer and Acanthocicer of subgenus Viciastrum consist of 23 and 9 perennial species, respectively (Sharma et al., 2013). The distribution of different *Cicer* species is given in Table 2.2.

2.4 Gene pools

The concept of gene pools and their uses in crop improvement were proposed by Harlan and de Wet (1971). The *Cicer* genus is bestowed with rich germplasm in

Table 2.1 Prehistoric sites containing chickpea or unidentified *Cicer* species listed in chronological order.

Remains	Site	Country	Period	Age	References
C. arietinum	Tell el-Kerkh	Syria	Early Neolithic (9350–7900 uncal BP)	7260 BC	Tanno and Willcox (2006)
Cicer sp.	Nevali Cori	Turkey	Early Neolithic (9350–7900 uncal BP)	7250 BC	Pasternak (1998)
Cicer sp.	Çayönü	Turkey	Early Neolithic (9350–7900 uncal BP)	7150 BC	van Zeist and de Roller (1991, 1992)
Cicer sp.	Nevali Cori	Turkey	Early Neolithic (9350–7900 uncal BP)	7150 BC	Pasternak (1998)
Cicer sp.	Asikli Höyük	Turkey	Early Neolithic (9350–7900 uncal BP)	6700 BC	van Zeist and de Roller (1995)
Cicer sp.	Wadi el-Jilat	Jordan	Early Neolithic (9350–7900 uncal BP)	6600 BC	Garrad and Colledge (1996)
C. arietinum	Jericho	Palestine	Early Neolithic (9350–7900 uncal BP)	6500 BC	Hopf (1983)
C. arietinum	Ain Ghazal	Jordan	Early Neolithic (9350–7900 uncal BP)	6350 BC	Rollefson et al. (1985)
C. arietinum	Ghoraife	Syria	Early Neolithic (9350–7900 uncal BP)	6275 BC	van Zeist and Bakker-Heeres (1982)
Cicer sp.	Ramad	Syria	Early Neolithic (9350–7900 uncal BP)	6100 BC	van Zeist and Bakker-Heeres (1982)
C. arietinum	Abu Hureyra	Syria	Early Neolithic (9350–7900 uncal BP)	6020 BC	de Moulins (2000)

Table 2.1 Prehistoric sites containing chickpea or unidentified *Cicer* species listed in chronological order. *continued*

Remains	Site	Country	Period	Age	References
C. arietinum	Çatalhöyük	Turkey	Early Neolithic (9350–7900 uncal BP)	6000 BC	Fairbairn et al. (2002)
C. arietinum	Otzaki	Greece	Late Neolithic (5450–3500 BC)	5500 BC	Kroll (1981)
C. arietinum	Hacilar	Turkey	Late Neolithic (5450–3500 BC)	5450 BC	Helbaek (1970)
C. arietinum	Dimini	Greece	Late Neolithic (5450–3500 BC)	3500 BC	Kroll (1979)
C. arietinum	Tel Arad	Israel	Bronze Age (2800–1300 BC)	2800 BC	Hopf (1978)
C. arietinum	Bab edh-Dhra	Jordan	Bronze Age (2800–1300 BC)	2800 BC	McCreery (1979)
C. arietinum	Ur	Iraq	Bronze Age (2800–1300 BC)	2500 BC	Ellison et al. (1978)
C. arietinum	Lachish	Israel	Bronze Age (2800–1300 BC)	2300 BC	van der Maesen (1987)
C. arietinum	Harappa	Pakistan	Bronze Age (2800–1300 BC)	2000 BC	Fuller (2000)
C. arietinum	Mohenjo Daro	Pakistan	Bronze Age (2800–1300 BC)	2000 BC	Fuller (2000)
C. arietinum	Atranjikhera	India	Bronze Age (2800–1300 BC)	2000 BC	Chowdury et al. (1971)
C. arietinum	Kalibangan	India	Bronze Age (2800–1300 BC)	2000 BC	Vishnu-Mittre and Savithri (1982)
C. arietinum	Tell Bazmosian	Iraq	Bronze Age (2800–1300 BC)	1900 BC	Helbaek (1965)
C. arietinum	Lal Quila	India	Bronze Age (2800–1300 BC)	1750 BC	Saraswat (1992)

Continued

Table 2.1 Prehistoric sites containing chickpea or unidentified *Cicer* species listed in chronological order. *continued*

Remains	Site	Country	Period	Age	References
C. arietinum	Sanghol	India	Bronze Age (2800–1300 BC)	1700 BC	Saraswat (1992)
C. arietinum	Inamgaon	India	Bronze Age (2800–1300 BC)	1650 BC	Saraswat (1992)
C. arietinum	Nevasa	India	Bronze Age (2800–1300 BC)	1650 BC	Saraswat (1992)
C. arietinum	Hulas	India	Bronze Age (2800–1300 BC)	1600 BC	Vishnu-Mittre et al. (1985)
C. arietinum	Senuwar	India	Bronze Age (2800–1300 BC)	1500 BC	Saraswat (1992)
C. arietinum	Daimabad	India	Bronze Age (2800–1300 BC)	1500 BC	Saraswat (1992)
C. arietinum	Crete	Crete	Bronze Age (2800–1300 BC)		van der Maesen (1972)
C. arietinum	Deir el Medineh	Egypt	Bronze Age (2800–1300 BC)	1400 BC	Darby et al. (1977)
C. arietinum	Jorwe	India	Iron Age (1300–500 BC)	1150 BC	Kajale (1991)
C. arietinum	Narhan	India	Iron Age (1300–500 BC)	1150 BC	Saraswat (1992)
C. arietinum	Tell Megiddo	Israel	Iron Age (1300–500 BC)	1000 BC	van der Maesen (1987)
C. arietinum	Sringaverapura	India	Iron Age (1300–500 BC)	825 BC	Saraswat (1992)
C. arietinum	Nimrud	Iraq	Iron Age (1300–500 BC)	700 BC	Helbaek (1966)
C. arietinum	Salamis	Cyprus	Iron Age (1300–500 BC)	550 BC	Renfrew (1967)
C. arietinum	Lalibela cave	Ethiopia	Iron Age (1300–500 BC)	290 BC	Dombrowski (1970)

Table 2.1 Prehistoric sites containing chickpea or unidentified *Cicer* species listed in chronological order. *continued*

Remains	Site	Country	Period	Age	References
C. arietinum	Nevasa	India	Recent (300 BC–200 AD)	200 BC	van der Maesen (1987)
C. arietinum	Kolhapur	India	Recent (300 BC–200 AD)	50 BC	Kajale (1991)
C. arietinum	Tripuri	India	Recent (300 BC–200 AD)	0 AD	Kajale (1991)
C. arietinum	Bokardan	India	Recent (300 BC–200 AD)	0 AD	Kajale (1991)
C. arietinum	Ter	India	Recent (300 BC–200 AD)	25 AD	Kajale (1991)

Note: Most Neolithic dates are presented in the literature as uncalibrated years before present, so the conversion to BC is approximate.
Table adapted from Chapter History and Origin by Redden, R.J., Berger, J.D., 2007. History and origin of chickpea. In: Yadav, S.S., Redden, R.J., Chen, W., Sharma, B. (Eds.), Chickpea Breeding and Management, first ed. CAB International, Wallingford, UK, pp. 1–13 in Chickpea).

the form of wild species. It is classified into three gene pools, primary, secondary, and tertiary, based on their crossability with cultigen. The wild species has been categorized based on their usefulness for improving the cultigen. The classification proposed by Harlan and de Wet (1971) was further modified by placing the species *C. reticulatum* in the secondary gene pool (van der Maesen et al., 2007). The recent proposed classification is almost like the previous gene pool. The primary gene pool consists of cultivated species and landraces. The secondary gene pool consists of the progenitor species, *C. reticulatum* and *C. echinospermum*, the species that are crossable with *C. arietinum* but with reduced fertility of the resulting hybrids and progenies; nevertheless, both are cross-compatible with the cultigen and do not need in vitro interventions to produce hybrids. The tertiary gene pool consists of all the annual and perennial *Cicer* species that are not crossable with cultivated species. All the perennial *Cicer* species are in the tertiary gene pool, as none of the species of this group is known to cross readily with the cultivated species and produce mature seeds (Mallikarjuna et al., 2011) (Table 2.3).

2.4.1 Primary gene pool

The primary gene pool comprises the biological species that includes the cultigen itself and any wild forms that breeding tests show to be conspecific with it. This pool consists of cultivated species and vast collection of landraces. Species within the primary gene pool are readily intercrossed and produce fertile progenies (Upadhyaya et al., 2008). Therefore, conventional breeding is used for gene flow between the

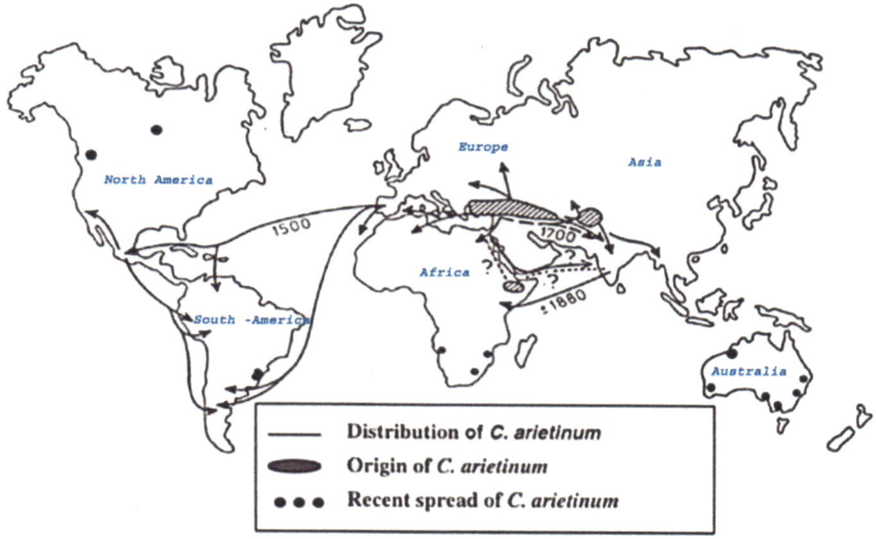

FIGURE 2.3

Areas of origin of cultivated chickpea and subsequent spread around the world.

Modified from van der Maesen, L.J.G., 1972. Cicer *L., a Monograph of the Genus, With Special Reference to the Chickpea (*Cicer arietinum *L.), Its Ecology and Distribution. Mendelingen Landbouwhogeschool Wageningen, Wageningen, The Netherlands, pp. 1–341; Figure adapted from Croser, J.S., Ahmad, C.F., Clarke, H.J., Siddique, K.H.M., 2003. Utilisation of wild Cicer in chickpea improvement progress, constraints, and prospects. Aust. J. Agric. Res. 54, 429–444.*

species of the primary gene pool. India has released 126 cultivars between 1967 and 2003. Pedigree analysis of 86 cultivars developed from crosses has revealed that although 95 progenitors were involved, just 10 of these contributed 35% of the genetic base (Shivkumar et al., 2004). This suggests that many cultivars share a narrow genetic base. It is predicted that a global composite collection would enhance the utilization of germplasm in improvement of chickpea. Nevertheless, due to narrow ancestry and numerous bottlenecks, chickpea has limited genetic variation in the primary gene pool. Despite constraints, extensive international breeding efforts have led to the development of over 300 improved varieties (Gowda and Gaur, 2004).

2.4.2 Secondary gene pool

The secondary gene pool comprises all those biological species among which gene flow is possible via interspecific hybrids. Species distant from the cultigen are placed in the secondary gene pool (Ladizinsky and Adler, 1976). Therefore, gene flow between the species that belong to the secondary gene pool is more difficult, and the progenies produced have substantial degrees of sterility (Harlan and de Wet, 1971).

Table 2.2 Distribution of different annual and perennial *Cicer* species.

S. No	Species	Distribution
Annual cultivated *Cicer*		
1	*Cicer arietinum* (Chickpea)	Mediterranean region of Burma, Ethiopia, Mexico, Chile
Annual wild *Cicer* species		
1	*C. reticulatum*	E Turkey
2	*C. echinospermum*	Turkey, E Anatolia, N Iraq
3	*C. judaicum*	Palestine, Lebanon
4	*C. bijugum*	SE Turkey, N Syria, M Iraq
5	*C. pinnatifidum*	Cyprus, N Iraq, Syria, Turkey, USSR
6	*C. chorassanicum*	N and C Afghanistan, N and NE Iraq
7	*C. cuneatum*	Ethiopia, SE Egypt, NE Sudan, Saudi Arabia
8	*C. yamashitae*	Afghanistan
Perennial wild *Cicer* species		
1	*C. acanthophyllum*	Afghanistan, Pakistan, USSR
2	*C. anatolicum*	Turkey, Iran, Iraq, Armenia
3	*C. atlanticum*	Morocco
4	*C. balcaricum*	Caucasus (USSR)
5	*C. baldshuanicum*	USSR
6	*C. canariense*	Canary islands, Tenerife, La Palma
7	*C. fedtschenkoi*	USSR, N and NE Afghanistan
8	*C. flexuosum*	USSR
9	*C. floribundum*	Turkey
10	*C. graecum*	Greece
11	*C. grande*	USSR
12	*C. heterophyllum*	Turkey
13	*C. incanum*	USSR
14	*C. incisum*	Greece, Turkey, Tran, Lebanon, USSR
15	*C. isauricum*	Turkey
16	*C. kermanense*	SE Iran
17	*C. korshinskyi*	USSR
18	*C. laetum*	USSR
19	*C. luteum*	USSR-Tadzhik SSR
20	*C. maracanthum*	Afghanistan, India, Pakistan, USSR
12	*C. microphyllum*	E Afghanistan, Tibet, India, Pakistan, USSR
22	*C. mogolatvicum*	USSR
23	*C. montbretii*	Albania, Bulgaria, Turkey
24	*C. multijugum*	Afghanistan

Continued

Table 2.2 Distribution of different annual and perennial *Cicer* species. *continued*

S. No	Species	Distribution
25	*C. nuristanicum*	Afghanistan, India, Pakistan
26	*C. oxyodon*	Iran, Afghanistan, N Iraq
27	*C. paucijugum*	USSR
28	*C. pungens*	Afghanistan, USSR
29	*C. rassuloviae*	USSR
30	*C. rechingeri*	Afghanistan
31	*C. songaricum*	USSR
32	*C. spyroceras*	Iran
33	*C. stapfianum*	Iran
34	*C. subaphyllum*	Iran
35	*C. tragacanthoides*	Iran, USSR

Table 2.3 Gene pools of different *Cicer* species.

Primary	Secondary	Tertiary
Landraces and cultivars of *C. arietinum*	*C. reticulatum*, *C. echinospermum*	*C. bijugum*, *C. judaicum*, *C. pinnatifidum*, *C. chorassanicum*, *C. yamashitae*, *C. cuneatum*, *C. atlanticum*, *C. incisum*, *C. floribundum*, *C. graecum*, *C. heterophyllum*, *C. uludereensis*, *C. isauricum*, *C. montbretii*, *C. acanthophyllum*, *C. anatolicum*, *C. balcaricum*, *C. baldshuanicum*, *C. flexuosum*, *C. grande*, *C. incanum*, *C. korshinskyi*, *C. laetum*, *C. luteum*, *C. macracanthum*, *C. microphyllum*, *C. multijugum*, *C. nuristanicum*, *C. paucijugum*, *C. pungens*, *C. rassuloviae*, *C. rechingeri*, *C. songaricum*, *C. stapfianum*, *C. subaphyllum*, *C. tragacanthoides*, *C. kermanense*, *C. mogoltavicum*, *C. oxyodon*, *C. spiroceras*, *C. canariense*

Chromosomal rearrangement is the underlying mechanism for sterile progeny. According to van der Maesen et al. (2007), the secondary gene pool consists of two wild species *C. reticulatum* and *C. echinospermum*. The species *C. reticulatum* often produces fertile hybrids and progenies when crossed with *C. arietinum*. Studies based on morphological and cytological similarities as well as crossability (Ladizinsky and Adler, 1976) and molecular analysis suggest that the species *C. reticulatum* is a wild progenitor of *C. arietinum*. Utilization of useful traits from various accessions of

C. reticulatum into cultivated species has been reported and they represent a potential source for broadening the genetic base (ascochyta blight resistance—Collard et al., 2003; *Helicoverpa armigera* resistance—Mallikarjuna et al., 2007a; nematode resistance—Malhotra et al., 2002; high yield—Singh and Ocampo, 1997). Irula et al. (2002) from ICARDA reported natural hybrids between chickpea and *C. reticulatum*; and between *C. echinospermum* and *C. reticulatum*. This strongly suggests compatibility among these three species. Ladizinsky and Adler (1976) reported that *C. reticulatum* and *C. echinospermum* are distinct from each other by major reciprocal translocation and their hybrid was completely sterile. However, Singh and Ocampo (1997) conveyed prospect to obtain completely fertile selections with some accessions of *C. echinospermum*. Foundation of placing both *C. reticulatum* and *C. echinospermum* in the secondary gene pool is their crossability with the cultigen.

2.4.3 Tertiary gene pool

The tertiary gene pool embraces those taxa among which hybridization is possible, but the hybrids themselves are infertile, inviable, or anomalous and do not permit gene flow by normal introgressive processes. This gene pool consists of remaining annual and perennial species. The annual *Cicer* species belonging to the tertiary gene pool have great importance. These species have important traits/genes for improvement of chickpea, such as drought resistance *C. pinnatifidum* (Bhattarai and Fettig, 2005); ascochyta blight resistance in *C. judaicum*, *C. bijugum*, and *C. pinnatifidum* (Pande et al., 2006); and botrytis gray mold resistance in *C. judaicum* (Pande et al., 2006). Hybridization between the species related to tertiary gene pool is not possible or hybrids are completely sterile. Numerous reports revealed that wild species in the tertiary gene pool are distantly related to cultivated chickpea; however, techniques are being improvised for making wide crosses in chickpea using tertiary gene pool species (Mallikarjuna, 1999; Mallikarjuna et al., 2007b; Mallikarjuna and Jadhav, 2008). They require specialized techniques, like hormone applications, embryo rescue, and *in-vitro* attempts to root hybrid shoot/plants are used for transfer of gene into the cultivated background. Therefore, all the perennial species are present in tertiary gene pool as none of them are known to be able to cross readily with the cultivated species and produce mature seeds. Although hybrid shoots/plants can be obtained due to special efforts, they have proven to be fragile and have not withstood transfer to glasshouse/field and hence are not yet available for chickpea improvement.

2.5 Barriers in interspecific hybridization and crossability groups

Ahmad et al. (1988) have examined the barrier to interspecific hybridization within and between the crossability groups of chickpea. Their studies revealed the normal

Table 2.4 Crossability groups of chickpea based on hybridization and fertility of hybrids (Muehlbauer et al., 1994).

Crossability group	Species
Group I	*C. arietinum*, *C. reticulatum*, *C. echinospermum*
Group II	*C. bijugum*, *C. pinnatifidum*, *C. judaicum*, *C. yamashitae*
Group III	*C. chorassanicum*
Group IV	*C. cuneatum*

pollination and fertilization in many interspecific cross combinations. The genetic disharmony between the maternal and paternal genomes has been attributed for sterility. Based on crossability and fertility of hybrids in interspecific crosses, annual members of *Cicer* have been classified into four crossability groups (Ladizinsky and Adler, 1976). The first group constitutes of three species namely *C. arietinum*, *C. reticulatum*, and *C. echinospermum*. The *C. reticulatum* has been proposed as the putative progenitor of the cultivated *C. arietinum* (Ladizinsky, 1975; Kabir and Singh, 1988) and recognized as a subspecies (Moreno and Cubero, 1978). The second crossability group includes *C. bijugum*, *C. pinnatifidum*, *C. yamashitae*, and *C. judaicum*. The *C. chorassanicum* and *C. cuneatum*, which can be crossed neither with any other annual species nor with each other, form the third and fourth groups, respectively (Muehlbauer et al., 1994) (Table 2.4).

References

Abbo, S., Molina, C., Jungmann, R., Grusak, A., Berkovitch, Z., Reifen, R., et al., 2005. Quantitative trait loci governing carotenoid concentration and weight in seeds of chickpea (*Cicer arietinum* L.). Theor. Appl. Genet. 111, 185–195.

Abbo, S., Shtienberg, D., Lichtenzveig, J., Lev-Yadun, S., Gopher, A., 2003. The chickpea, summer cropping, and a new model for pulse domestication in the ancient near east. Q. Rev. Biol. 78, 435–448.

Ahmad, F., Slinkard, A.E., Scoles, G.J., 1988. Investigations into the barrier(s) to interspecific hybridization between *Cicer arietinum* L. and eight other annual *Cicer* species. Plant Breed. 100, 193–198.

Bar-Yosef, O., 1998. The Natufian culture in the Levant, threshold to the origins of agriculture. Evol. Anthropol. 16, 159–177.

Bhattarai, T., Fettig, S., 2005. Isolation and characterization of a dehydrin gene from *Cicer pinnatifidum*, a drought-resistant wild relative of chickpea. Plant Physiol. 123, 452–458.

Chowdury, K.A., Saraswar, K.S., Hasam, S.M., Gaur, R.G., 1971. 4000–3500 year old barley, rice and pulses from Atranjikhera. Sci. Cult. 37, 531–533.

Collard, B.C.Y., Pang, E.C.K., Ades, P.K., Taylor, P.W.J., 2003. Preliminary investigation of QTLs associated with seedling resistance to ascochyta blight from *Cicer echinospermum*, a wild relative of chickpea. Theor. Appl. Genet. 107, 719–729.

Croser, J.S., Clarke, H.J., Siddique, K.H.M., Khan, T.N., 2003. Low-temperature stress: implications for chickpea (*Cicer arietinum* L) improvement. Crit. Rev. Plant Sci. 22 (2), 185–219.
Cubero, J.I., 1975. The research on chickpea (*Cicer arietinum*) in Spain. In: Proceedings of the International Workshop on Grain Legumes, 13–16 January 1975. International Crops Research Institute for the Semi-arid Tropics, Hyderabad, India, pp. 17–122.
Darby, W.J., Ghaloungui, P., Grivetti, L., 1977. Food: The Gift of Osiris. Academic Press, London.
de Moulins, D., 2000. Abu hureyra 2: plant remains from the neolithic. In: Moore, A.M.T., Hillman, G.C., Legge, A.J. (Eds.), Village on the Euphrates: From Foraging to Farming at Abu Hureyra. Oxford University Press, Oxford, pp. 399–416.
Dombrowski, J., 1970. Preliminary report on excavation in Lalibela and Natchabiet caves, Begemeder. Ann. Ethiopie 8, 21–29.
Duke, J.A., 1981. Handbook of Legumes of World Economic Importance. Plenum, New York, USA, pp. 52–57.
Ellison, R., Renfrew, J., Brothwell, D., Seelwy, N., 1978. Some food offerings from Ur, excavated by Sir Leonard Wooley, and previously unpublished. J. Archeol. Sci. 5, 167–177.
Fairbairn, A., Asouti, E., Near, J., Martinoli, D., 2002. Macro-botanical evidence for plant use at Neolithic Çatalhöyük, south-central Anatolia, Turkey. Veg. Hist. Archaeobotany 11, 41–54.
FAO, 2019. Food and Agriculture Organization of the United Nations. http://www.fao.org/faostat/en/#data/QC.
Fuller, D.Q., 2000. Fifty years of archeobotanical studies in India: laying a solid foundation. In: Settar, S., Korisettar, R. (Eds.), Indian Archeology in Retrospect. Archeology and Interactive Disciplines. Manohar, New Delhi, pp. 247–363.
Garrad, A., Colledge, S.L.M., 1996. The emergence of crop cultivation and caprine herding in the "Marginal Zone" of the Southern Levant. In: Harris, D.R. (Ed.), Origins and Spread of Agriculture and Pastoralism in Eurasia. Smithsonian Institution Press, Washington, DC, USA, pp. 204–226.
Gowda, C.L.L., Gaur, P.M., 2004. Global scenario of chickpea research – present status and future thrusts. In: Ali, M., Singh, B.B., Shivkumar, S., Dhar, V. (Eds.), Pulses in New Perspective. Indian Society of Pulses Research and Development, Indian Institute of Pulses Research, Kanpur, Uttar Pradesh, India, pp. 1–22.
Harlan, J.R., de Wet, M.J., 1971. Towards a rational classification of crop plant. Taxon 20, 509–517.
Helbaek, H., 1965. Isin-Larson and Horian food remains at Tell Bazmosian in the Dokan valley. Sumer 19, 27–35.
Helbaek, H., 1966. The plant remains. In: Mallowan, M.E.L. (Ed.), Nimrud and Its Remains. Collins, London, pp. 613–620.
Helbaek, H., 1970. The plant husbandry at Hacilar. In: Mellaart, J. (Ed.), Excavation at Hacilar. Edinburgh University Press, Edinburgh, pp. 189–244.
Hopf, M., 1978. Plant remains, strata V-I. In: Amiran, R. (Ed.), Early Arad: The Chalcothic Settlement and the Early Bronze Age City. Israel Exploration Society, Jerusalem, pp. 64–82.
Hopf, M., 1983. The pottery phases of the Tell and other finds. In: Kenyon, K.M., Holland, T.A. (Eds.), Excavation at Jericho. British School of Archeology in Jerusalem, London, pp. 576–621.
Iruela, M., Rubio, J., Cubero, J.I., Gil, J., Millan, T., 2002. Phylogenetic analysis in the genus *Cicer* and cultivated chickpea using RAPD and ISSR markers. Theor. Appl. Genet. 104, 643–651.

Jukanti, A.K., Gaur, P.M., Gowda, C.L., Chibbar, R.N., 2012. Nutritional quality and health benefits of chickpea (*Cicer arietinum* L.): a review. Br. J. Nutr. 108, 11–26.

Kabir, G., Singh, R.M., 1988. Seed protein electrophoresis in six species and two F_1s of *Cicer*. Proc. Indian Acad. Sci. Plant Sci. 98, 183–189.

Kajale, M.D., 1991. Current status of Indian palaeoethnobotany: introduced and indigenous food plants with a discussion of the historical and evolutionary development of Indian agriculture and agricultural systems in general. In: Renfrew, J.M. (Ed.), New Light on Early Farming – Recent Developments in Palaeoethnobotany. Edinburgh University Press, Edinburgh, pp. 155–189.

Kazan, K., Muehlbauer, F.J., 1991. Allozyme variation and phylogeny in annual species of *Cicer* (*Leguminosae*). Plant Syst. Evol. 175, 11–21.

Kroll, H., 1979. Kulturpflanzen aus Dimini. Archaeo-Physika 8, 173–189.

Kroll, H., 1981. Thessalische kulturpflanzen. Z. Archaol. 15, 97–103.

Kupicha, F.K., 1981. Vicieae. In: Polhill, R.M., Raven, P.H. (Eds.), Advances in Legume System. Royal Botanic Gardens, Kew, pp. 377–381.

Ladizinsky, G., Adler, A., 1976. The origin of chickpea (*Cicer arietinum* L.). Euphytica 25, 211–217.

Ladizinsky, G., 1975. A new *Cicer* from Turkey. Notes R. Bot. Gard. Edinb. 34, 201–202.

Malhotra, R.S., Singh, K.B., Di Vito, M., Greco, N., Saxena, M.C., 2002. Registration of ILC 10765 and ILC 10766 chickpea germplasm lines resistant to cyst nematode. Crop Sci. 42, 1756.

Mallikarjuna, N., 1999. Ovule and embryo culture to obtain hybrids from interspecific incompatible pollinations in chickpea. Euphytica 110, 1–6.

Mallikarjuna, N., Coyne, C., Cho, S., Rynearson, S., Rajesh, P.N., Jadhav, D.R., Muehlbaur, F., 2011. Cicer. In: Kole, C. (Ed.), Wild Crop Relatives: Genomic and Breeding Resources, Legume Crops and Forages, first ed. Springer-Verlag, Berlin Heidelberg, pp. 63–82.

Mallikarjuna, N., Jadhav, D., Vakiti, N., Amudhavalli, C., Chandra, S., Hoisington, D., 2007b. Progress in the interspecific hybridization between *Cicer arietinum* and the wild species *C. bijugum*. SAT eJ. 5 (1). ejournal.icrisat.org.

Mallikarjuna, N., Jadhav, D.R., 2008. Techniques to produce hybrid between *Cicer arietinum* L. × *C. pinnatifidum* Jaub. Indian J. Genet. 68, 1–8.

Mallikarjuna, N., Sharma, H.C., Upadhyaya, H.D., 2007a. Exploitation of wild relatives of pigeonpea and chickpea for resistance to *Helicoverpa armigera*. eJ. SAT Agric. Res. Crop. Improv. 3, 4–7.

McCreery, D., 1979. Flotation of the Bab edh-Dhra and Numeira plant remains. Ann. Am. School Orient Res. 46, 165–169.

McIntosh, M., Miller, C., 2001. A diet containing food rich in soluble and insoluble fiber improves glycemic control and reduces hyperlipidemia among patients with type 2 diabetes mellitus. Nutr. Rev. 59, 52–55.

Moreno, M., Cubero, J.I., 1978. Variation in *Cicer arietinum* L. Euphytica 27, 465–485.

Muehlbauer, F.J., 1993. Use of wild species as a source of resistance in cool-season food legume crops. In: Singh, K.B., Saxena, M.C. (Eds.), Breeding for stress tolerance in cool-season food legumes. John Wiley & Sons, Chichester, UK, pp. 359–372.

Muehlbauer, F.J., Kaiser, W.J., Simon, C.J., 1994. Potential for wild species in cold season food legume breeding. Euphytica 73, 109–114.

Pande, S., Ramgopal, D., Kishore, G.K., Mallikarjuna, N., Sharma, M., et al., 2006. Evaluation of wild Cicer species for resistance to Ascochyta blight and Botrytis Gray Mold in controlled environment at ICRISAT, Patancheru, India. Int. Chickpea Pigeonpea Newsl. 13, 25–26.

Pasternak, R., 1998. Investigations of botanical remains from Nevali Cori PPNB, Turkey: short interim report. In: Damania, A.B., Valkoun, J., Wilcox, G., Qualset, C.O. (Eds.), Origins of Agriculture and Crop Domestication. ICARDA, Aleppo, Syria, pp. 170–177.

Pittaway, J.K., Ahuja, K.D., Cehun, M., Chronopoulos, A., Robertson, I.K., Nestel, P.J., et al., 2006. Dietary supplementation with chickpeas for at least 5 weeks results in small but significant reductions in serum total and low-density lipoprotein cholesterols in adult women and men. Ann. Nutr. Metab. 50, 512–518.

Redden, R.J., Berger, J.D., 2007. History and origin of chickpea. In: Yadav, S.S., Redden, R.J., Chen, W., Sharma, B. (Eds.), Chickpea Breeding and Management, first ed. CAB International, Wallingford, UK, pp. 1–13.

Renfrew, J.M., 1967. Carbonized seeds and fruits from the funeral pyres of Salamis, 6th–5th century BC. In: Kargeorghis, V. (Ed.), Excavations in the Necropolis of Salamis. Harrison and Sons, UK.

Rollefson, G.O., Simmonds, A., Donaldson, M.L., Gillespie, W., Kafafi, Z., et al., 1985. Excavation at the Pre-Pottery Neolithic B village of 'Ain Ghazal' (Jordan). Mittl. Dtsch. Orient-Gesellschaft Berl. 117, 69–116.

Saraf, C.S., Rupela, O.P., Hegde, D.M., Yadav, R.L., Shivkumar, B.G., et al., 1998. Biological nitrogen fixation and residual effects of winter grain legumes in rice and wheat cropping systems of the Indo-Gangetic plain. In: Kumar, J.V.D.K., Johansen, C., Rego, T.J. (Eds.), Residual Effects of Legumes in Rice and Wheat Cropping Systems of the Indo-Gangetic Plain. Oxford/IBH Publishing, New Delhi, pp. 14–30.

Saraswat, K.S., 1992. Archaeobotanical remains in ancient cultural and socio-economical dynamics of the Indian sub-continent. Palaeobotanist 40, 514–545.

Sharma, S., Upadhyaya, H.D., Roorkiwal, M., Varshney, R.K., Gowda, C.L.L., 2013. Chickpea. In: Singh, M., Upadhyaya, H.D., Bisht, I.S. (Eds.), Genetic and Genomic Resources of Grain Legume Improvement, first ed. Elsevier, London, pp. 81–111.

Shivkumar, S., Gupta, S., Chandra, S., Singh, B.B., 2004. How wide is the genetic base of pulse crops. In: Ali, M., Singh, B.B., Shivkumar, S., Dhar, V. (Eds.), Pulses in New Perspective. Proceedings of National Symposium on Crop Diversification and Natural Resource Management. Indian Institute of Pulses Reasearch, Kanpur, Uttar Pradesh, India, pp. 211–221.

Singh, K.B., 1997. Chickpea (*Cicer arietinum* L.). Field Crop. Res. 53, 161–170.

Singh, K.B., Ocampo, B., 1997. Exploitation of wild Cicer species for yield improvement in chickpea. Theor. Appl. Genet. 95, 418–423.

Singh, K.B., Pundir, R.P.S., Robertson, L.D., van Rheene, H.A., Singh, U., Kelley, T.J., et al., 1997. Chickpea. In: Fuccillo, D., Sears, L., Stapleton, P. (Eds.), Biodiversity in Trust. Cambridge University Press, Cambridge, pp. 100–113.

Singh, S., Singh, I., Kapoor, K., Gaur, P.M., Chaturvedi, S.K., Singh, N.P., Sandhu, J.S., 2014. Chickpea. In: Singh, M., Bisht, I.S., Dutta, M. (Eds.), Broadening the Genetic Base of Grain Legumes, first ed. Springer, London, pp. 51–73.

Tanno, K., Willcox, G., 2006. The origins of cultivation of *Cicer arietinum* L. and *Vicia faba* L.: early finds from north west Syria (Tell el-Kerkh, late 10th millennium BP). Veg. Hist. Archaeobotany 15, 197–204.

Upadhyaya, H.D., Dwivedi, S.L., Baum, M., Varshney, R.K., Udupa, S.M., et al., 2008. Genetic structure, diversity and allelic richness in composite collection and reference set in chickpea (*Cicer arietinum* L.). BMC Plant Biol. 8, 106–118.

van der Maesen, L.J.G., 1972. *Cicer* L., a Monograph of the Genus, With Special Reference to the Chickpea (*Cicer arietinum* L.), Its Ecology and Distribution. Mendelingen Landbouwhogeschool Wageningen, Wageningen, The Netherlands, pp. 1–341.

van der Maesen, L.J.G., 1987. Origin, history and taxonomy of chickpea. In: Saxena, M.C., Singh, K.B. (Eds.), The Chickpea. CAB International, Cambridge, pp. 11–34.

van der Maesen, L.J.G., Maxted, N., Javadi, F., Coles, S., Davies, A.M.R., 2007. Taxonomy of the genus *Cicer* revisited. In: Yadav, S.S., Redden, R.J., Chen, W., Sharma, B. (Eds.), Chickpea Breeding and Management. CABI, Wallingford, pp. 14–47.

van Zeist, W., Bakker-Heeres, J., 1982. Archeobotanical studies in the the Levant 1. Neolithic sites in the Damascus Basin: Aswad, Ghoraife, and Ramad. Palaeohistoria 24, 165–256.

van Zeist, W., de Roller, G., 1991, 1992. The plant husbandry of aceramic Çayönü, SE Turkey. Palaeohistoria 33/34, 65–96.

van Zeist, W., de Roller, G., 1995. Plant remains from Asikli Höyük, a pre-pottery Neolithic site in central Anatolia. Veg. Hist. Archaeobotany 4, 179–185.

Varshney, R.K., Song, C., Saxena, R.K., Azam, S., Yu, S., Sharpe, A.G., et al., 2013. Draft genome sequence of chickpea (*Cicer arietinum*) provides a resource for trait improvement. Nat. Biotechnol. 31, 240–246. Epub 2013 Jan 27. http://www.nature.com/nbt/journal/vaop/ncurrent/full/nbt.2491.html.

Vavilov, N.I., 1951. The origin, variation, immunity and breeding of cultivated plants. Chron. Bot. 13, 1–366.

Vishnu-Mittre, Savithri, R., 1982. Food economy of the Harappans. In: Possehl, G.L. (Ed.), Harappan Civilization: A Contemporary Perspective. Oxford and IBH, New Delhi, pp. 205–221.

Vishnu-Mittre, Saraswat, K.S., Sharma, A., Chanchala, 1985. Indian Archeology: A Review - (1982-83). Archeological Survey of India, New Delhi, pp. 146–150.

Zohary, D., Hopf, M., 1973. Domestication of pulses in the old world. Science 182, 887–894.

Zohary, D., Hopf, M., 2000. Pulses. In: Domestication of Plants in the Old World: The Origin and Spread of Cultivated Plants in West Asia, Europe, and the Nile Valley, third ed. Oxford University Press, New York, pp. 108–111.

CHAPTER

Chickpea genetic resources: collection, conservation, characterization, and maintenance

3

Rahul Chandora[1], Gayacharan[2], Neelam Shekhawat[3], Nikhil Malhotra[1]
[1]*ICAR-National Bureau of Plant Genetic Resources, Shimla, Himachal Pradesh, India;* [2]*ICAR-National Bureau of Plant Genetic Resources, New Delhi, Delhi, India;* [3]*ICAR-National Bureau of Plant Genetic Resources, Jodhpur, Rajasthan, India*

3.1 Introduction

Chickpea (*Cicer arietinum* L.) is a self-pollinated true diploid ($2n = 2x = 16$) with a moderately sized genome of around 750 Mbp (Varshney et al., 2013) evolved from its immediate wild progenitor, *C. reticulatum*. It is a cool season leguminous crop belonging to Fabaceae, cultivated in 59 countries. Major chickpea producing countries are India, Pakistan, Turkey, Myanmar, Australia, Iran, Ethiopia, Mexico, Canada, and the United States. Chickpea contributes to global nutritional security owing to its 23% protein content (Singh and Pratap, 2016) and it contains essential amino acids like tryptophan, methionine, and cysteine (Williams and Singh, 1987). Chickpea is one of the founder crops of modern agriculture (Zohary and Hopf, 1973). It ranks second worldwide as a food legume crop and is primarily a crop of developing countries, contributing to a large part of human food and animal feed in these areas. In Indian context, the crop is very important, as a major portion of the population is vegetarian, where pulses are the major source of protein. Chickpea is grown mostly in South Asia and Sub-Saharan Africa, which accounts for more than 75% of the world chickpea area. The present global production of chickpea is more than 14.7 million tons, and the productivity is about 1014.6 kg/ha (http://www.fao.org/faostat/en/#data/QC.). During the last 20 years, the global area, production, and yield of chickpea have increased by 8%, 40%, and 27% respectively. India is the largest producer of chickpea in the world and accounts for more than 70% of the total annual pulse production with 11.16 million tons of chickpea harvested from 10.76 million ha during 2017–18 (Dixit, 2018), which is far below than the actual potential of this crop. Chickpea is one of the ancient grain legume crops which was domesticated in the Old World (Van der Maesen, 1972) and most probably originated in Turkey and adjoining areas of Syria. From there, it

Chickpea: Crop Wild Relatives for Enhancing Genetic Gains. https://doi.org/10.1016/B978-0-12-818299-4.00003-8
Copyright © 2020 Elsevier Inc. All rights reserved.

had spread toward the West and the South with human migration via the Silk Route (Singh et al., 1997). Primary centers of origin have been identified in the Mediterranean, southwest Asia, and a secondary center of origin in Ethiopia (Vavilov, 1951). In the 1970s, some explorers found wild population of chickpea in southeast Turkey. Its similarity to the cultivated chickpea in terms of cross fertility and morphology was found to be so close that it was thought to be the modern crop's wild ancestor and classified as a separate species, *C. reticulatum*.

Cultivated chickpea is different from its wild species in its erect plant growth habit, altered phenology, seed coat texture, and reduced seed dormancy. This is due to the "domestication syndrome" (Hammer, 1984). Furthermore, variation exists in the cultivated species such as seed shape, size, and color based on which, cultivated chickpea is classified into two types (Cubero, 1975): *desi* type (microsperma), predominantly grown in semiarid tropics (Muehlbauer and Singh, 1987) characterized by small seed type, bushy growth habit, pink flowers, anthocyanin pigment, angular seed shape with dark color, and rough seed coats as well as high fiber content; and *kabuli* type (macrosperma), majorly found in the temperate regions, in Mediterranean countries, having white flowers with an owl-shaped head, anthocyanin pigment absent, white or beige seed coat color with smooth surface, erect growth habit, and large seeds with low content of fiber. *Desi* type is considered as an early domesticated form of the cultivated chickpea and subsequently *kabuli* is the derived form, because seeds of the *desi* type are more similar to those of wild *C. reticulatum* as compared to the *kabuli* type (Ladizinsky and Adler, 1976). Both types are grouped into two interconnected clusters within cultivated chickpea (Smithson et al., 1985; Smartt, 1990) representing two diverse gene pools. This has been supported by DNA markers (Iruela et al., 2002; Upadhyaya et al., 2008; Roorkiwal et al., 2014). *Desi* type predominates in chickpea production and accounts for about 85% while *kabuli* accounts for about 15% of the world chickpea production. Cultivated chickpea gene pool possesses narrow genetic base due to domestication bottleneck (Spillane and Gepts, 2001), genetic drift (Cooper et al., 2001), migration, and lesser use of genetic resources in chickpea breeding (Kumar et al., 2004). A key factor contributing to low grain yield is the loss of invaluable genes linked to higher productivity including biotic and abiotic stress tolerance in the cultivated types and a more general narrowing down of their genetic base starting with its domestication (Varshney et al., 2013; Ladizinsky and Adler, 1975; Tuwafe et al., 1988; Kazan et al., 1991; Ahmad et al., 1992a,b; Labdi et al., 1996; Tayyar et al., 1996; Simon and Muehlbauer, 1997; Siddique et al., 2000). In the past, crop wild relatives (CWRs) have contributed substantially to the genetic improvement of many useful characters including seed yield in different annual crop plants (Stalker, 1980). Similarly in chickpea, the productivity enhancement–related traits have been introgressed from *C. reticulatum*, *C. echinospermum* (Sandhu et al., 2006; Singh et al., 2005), and *C. pinnatifidum* species (Sandhu et al., 2006). CWRs, however, provide the broadest range of genetic diversity and represent additional novel gene pool of genetic resources in grain legumes, including chickpea, and have the ability to provide a wide possible range of germplasm resources for introgression of various traits of agro-economic importance (Singh et al., 2014). This is because the effectiveness of any breeding program depends on the extent and nature of phenotypic and genotypic variation. Therefore, narrow genetic base warrants systematic exploration and collection of chickpea from hotspot areas with high concentration of wild species, their characterization and evaluation, maintenance for effective conservation, and efficient utilization into adapted elite genetic background. This is of considerable significance in view of improving the chickpea production in the country which will help to underpin future food and nutritional security.

3.2 Species distribution and gene pools

Distribution pattern of wild relatives of crops provides critical information like where domestication may have occurred. Chickpea wild relatives are mainly found in Turkey, Syria, Lebanon, Afghanistan, India, Ethiopia, Pakistan, Israel, and Jordan (Fig. 3.1). *C. bijugum*, *C. echinospermum*, and *C. reticulatum*, which are closely related to chickpea, are widely distributed in southeastern Turkey and adjoining Syria (Ladizinsky and Adler, 1975; Ladizinsky, 1998). However, many *Cicer* species are found in specific areas like *C. anatolicum* in Armenia and Turkey, *C. bijugum* in Syria and Turkey, *C. microphyllum* in India and Pakistan, *C. macracanthum* in Pakistan, etc. Genesys database (www.genesys-pgr.org) shows that around 1500 wild *Cicer* species, including duplicates, are being ex situ maintained in different gene banks. *C. reticulatum* forms a major part of the collected germplasm, followed by *C. judaicum, C. pinnatifidum*, *C. echinospermum*, and *C. bijugum*.

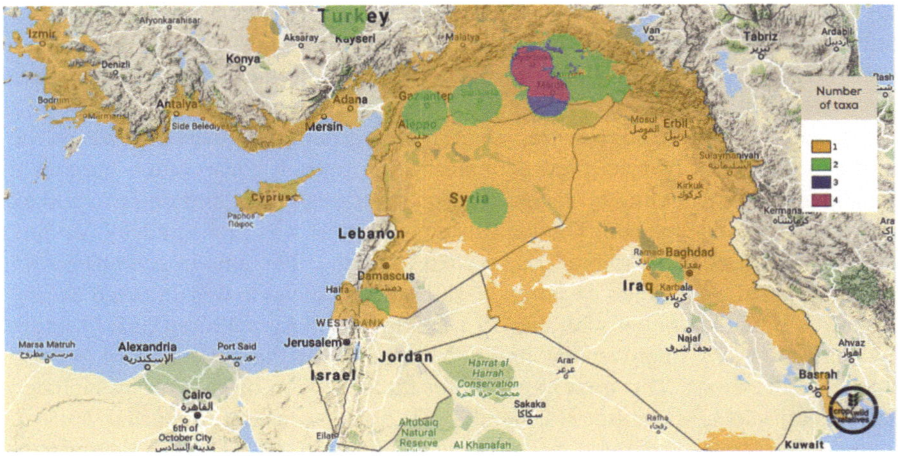

FIGURE 3.1

Species richness of *Cicer* genus.

Source: www.genesys-pgr.org.

Currently, *Cicer* genus comprises 44 species, including 35 wild perennials, 8 wild annuals, and the cultivated annual (Muehlbauer, 1993; Vander Maesen, 1972). According to crossing compatibility with cultivated chickpea, *Cicer* species are classified into three gene pools (Harlan and de Wet 1971). The primary gene pool consists of the cultivated species, *Cicer arietinum* and its immediate putative progenitor, *C. reticulatum*. They differ either by a reciprocal inversion, a paracentric inversion, or by location of chromosomal satellites (Ladizinsky, 1998). Both are easily crossable with each other and freely exchange their genes. Wild annual species *C. echinospermum* comes under the secondary gene pool and differs from the domestic species by a single reciprocal translocation (Ladizinsky, 1998), which is crossable with cultivated chickpea, but with reduced fertility level of the hybrids and their advanced progenies. The tertiary gene pool consists of the remaining 6 annual and 35 perennial species, which are genetically isolated from the cultivated one (Nguyen et al., 2004). These species are not easily crossable with *C. arietinum* and require specialized techniques for the transfer of genes into cultivated backgrounds of chickpea.

3.3 Germplasm conservation

Genetic resources provide basic material for selection and improvement, through breeding, to ensure food security needs of rapidly rising global human population which is expected to reach c.9 billion by 2050 from c.7 billion in the present. The constant population rise is a major cause for disturbance in the ecosystems and adverse climatic changes. To sustain and enhance agricultural production amid fast-changing crop growing conditions, in situ and ex situ conserved plant biodiversity plays a very important role. Crop improvement programs and existence of cultivated species depends upon the level of the variability existing in the gene pool. Hence, conservation becomes the foremost important activity to generate raw material of germplasm diversity for present and future use.

Sustainable and effective conservation of crop genetic resources depends on the active management of conserved germplasm. Germplasm conservation requires a holistic approach adopting both conservation strategies viz. in situ on-farm and ex situ in a balanced manner. The ex situ conservation was pioneered by Vavilov (1926) and subsequent explorations resulted in large collections in gene banks. In India, systematic chickpea exploration and collection work was initiated during the 1970s in collaboration with the US Department of Agriculture (USDA) under its Regional Pulse Improvement program, resulting in the collection of ~7000 chickpea accessions. The areas explored for the collection of germplasm included parts of Rajasthan, Maharashtra, Gujarat, Bihar, Odisha, eastern Arunachal Pradesh, southern Karnataka, and Tamil Nadu (Singh and Singh, 1997). The genetic diversity of wild *Cicer* species, particularly *C. microphyllum*, has been collected by several explorers of National Bureau of Plant Genetic Resources (NBPGR) from different ecological habitats, especially from Trilokinath, Kukumseri, Throat and Mayar

Valley in Lahaul and Lossar, and Tabo in Spiti region of Himachal Pradesh. Many other wild *Cicer* species like *C. nuristanicum, C. microphyllum, C. macrocanthum, C. judaicum, C. reticulatum, C. pinnatifidum, C. yamashitae,* etc. have been collected from Pakistan, Afghanistan, Turkey, Syria, and adjoining areas by International Crops Research Institute for Semi-Arid Tropics (ICRISAT) and International Center for Agricultural Research in the Dry Areas (ICARDA). Conservation of genetic diversity of cultivated and wild species and the utilization of available germplasm are essential for the future of chickpea production.

3.3.1 Ex situ conservation

Ex situ conservation means to maintain the genetic diversity and genetic integrity of collected species outside their natural habitats, to avoid their genetic erosion (loss or degeneration). Seeds of chickpea are orthodox in nature. They can be dried and stored for a long period at $5 \pm 2\%$ at $-18°C$ and 15% relative humidity (RH) with a minimum loss of seed viability. In ex situ conservation of seeds, working collections (active collections) of chickpea stored for short term (10–20 years) under relatively less stringent conditions at $+5°C$ and a controlled air humidity of $\pm 35\%$ RH is the basic source of distribution and utilization, whereas the base collection is maintained at $-18°C$ in vacuum-packed aluminum foil pouches at 3%–7% seed moisture content and is regularly monitored every 10 years to check seed viability. This serves as a backup for active collection. Seeds can be conserved for more than 50 years without any loss of viability (Usberti and Gomes, 1998).

Around 97,400 accessions of chickpea germplasm are conserved in more than 30 gene banks worldwide. Most of these accessions originated in India, Iran, Syria, and Turkey (Fig. 3.2).

ICRISAT holds the largest collection of chickpea, i.e., 20,764 accessions which comprises 20,456 of cultivated types and 308 of 18 wild *Cicer* species (Table 3.2) collected from 60 countries. It has a global responsibility of collecting, conserving, and distributing the chickpea germplasm, consisting of landraces, modern cultivars, genetic stocks, mutants, and wild species of chickpea. Majority of collections (16,114 accessions) were obtained through donations from 58 countries and remaining 4153 accessions were collected from 65 collection missions in 14 countries. Of the entire cultivated chickpea accessions, 75.2% are of *desi* small seed type, 23.1% are of *kabuli* large seed type, and 3.8% are of intermediate type (Upadhyaya et al., 2002). Similarly, ICARDA gene bank has collected and conserved 15,734 accessions including 540 accessions of wild *Cicer* species in its global germplasm repository from 61 countries across six continents (Asia, Africa, the Americas, Europe, and Australia). The Indian National Genebank at NBPGR, New Delhi, conserves 14,704 chickpea accessions which represent a large proportion of the cultivated diversity found in India and some genotypes introduced from other countries. Other gene banks like Australian Temperate Field Crops Collection (ATFCC), Victoria, Australia, and Western Regional Plant Introduction Station, USDA-ARS, USA, conserve 8655 and 8038 accessions, respectively. Gene

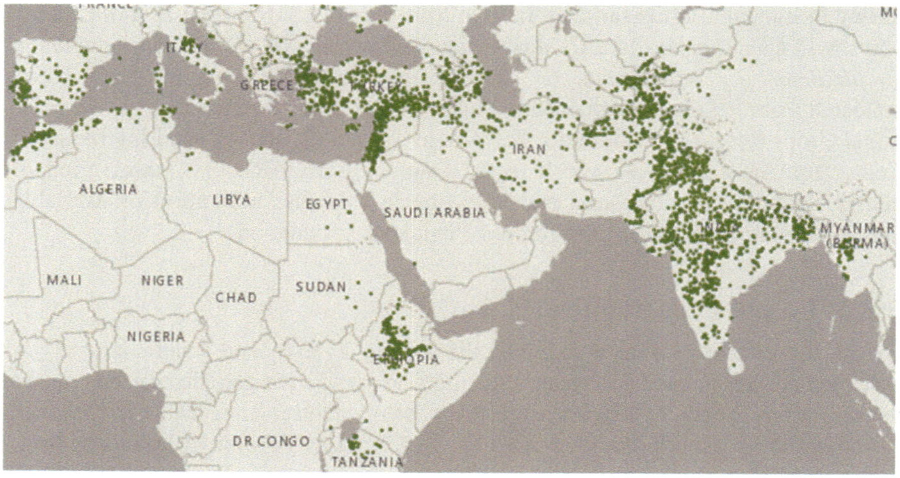

FIGURE 3.2

Representation of *Cicer* germplasm collection density from major chickpea growing areas.

Source: www.genesys-pgr.org.

bank databases around the world indicate that a substantial amount of chickpea collections are conserved. But this is an overestimation of numbers as these collections also include duplicates which generally happen in the process of collection, conservation, and exchange. Wild species of *Cicer* represents only less than 1% of the total accessions (conserved in about 10 gene banks worldwide). Due to the narrowed gene pool of cultivated chickpea, recent utilization of wild species through prebreeding programs has significantly increased. Therefore, now priority is being given to the collection and conservation of wild species. The whole summary of ex situ chickpea germplasm collection held in various major gene banks across the world is presented in Table 3.1.

3.3.2 In situ conservation

Conservation of species or landrace or population in their natural habitat is known as in situ conservation. Here, evolutionary processes continue to make it a dynamic system. Cultivated species are generally conserved through traditional seed conservation practices in traditional farming systems. However, in the current scenario, in situ conservation has become important in conserving wild as well as semidomesticated forms of chickpea. For example, a rare wild perennial species, *C. montbretii* is conserved in the protected areas of Ropotamo and Uzunbodzhak Reserves and Strandja Nature Park, Bulgaria (Petrova and Angelova, 2013). Strandja Mountain, in southern Bulgaria, is also considered as one of centers of origin of cultivated chickpea (Koinov, 1968; Kaiser et al., 1998). Other wild

Table 3.1 Ex situ *Cicer* collections gene banks.

Institute	Wild	Cultivated	Total
International Crop Research Institute for the Semi-Arid Tropics (ICRISAT), India	308	20,456	20,764
International Centre for Agricultural Research in the Dry Areas(ICARDA), Beirut, Lebanon	540	15,194	15,734
National Bureau of Plant Genetic Resources (NBPGR), New Delhi, India	69	14,635	14,704
Australian Temperate Field Crops Collection (ATFCC), Australia	246	8409	8655
Western Regional Plant Introduction Station, USDA-ARS, USA	194	7844	8038
National Plant Gene Bank of Iran, Seed and Plant Improvement Institute (NPGBI-SPII), Iran	–	5700	5700
N.I. Vavilov All-Russian Scientific Research Institute of Plant Industry (VIR), Russian Federation	–	2767	2767
Plant Genetic Resources Program (PGRP), Pakistan	89	2057	2146
Plant Genetic Resources Department, Aegean Agricultural Research Institute (AARI), Turkey	21	2054	2075
Institute of Plant Production nd. a. V. Ya. Yuryev of NAAS, Ukraine	–	1760	1760
Estación de Iguala, Instituto Nacional de Investigaciones Agrícolas, Iguala, Mexico	–	1600	1600
Institute of Biodiversity Conservation (IBC), Ethiopia	–	1173	1173
Institute for Agrobotany (RCA), Hungary	9	1161	1170
Uzbek Research Institute of Plant Industry (UzRIPI), Uzbekistan	–	1055	1055
Total	1476	85,865	87,341

Source: http://www.fao.org/wiews-archive/germplasm_query.htm?i_l=EN.

species, *C. reticulatum*, *C. bijugum*, *C. echinospermum*, and *C. pinnatifidum* in Turkey and Syria; *C. judaicum* in Israel, Jordan, and Lebanon; and *C. yamashitae* in Afghanistan are conserved in the wild. Another species, *C. microphyllum* is naturally conserved in the cold desert region of Lahaul and Spiti in northwestern Indian Trans-Himalayas. This wild chickpea has genes for large pod size (1.5–2 cm), more number of seeds (5–8), and tolerance to drought and cold as it remains under the snow for 6 months. These characteristics make this crop particularly important economically. Chickpea landraces (farmers' traditional cultivars) are generally being maintained in traditional farming systems by inheriting traditional seeds from one generation to the next generation. Recently, keeping in view the loss of landrace diversity due to their replacement with modern chickpea varieties, on-farm (Maxted et al., 2012) in situ conservation method is being promoted by conservationists. Here, custodian farmers are sensitized and incentivized to continue conserving their traditional cultivars by cultivating them. As these landraces face diverse kind of

Table 3.2 Status of wild *Cicer* species accessions conserved at ICRISAT gene bank.

Species	No. of accessions	Country of origin
Annual		
C. bijugum	49	Iraq, Syria, and Turkey
C. chorassanicum	4	Afghanistan
C. cuneatum	5	Ethiopia
C. echinospermum	18	Turkey
C. judaicum	70	Afghanistan Ethiopia, India, Israel, Jordan, Lebanon, Morocco, Syria, and Turkey
C. ppinnatifidum	42	Ethiopia, Israel, Lebanon, Syria, and Turkey
C. reticulatum	36	Turkey
C. yamashitae	7	Afghanistan
Perennial		
C. anatolicum	3	Turkey
C. canariense	1	Spain
C. floribundum	1	Turkey
C. macracanthum	5	Pakistan
C. microphyllum	52	India and Pakistan
C. montbretii	2	Turkey
C. multijugum	1	Russian Federation
C. nuristanicum	2	Pakistan
C. pungens	9	Afghanistan
C. rechingeri	1	Afghanistan
Total	308	

selection processes and cellular evolutionary processes, genetic material keeps reshaping itself as per the requirement of the growing environment. Therefore, these landraces become gold mines for the crop breeders/researchers. According to Erna Bennett, landraces will perform better under climatic changes as compared to the modern cultivars because they possess more plasticity traits and dynamic populations, which evolve themselves under heterogeneous environment. This is of utmost importance for the success of future on-farm conservation of species.

Ex situ conservation strategies should ideally be complemented with in situ conservation for conservation of the entire diversity of chickpea because the total gene pool of chickpea cannot be included in the ex situ collections. Therefore, establishing good linkage between gene banks and in situ conservation (community seed banks/on-farm) is the key to achieve successful conservation of the entire gene pool/genetic diversity (Engels et al., 2002).

3.3.3 Duplicates

One of the major concerns in germplasm conservation for all the gene banks is the presence of duplicate accessions. It is absolutely necessary to identify duplicates for the rationalization of the collection. Various tools are being utilized for PGR (Plant Genetic Resources) systematization. One such tool, known as "PGRdup," has been developed by NBPGR, New Delhi, India. It is software that uses passport databases for the identification of possible duplicates in gene bank collections (Aravind et al., 2015).

3.3.4 Safety duplication

Safety duplication is vital to ensure that a genetically identical subsample of an accession is available. This is necessary to alleviate the prospect of partial or total loss of germplasm that can occur as a result of natural or man-made calamity. Safety duplicates are genetically identical to the long-term collection and are also known as the secondary most-original-sample (Engels and Visser, 2003). The process involves duplication of the material as well as its related information, including database backup. A safety duplicate sample for every original accession should be stored at a geographically distant area, under the same or better conditions than those in the original gene bank. To minimize the risks that can happen in an individual country, safety duplication should ideally take place outside that country. The chickpea germplasm collection of ICARDA is safely duplicated at ICRISAT. As a safety backup, 16,996 chickpea accessions were deposited in Svalbard Global Seed Vault, Norway. Safety duplication is undertaken under a "black-box" approach. This means that the repository gene bank has no authorization to the use and distribution of the duplicates. The depositor is responsible for the material's quality, viability, and regeneration from base collection when the material begins to lose viability. The deposit can be returned when the original collection is lost or destroyed or can be recalled when replaced with newly regenerated germplasm. All original seeds collected by a gene bank should undergo safety duplication. The fact that an accession is present in another collection does not immediately signify that the accession is safely duplicated in long-term conservation conditions. It is apparent that chickpea collections are inadequately duplicated and are thus at risk. Hence, safety duplication requires a formal arrangement.

3.4 Germplasm characterization and evaluation

Systematic characterization and evaluation is necessary to facilitate effective utilization of available genetic resources in crop breeding programs and for efficient management of germplasm. Since the 1970s, a large number of chickpea germplasm accessions have been characterized and evaluated, in batches, for morphological and agronomic traits (Singh et al., 2016). About 99% of germplasm accessions of chickpea at ICRISAT have been characterized and evaluated for agro-morphological traits (Sharma et al., 2005; Upadhyaya et al., 2006a). The first large-scale evaluation of 5477 accessions was carried out by Narayan and Macefield (1976), who observed substantial amount of genetic variability for plant type, flower color, days to flowering, days to seed maturity, pod size, number of seeds per pod, weight of 100 seeds, and seed color from different geographical locations. This was followed by various joint evaluations by collaborative research at several national and international organizations (Muehlbauer and Sarker, 2017).

In India, systematic research in chickpea started with the inception of All India Coordinated Pulse Improvement Program (AICPIP) during 1967 (http://www.aicrpchickpea.res.in/) followed by the establishment of a separate All India Coordinated Research Project (AICRP) on chickpea during 1993. However, characterization of Indian chickpea collections was mainly taken care by NBPGR, which is also the host institute for the Indian National Genebank. Along with characterization, NBPGR is also actively carrying out evaluation of conserved chickpea germplasm (Fig. 3.3).

For assessing the level of existing variability and identification of trait-specific accessions, the entire collection of cultivated chickpea and wild species, majority of which were indigenous collections, were characterized and substantial amount of variability was observed for both qualitative and quantitative traits (Archak et al., 2016). Since 2004–18, around 4500 accessions have been evaluated for important agronomic traits as well as biotic and abiotic stresses, through multilocation trials (unpublished data). A set of 65 large seeded *kabuli* type accessions

FIGURE 3.3

Field view: characterization and preliminary evaluation of chickpea germplasm conserved in National Gene Bank of India.

were evaluated across 13 environments for various quantitative and qualitative traits. Two germplasm lines, ICC-17109 and ICC-17452 exhibited high yield and stable performance across the environments (Gowda et al., 2011). These mega evaluation and characterization studies revealed the availability of considerable amount of genetic variability in cultivated chickpea gene pool for agro-morphological traits over various geographical locations (Upadhyaya et al., 2001, 2006b; Sewak et al., 2012).

A source of resistance in cultivated chickpea represents a narrow genetic variability, and wild *Cicer* species have been used to widen the genetic base of cultivated gene pool by incorporation of various adapted genes (Singh et al., 2014). Chickpea is attacked by a wide range of pathogens and pests, causing significant economic losses. Out of 47 diseases, fusarium wilt (*Fusarium oxysporum*) and ascochyta blight (*Ascochyta rabiei*) are the most destructive diseases in chickpea while other diseases like dry root rot, botrytis gray mold, and phytophthora root rot have regional importance. One of the major difficulties in modern cultivars is their vulnerability to wilt and blight disease due to emergence of new races of pathogens, which requires identification of durable genes for resistance and incorporation of those resistance genes into commercial varieties.

Cicer species have been extensively screened and several of them were reported promising sources of resistance to diseases caused by various fungi, insects, and nematodes. Shah et al. (2005) identified a high level of resistance to ascochyta blight after evaluation of 44 wild *Cicer* germplasm accessions. Similarly, 25 wild *Cicer* species were screened for resistance to ascochyta blight by artificial inoculation of the germplasm under glasshouse (Benzohra et al., 2014). In an evaluation trial of 13,500 chickpea accessions, 150 *desi* type and 10 *kabuli* type chickpea accessions were found to be resistant against fusarium wilt (Haware et al., 1992). Pande et al. (2004) reported one germplasm line namely ICC-14395, a cultivar, ICCV-2, and one advanced breeding line (ICC X 830203-BH-BH-11H) resistant to dry root rot. A single variety (PBG-5) with high yield and resistance to biotic stresses was developed using germplasm lines BG-257 and E100Y (Sandhu et al., 2004). Among 48 studied lines, 10 accessions had showed resistance reaction against chickpea blight (Ahmad et al., 2013). Likewise, sources of resistant genes against wilt along with variability in morphological and physiological traits were identified in 110 lines of chickpea (Halila and Strange, 1997). Resistance to multiple diseases were performed by various accessions against botrytis gray mold and root knot nematode suggesting introgression of these useful genes from wild to cultivated gene pool for creating additional variability (Singh et al., 2014).

Wild species of chickpea viz., *C. judaicum, C, pinnatifidum, C. bijugum* have given various genes for resistance to ascochyta blight, botrytis gray mold, and fusarium wilt (Singh et al., 1982). Similarly, *C. bijugum, C. judaicum, C. pinnatifidum, C. reticulatum* are extensively used wild species as sources of resistance to ascochyta blight, fusarium wilt, leaf minor, bruchids, cyst nematodes, gray mold, and phytophthora root rot (Singh et al., 2008). *C. microphyllum* has also been identified with having resistance to pod borer (Sharma et al., 2005).

Because chickpea is a winter season crop, cold temperatures seriously affect the yield in chickpea. Introduction of cold tolerance genes into cultivated high-yielding cultivars is necessary to obtain high yield during winters. The highly variable genetic resources for cold tolerance were identified in an evaluation trial of 3276 lines. Significant levels of cold tolerance were also present in *C. bijugum, C. reticulatum, C. echinospermum*, and *C. pinnatifidum* (Singh et al., 1990). Toker (2005) identified 3 accessions out of 43 accessions of wild *Cicer* species that were highly cold tolerant. Wild species were almost insensitive to the cold temperature, and interspecific hybrids between cultivated and wild were significantly more cold tolerant than cultivated varieties. The genes responsible for vernalization are available in *C. reticulatum* accessions, which were absent in cultigens (Berger et al., 2005). It was found that cold tolerance is controlled by at least five genes with both additive and nonadditive gene effects and was dominant over susceptibility (Malhotra and Singh, 1991).

The breeding for drought tolerance is very slow due to various factors like variability in available moisture across the years, complex genetic basis of the character, root traits, etc. (Turner et al., 2001). Canci and Toker (2009) evaluated eight wild annual *Cicer* species and reported four accessions of *C. reticulatum* and one accession of *C. pinnatifidum* that were resistant to drought and heat stress. One high-yielding early-maturing and bold seeded *desi* type chickpea accession, ICC-13124 has been identified as a source of genes responsible for drought tolerance (Parameshwarappa et al., 2012). Another accession, ICC-4958 is identified as a potential source of drought resistance due to better SCMR and a strong root system. Germplasm accession, ICC-16374 could also be utilized as a source of drought resistance (Kashiwagi et al., 2006a). Yield is a collective function of transpiration and transpiration efficiency. The improvement in transpiration efficiency is one of the important components for improving the drought resistance trait (Kashiwagi et al., 2006b). Several traits such as biomass, length density, and depth of root system are some important traits which significantly contribute in drought avoidance mechanism and increase yield. Kashiwagi et al. (2006) characterized and evaluated 216 germplasm lines for root traits and observed ample amount of variability for root traits length density (RLD). There is a need for further germplasm exploration to identify more landraces having RLD traits to progress the selection and breeding for drought in chickpea. A set of 37 landraces from 14 countries were evaluated for drought tolerance by recoding physiological parameters like relative water content, membrane stability index, etc. Kumar et al. (2015) have identified promising donors viz. IG 5856 (Jordan) and IG5904 (Iraq) for terminal drought tolerance in chickpea. Salinity adaptation of chickpea breeding programs could be enhanced by evaluation of whole germplasm for capturing nearly the whole range of variations responsible for salinity stress. Studies have showed that *kabuli* type accessions are more tolerant toward salinity stress as compared to *desi* type of germplasm (Serraj et al., 2004).

In chickpea, CWRs have the largest range of genetic diversity. They provide a wide range of germplasm resources for introgression of various traits of agro-economic importance (Singh et al., 2014). *C. reticulatum* represents a potential source of genes for broadening the genetic base of cultivated gene pool because it has been identified with having useful genetic variations. *C. echinospermum* is another potential source of increased genetic variation. The summary of evaluation of wild species against important agro-morphological traits including major biotic and abiotic stresses is presented in Table 3.3 and Fig. 3.4.

There is always a need to explore and identify resistant germplasm for biotic and abiotic stresses because of appearance of the new physiological races of previously released resistant cultivars. Interspecific/wide hybridization between wild and cultivated species is used for increasing the genetic variation and introduction of resistance to various biotic and abiotic stresses. The durability of resistance in commercial varieties could be enhanced by pyramiding of resistance genes.

3.4.1 Genomic resources and their utilization, characterization and evaluation

Genomic tools and molecular markers have become an integrated part of any crop improvement program. Molecular tools and techniques play an important role in trait identification, trait introgression, marker-assisted selection, identification of diverse lines, DNA fingerprinting of varieties, etc. Whole genome of cultivated chickpea (*C. arietinum*) and its immediate wild progenitor species (*C. reticulatum*) have been sequenced by different research groups. The whole genome of *desi* type chickpea accession, ICC-4958 was sequenced using clone by clone approach in combination with next-generation sequencing technology, bacterial artificial chromosome, and genetic map (Jain et al., 2013). Furthermore, an advanced version of this genome with more than 94% gene spacing was submitted using additional sequence data and an improved genetic map (Parween et al., 2015). Identification of about 2 million SNPs (single nucleotide polymorphisms) was achieved as a result of whole genome resequencing of parental chickpea lines (Thudi et al., 2016). *Kabuli* type chickpea variety, CDC Frontier was sequenced utilizing whole genome shotgun sequencing method (Varshney et al., 2013). Whole genome of *C. reticulatum* (accession PI-489777), the wild progenitor of cultivated chickpea, was also sequenced using shotgun sequencing approach (Gupta et al., 2016). The complete plastid genome of *C. arietinum,* which is 125,319 bp in size, has also been sequenced (Jansen et al., 2008), which could be useful in understanding gene regulations and functions of plastid genes, phylogenetic studies, photosynthetic efficiency, etc. The genomic resources of chickpea increase the precision and accuracy of chickpea improvement programs, particularly in marker-assisted foreground and background selection, QTL identification, and trait-associated gene discovery.

Genetic diversity assessment and tagging of economically important genes can be efficiently done by using molecular markers. Due to high cost and limited availability of genomic resources, the approaches of genomic-assisted breeding in chickpea remained low until 2005. ICRISAT played an important role in generation of molecular markers and genomic resources including whole genome sequencing. At present, more than 2000 SSR markers are available in chickpea (Sharma et al.,

Table 3.3 Sources of useful traits identified in chickpea landraces and their wild relatives.

Trait of interest	Wild relatives/species	Reference(s)
Cold hardiness	*C. bijugum, C. echinospermum, C. judaicum, C. pinnatifidum, C. reticulatum*	Van der Maesen and Pundir (1984); Singh et al. (1998); Sandhu (2004); Toker (2005); Berger et al. (2005); Saeed and Darvishzadeh (2017)
Drought tolerance	*C. anatolicum, C. echinospermum, C. microphyllum, C. montbretii, C. oxyodon, C. pinnatifidum, C. reticulatum, C. songaricum*	Toker et al. (2007); Canci and Toker (2009); Maqbool et al. (2017)
Heat resistance	*C. pinnatifidum, C. reticulatum*	Canci and Toker (2009); Devasirvatham et al. (2012)
Salinity resilience	*C. microphyllum*	Srivastava et al. (2016)
Ascochyta blight resistance	*C. arietinum, C. bijugum, C. cuneatum, C. echinospermum, C. judaicum, C. montbretii, C. pinnatifidum, C. reticulatum*	Shah et al. (2005); Pande et al. (2010); Ahmad et al. (2013); Kaur et al. (2013); Singh et al. (2014); Li et al. (2017)
Botrytis gray mold resistance	*C. bijugum, C. judaicum, C. pinnatifidum, C. reticulatum*	Basandrai et al. (2008); Kaur et al. (2013); Singh et al. (2014); Hegde et al. (2018)
Bruchids resistance	*C. reticulatum*	Eker et al. (2018)
Fusarium wilt resistance	*C. bijugum, C. cuneatum, C. echinospermum, C. judaicum, C. pinnatifidum, C. reticulatum*	Nguyen et al. (2004); Singh et al. (2005); Ahmad et al. (2013)
Helicoverpa pod borer resistance	*C. bijugum, C. cuneatum, C. echinospermum, C. microphyllum, C. pinnatifidum, C. reticulatum*	Sharma et al. (2006); Golla et al. (2018)
Phytophthora root rot resistance	*C. bijugum, C. echinospermum, C. pinnatifidum, C. reticulatum*	Knights et al. (2008)
Root knot nematode resistance	*C. bijugum, C. echinospermum, C. judaicum, C. pinnatifidum, C. reticulatum*	Singh et al. (2014)
Root lesion nematode resistance	*C. echinospermum, C. reticulatum*	Thompson et al. (2011)
Seed beetle	*C. cuneatum, C. echinospermum, C. judaicum, C. reticulatum*	Gupta and Parihar (2015)
High no. of seeds per plant	*C. cuneatum, C. montbretii*	Singh et al. (2014), Gupta et al. (2016)

Table 3.3 Sources of useful traits identified in chickpea landraces and their wild relatives.—*cont'd*

Trait of interest	Wild relatives/species	Reference(s)
Yield components	*C. pinnatifidum, C. reticulatum*	Singh et al. (2005); Singh et al. (2012); Singh et al. (2014), Gupta et al. (2016)

2013). Recently, hybridization-based diversity arrays technology (DArT) and sequence-based single base nucleotide polymorphism (SNP) markers have also been introduced in chickpea (Sharma et al., 2013).

Molecular markers are utilized in each every aspect of crop improvement programs. To understand population structure and the level of diversity among genotypes, some of the most extensively used tools by molecular breeders are DNA-based analysis (Buhariwalla, 2006; Millan et al., 2006; Bharadwaj et al., 2011; Choudhary et al., 2012; Kujur et al., 2015), genome-wide and candidate gene–based association analysis (Thudi et al., 2014; Kujur et al., 2015; Diapari et al., 2014; Jadhav et al., 2015; Upadhyaya et al., 2016), trait identification, tagging and introgression (Varshney et al., 2013; Kujur et al., 2015; Jaganathan et al., 2015), genetic map construction (Pfaff and Kahl, 2003; Sharma et al., 2004; Gujaria et al., 2011; Deokar et al., 2014), etc.

3.4.2 Development of core and mini-core collection

Selection of a single or few trait-specific accessions from gene banks where thousands of accessions are conserved becomes a difficult task. Therefore, the concept of core collection was realized (Frankel and Brown, 1984), to enhance the selection efficiency and the use of genetic resources in the crop improvement programs. In general, a core collection comprises c.10% of the accessions from the entire collection that represents a maximum available genetic diversity (Brown, 1989a). Furthermore, the core set becomes amenable for extensive evaluation of agro-morphological traits, biotic stresses, abiotic stresses, etc. (Brown, 1989b). However, in the case of many crops, where gene banks contain several thousands of accessions, even handling a core set of c.10% of total collections remains a difficult task. Therefore, a two-stage strategy, i.e., core of core, known as "mini-core," was introduced in chickpea (Upadhyaya and Ortiz, 2001). Mini-core becomes c.1% of the entire collections of the species.

In chickpea, the first core was developed utilizing the entire chickpea collection (16,991 accessions back then) conserved at ICRISAT Genebank (Upadhyaya et al., 2001). To develop the core, germplasm accessions were stratified based on their geographical origin followed by their clustering attributes utilizing data on 13 quantitative traits. A core subset of 1956 accessions was developed, which was further

FIGURE 3.4

Promising accessions of wild *Cicer* species carrying important traits of interest.

subjected for mini-core development. This mini-core set of 211 accessions was developed based on detailed data of 22 morphological and agronomical traits. Likewise, a core collection comprising of 505 accessions, selected from the entire collection of 3350 accessions, was developed by Hannan et al. (1994) at the Western Regional Plant Introduction Station (WRPIS) in Pullman, Kibret (2011) also developed core set of Ethiopian chickpea germplasm containing 158 accessions representing the entire Ethiopian chickpea genetic diversity. These core collections were evaluated for various traits of interest over a period of time. Similarly another chickpea core of 1103 accessions was developed by NBPGR from 14561 chickpea accessions based on the information about geographical origin and characterization data on 20 descriptors (8 quantitative and 12 qualitative). Around 70.0% of the accessions of core originated from the Indian subcontinent (Archak et al., 2016).

The development of core subsets has reflected an enhanced utilization of chickpea germplasm and identification of several trait-specific accessions for various important characteristics. The chickpea mini-core developed by Upadhyaya et al. (2001) was extensively evaluated by various workers for different traits like drought, salinity, heat, fusarium wilt, dry root rot, ascochyta blight, botrytis gray mold, pod borer, and herbicide resistance (Upadhyaya et al., 2013). The core was also used to identify accessions with superior breeding values (Meena et al., 2010), to utilize them in chickpea improvement programs. The mini-core was also utilized to understand the nutritional variability available in chickpea by making use of 10 important parameters, viz., total protein, total free amino acids, phytic acid, tannin, total phenolics, total flavonoids, lectin, DPPH radical scavenging activity, in vitro digestibility of protein and starch (Bhagyawant et al., 2018). Similarly, another core set of chickpea consisting of 1103 accessions (Archak et al., 2016) is being extensively screened for biotic and abiotic stresses. A large number of promising accessions resistant/tolerant to ascochyta blight (3), botrytis gray mold (55), wilt (67), dry root rot (5), helicoverpa resistance genes (15), drought avoidance root traits (18), and salinity tolerance (29) have been noted by evaluating the core and mini-core set of chickpea collection (Upadhyaya et al., 2001; Serraj et al., 2004; Pande et al., 2006; Kashiwagi et al., 2006a,b; Upadhyaya et al., 2008; Ahmad et al., 2013; and Singh et al., 2015).

3.5 Future challenges and prospects

Chickpea is the second most important legume crop, contributing significantly in securing nutritional and food security particularly in poor and developing countries. However, it could not get the due attention from researchers and breeders across the world. Despite research and breeding activities of last five decades in chickpea improvement, yield increase is relatively not satisfactory. Major constraints in chickpea improvement are its plant type, which is less amenable for mechanization, devastating pests and diseases, and its highly sensitive nature to heat and salinity stress. Recently, climatic changes have worsened the situation. Sudden temperature rises, short spells of winter, and erratic rains have become a common phenomenon which may reduce chickpea production in the future. Therefore, focused breeding programs on traits like heat, drought, and salinity tolerance are highly warranted. To some extent, niche-specific breeding programs combined with molecular marker technology, genomics, and advanced phonemics tools may sustain chickpea yield. Utilization of diverse chickpea germplasm particularly landraces and CWRs in identification and introgression of desired gene will help in improvement of chickpea cultivars.

References

Ahmad, F., Gaur, P.M., Slinkard, A.E., 1992a. Isozyme polymorphism and phylogenetic interpretations in the genus *Cicer* L. Theor. Appl. Genet. 83 (5), 620–627.

Ahmad, S., Khan, M.A., Sahi, S.T., Ahmad, R., 2013. Evaluation of chickpea germplasm against *Ascochyta rabiei* (pass) lab. J. Animal Plant Sci. 23, 440–443.

Ahmed, A., Iqbal, M., Javaid, M., Ahmad, Z., 1992b. Studies on comparative yield potential of some newly evolved small and bold-seeded mung bean genotypes. J. Anim. Plant Sci. 2, 139–140.

Aravind, J., Radhamani, J., Srinivasan, K., Subhash, B.A., Tyagi, R.K., 2015. An Introduction to PGR Dup Package.

Archak, S., Tyagi, R.K., Harer, P.N., Mahase, L.B., Singh, N., Dahiya, O.P., Nizar, M.A., Singh, M., Tilekar, V., Kumar, V., Dutta, M., 2016. Characterization of chickpea germplasm conserved in the Indian National Genebank and development of a core set using qualitative and quantitative trait data. Crop J. 4 (5), 417–424.

Basandrai, A.K., Pande, S., Basandrai, D., 2008. Recent advance in host-parasite interaction, chickpea *Ascochyta rabaie* and *Botrytis cinerea* systems. In: Setia, R.C., Nayyar, H., Setia, N. (Eds.), Crop Improvement, Strategies and Applications. IK International, New Delhi India, pp. 265–301.

Benzohra, I.E., Bendahmane, B.S., Benkada, M.Y., Labdi, M., 2014. Evaluation of wild *Cicer* species accessions for resistance to three pathotypes of *Ascochyta rabiei* (Pass.) Labr. in Algeria. Afr. J. Microbiol. Res. 8 (20), 2022–2029.

Berger, J.D., Buck, R., Henzell, J.M., Turner, N.C., 2005. Evolution in the genus *Cicer*-vernalisation response and low temperature pod set in chickpea (*C. arietinum* L.) and its annual wild relatives. Aust. J. Agric. Res. 56 (11), 1191–1200.

Bhagyawant, S.S., Gautam, A.K., Narvekar, D.T., Gupta, N., Bhadkaria, A., Srivastava, N., Upadhyaya, H.D., 2018. Biochemical diversity evaluation in chickpea accessions employing mini-core collection. Physiol. Mol. Biol. Plants 24 (6), 1165–1183.

Bharadwaj, C., Chauhan, S.K., Yadav, S., Tara Satyavathi, C., Singh, R., Kumar, J., Srivastava, R., Gayatri, R., 2011. Molecular marker-based linkage map of chickpea (*Cicer arietinum*) developed from desi × kabuli cross. Indian J. Agric. Sci. 81 (2), 116–118.

Brown, A.H.D., 1989a. Core collections: a practical approach to genetic resources management. Genome 31 (2), 818–824.

Brown, A.H.D., 1989b. The Case for Core Collections. The Use of Plant Genetic Resources, pp. 136–156.

Buhariwalla, H.K., 2006. ESTs from chickpea roots with putative roles indrought tolerance. Bio. Med. Central Plant Biol. 5, 16.

Canci, H., Toker, C., 2009. Evaluation of annual wild *Cicer* species for drought and heat resistance under field conditions. Genet. Resour. Crop Evol. 56 (1), 1.

Choudhary, P., Khanna, S.M., Jain, P.K., 2012. Genetic structure and diversity analysis of the primary gene pool of chickpea using SSR markers. Genet. Mol. Res. 11 (2), 891–905.

Cooper, C.L., Cooper, C.P., Dewe, P.J., O'Driscoll, M.P., O'Driscoll, M.P., 2001. Organizational Stress: A Review and Critique of Theory, Research, and Applications. Sage.

Cubero, J.I., 1975. Research on the chickpea (*Cicer arietinum*) in Spain. In: International Workshop on Grain Legumes. International Crops Research Institute for the Semi-Arid Tropics.

Deokar, A.A., Ramsay, L., Sharpe, A.G., Diapari, M., Sindhu, A., Bett, K., Warkentin, T.D., Tar'an, B., 2014. Genome wide SNP identification in chickpea for use in development of a high density genetic map and improvement of chickpea reference genome assembly. BMC Genomics 15 (1), 708.

Devasirvatham, V., Gaur, P.M., Mallikarjuna, N., Tokachichu, R.N., Trethowan, R.M., Tan, D.K., 2012. Effect of high temperature on the reproductive development of chickpea genotypes under controlled environments. Funct. Plant Biol. 39 (12), 1009–1018.

Diapari, M., Sindhu, A., Bett, K., Deokar, A., Warkentin, T.D., Taran, B., 2014. Genetic diversity and association mapping of iron and zinc concentrations in chickpea (*Cicer arietinum* L.). Genome 57 (8), 459–468.

Dixit, G.P., 2018. Project Coordinator's Report of All India Coordinated Research Project on Chickpea. Indian Institute of Pulses Research, Kanpur.

Eker, T., Erler, F., Adak, A., Imrek, B., Guven, H., Tosun, H.S., Sari, D., Sari, H., Upadhyaya, H.D., Toker, C., Ikten, C., 2018. Screening of chickpea accessions for resistance against the pulse beetle, *Callosobruchuschinensis* L. (Coleoptera: bruchidae). J. Stored Prod. Res. 76, 51–57.

IPGRI Handbooks for Genebanks No. 6. In: Engels, J.M.M., Visser, L. (Eds.), 2003. A Guide to Effective Management of Germplasm Collections. IPGRI, Rome, Italy.

Engels, J.M., Ramanatha, R., Brown, A.H.D., Jackson, M.T., 2002. Managing plant genetic diversity. In: Proceedings of an International Conference, Kuala Lumpur, Malaysia, 12–16 June 2000. CABI Publishing.

Frankel, O.H., Brown, A.H.D., 1984. Plant genetic resources today: a critical appraisal. In: Holden, J.H.W., Williams, J.T. (Eds.), Crop Genetic Resources: Conservation & Evaluation. George Allen & Unwin.

Golla, S.K., Rajasekhar, P., Akbar, S.M.D., Sharma, H.C., 2018. Proteolytic activity in the midgut of *Helicoverpa armigera* (Noctuidae: Lepidoptera) larvae fed on wild relatives of chickpea, *Cicer arietinum*. J. Econ. Entomol. 111 (5), 2409–2415.

Gowda, C.L.L., Upadhyaya, H.D., Dronavalli, N., Singh, S., 2011. Identification of large-seeded high-yielding stable kabuli chickpea germplasm lines for use in crop improvement. Crop Sci. 51 (1), 198–209.

Gujaria, N., Kumar, A., Dauthal, P., Dubey, A., Hiremath, P., Prakash, A.B., Farmer, A., Bhide, M., Shah, T., Gaur, P.M., Upadhyaya, H.D., 2011. Development and use of genic molecular markers (GMMs) for construction of a transcript map of chickpea (*Cicer arietinum* L.). Theor. Appl. Genet. 122 (8), 1577–1589.

Gupta, S., Nawaz, K., Parween, S., Roy, R., Sahu, K., Kumar Pole, A., Khandal, H., Srivastava, R., Kumar Parida, S., Chattopadhyay, D., 2016. Draft genome sequence of *Cicer reticulatum* L., the wild progenitor of chickpea provides a resource for agronomic trait improvement. DNA Res. 24 (1), 1–10.

Gupta, S., Parihar, A.K., 2015. Broadening the genetic base of pulse crops. In: Pulses: Challenges and Opportunities. Indian Society of Pulses Research & Development, IIPR, Kanpur, p. 86.

Halila, M.H., Strange, R.N., 1997. Screening of Kabuli chickpea germplasm for resistance to Fusarium wilt. Euphytica 96 (2), 273–279.

Hammer, K., 1984. Das domestikationssyndrom. Die Kulturpflanze 32 (1), 11–34.

Hannan, R.M., Kaiser, W.J., Muehlbauer, F.J., 1994. Development and utilization of the USDA chickpea germplasm core collection. In: Agronomy Abstracts, vol. 217. ASA, Madison, WI.

Harlan, J.R., de Wet, J.M., 1971. Toward a rational classification of cultivated plants. Taxon 20, 509–517.

Haware, M.P., Nene, Y.L., Pundir, R.P.S., Narayana Rao, J., 1992. Screening of world chickpea germplasm for resistance to fusarium wilt. Field Crops Res. 30 (1–2), 147–154.

Hegde, V.S., Tripathi, S., Bharadwaj, C., Agrawal, P.K., Choudhary, A.K., 2018. Genetics and genomics approaches to enhance adaptation and yield of chickpea (*Cicer arietinum* L.) in semi-arid environments. Sabrao J. Breed. Genet. 50 (2). http://www.fao.org/wiews-archive/germplasm_query.htm?i_l=EN.

Iruela, M., Rubio, J., Cubero, J.I., Gil, J., Millan, T., 2002. Phylogenetic analysis in the genus *Cicer* and cultivated chickpea using RAPD and ISSR markers. Theor. Appl. Genet. 104 (4), 643–651.

Jadhav, A.A., Rayate, S.J., Mhase, L.B., Thudi, M., Chitikineni, A., Harer, P.N., Jadhav, A.S., Varshney, R.K., Kulwal, P.L., 2015. Marker-trait association study for protein content in chickpea (*Cicer arietinum* L.). J. Genet. 94 (2), 279–286.

Jaganathan, D., Thudi, M., Kale, S., Azam, S., Roorkiwal, M., Gaur, P.M., Kishor, P.K., Nguyen, H., Sutton, T., Varshney, R.K., 2015. Genotyping-by-sequencing based intra-specific genetic map refines a "QTL-hotspot" region for drought tolerance in chickpea. Mol. Genet. Genom. 290 (2), 559–571.

Jain, M., Misra, G., Patel, R.K., Priya, P., Jhanwar, S., Khan, A.W., Shah, N., Singh, V.K., Garg, R., Jeena, G., Yadav, M., Kant, C., Sharma, P., Yadav, G., Bhatia, S., Tyagi, A.K., Chattopadhyay, D., 2013. A draft genome sequence of the pulse crop chickpea (*Cicer arietinum* L.). Plant J. 74 (5), 715–729.

Jansen, R.K., Wojciechowski, M.F., Sanniyasi, E., Lee, S.B., Daniell, H., 2008. Complete plastid genome sequence of the chickpea (*Cicer arietinum*) and the phylogenetic distribution of rps12 and clpP intron losses among legumes (Leguminosae). Mol. Phylogenetics Evol. 48 (3), 1204–1217.

Kaiser, W.J., Hannan, R.M., Muehlbauer, F.J., Mihov, M., 1998. First report of ascochyta blight of *Cicer montbretii*, a wild perennial chickpea in Bulgaria. Plant Dis. 82 (7), 830.

Kashiwagi, J., Krishnamurthy, L., Singh, S., Gaur, P.M., Upadhyaya, H.D., Panwar, J.D.S., Basu, P.S., Ito, O., Tobita, S., 2006b. Relationships between transpiration efficiency and carbon isotope discrimination in chickpea (*C. arietinum* L). J. SAT Agric. Res. 2 (1), 1–3.

Kashiwagi, J., Krishnamurthy, L., Singh, S., Upadhyaya, H.D., 2006a. Variation of SPAD chlorophyll meter readings (SCMR) in the mini-core germplasm collection of chickpea. J. SAT Agric. Res. 2 (1), 1–3.

Kashiwagi, J., Krishnamurthy, L., Upadhyaya, H.D., Krishna, H., Chandra, S., Vadez, V., Serraj, R., 2006. Genetic variability of drought-avoidance root traits in the mini-core germplasm collection of chickpea (*Cicer arietinum* L.). Euphytica 146 (3), 213–222.

Kaur, L., Sirari, A., Kumar, D., Singh, J.S., Singh, S., Kapoor, K., Singh, I., Gowda, C.L., Pande, S., Gaur, P., Sharma, M., 2013. Combining Ascochyta blight and Botrytis grey mould resistance in chickpea through interspecific hybridization. Phytopathol. Mediterr. 157–165.

Kazan, K., Kusmenoglu, I., Muehlbauer, F.J., Weeden, N.F., 1991. Duplication of aldolase and esterase loci in *cicer* (*cicereae* alef.). J. Hered. 82 (1), 58–63.

Kibret, K.T., 2011. Development and Utilization of Genetic Diversity Based Ethiopian Chickpea (*Cicer arietinum* L.) Germplasm Core Collection for Association Mapping. Doctoral dissertation. International Crops Research Institute for the Semi-Arid Tropics.

Knights, E.J., Southwell, R.J., Schwinghamer, M.W., Harden, S., 2008. Resistance to *Phytophthora medicaginis* Hansen and Maxwell in wild *Cicer* species and its use in breeding root rot resistant chickpea (*Cicer arietinum* L.). Aust. J. Agric. Res. 59 (4), 383–387.

Koinov, G., 1968. Chickpea (*Cicer arietnum* L.). Academy of Agricultural Sciences, sofia. 17.

Kujur, A., Upadhyaya, H.D., Shree, T., Bajaj, D., Das, S., Saxena, M.S., Badoni, S., Kumar, V., Tripathi, S., Gowda, C.L.L., Sharma, S., 2015. Ultra-high density intra-specific genetic linkage maps accelerate identification of functionally relevant molecular tags governing important agronomic traits in chickpea. Sci. Rep. 5, 9468.

Kumar, T., Bharadwaj, C., Rizvi, A.H., Sarker, A., Tripathi, S., Alam, A., Chauhan, S.K., 2015. Chickpea landraces: a valuable and divergent source for drought tolerance. IJTA (Int. J. Trop. Agric.) 33 (2 (Part II)), 633–638.

Kumar, S., Gupta, S., Singh, B.B., 2004. How wide the genetic base of pulse crops is: pulses in new perspective. In: Ali, M., Singh, B.B., Kumar, S., Dhar, V. (Eds.), Proceedings of the

National Symposium on Crop Diversification and Natural Resources Management. Indian Society of Pulses Research and Development, Kanpur, India, pp. 221–217.

Labdi, M., Robertson, L.D., Singh, K.B., Charrier, A., 1996. Genetic diversity and phylogenetic relationships among the annual *Cicer* species as revealed by isozyme polymorphism. Euphytica 88 (3), 181–188.

Ladizinsky, G., 1998. Plant Evolution under Domestication. Springer Science & Business Media.

Ladizinsky, G., Adler, A., 1975. Origin of chickpea as indicated by seed protein electrophoresis. Isr. J. Bot. 24.

Ladizinsky, G., Adler, A., 1976. Genetic relationships among the annual species of *Cicer* L. Theor. Appl. Genet. 48 (4), 197–203.

Li, Y., Ruperao, P., Batley, J., Edwards, D., Davidson, J., Hobson, K., Sutton, T., 2017. Genome analysis identified novel candidate genes for ascochyta blight resistance in chickpea using whole genome re-sequencing data. Front. Plant Sci. 8, 359.

Malhotra, R.S., Singh, K.B., 1991. Gene action for cold tolerance in chickpea. Theor. Appl. Genet. 82 (5), 598–601.

Maqbool, M.A., Aslam, M., Ali, H., 2017. Breeding for improved drought tolerance in Chickpea (*Cicer arietinum* L.). Plant Breed. https://doi.org/10.1111/pbr.12477.

Maxted, N., Castañeda Álvarez, N.P., Vincent, H.A., MagosBrehm, J., 2012. Gap Analysis: A Tool for Genetic Conservation. In: Collecting Plant Genetic Diversity: Technical Guidelines.

Meena, H.P., Kumar, J., Upadhyaya, H.D., Bharadwaj, C., Chauhan, S.K., Verma, A.K., Rizvi, A.H., 2010. Chickpea mini core germplasm collection as rich sources of diversity for crop improvement. J. SAT Agric. Res. 8, 1–5.

Millan, T., Clarke, H.J., Siddique, K.H., Buhariwalla, H.K., Gaur, P.M., Kumar, J., Gil, J., Kahl, G., Winter, P., 2006. Chickpea molecular breeding: new tools and concepts. Euphytica 147 (1–2), 81–103.

Muehlbauer, F.J., Singh, K.B., 1987. Genetics of Chickpea [*Cicer arietinum*].

Muehlbauer, F.J., 1993. Use of Wild Species as a Source of Resistance in Cool-Season Food Legume Crops.

Muehlbauer, F.J., Sarker, A., 2017. Economic Importance of Chickpea: Production, Value, and World Trade. Springer International Publishing, pp. 5–12.

Narayan, R.K.J., Macefield, A.J., 1976. Adaptive responses and genetic divergence in a world germplasm collection of chickpea (*Cicer arietinum* L.). Theor. Appl. Genet. 47 (4), 179–187.

Nguyen, T.T., Taylor, P.W.J., Redden, R.J., Ford, R., 2004. Genetic diversity estimates in *Cicer* using AFLP analysis. Plant Breed. 123 (2), 173–179.

Pande, S., Kishore, G.K., Rao, J.N., 2004. Evaluation of chickpea lines for resistance to dry root rot caused by Rhizoctonia. Int. Chickpea Pigeonpea Newsl. 11 (2), 37–38.

Pande, S., Kishore, G.K., Upadhyaya, H.D., Rao, J.N., 2006. Identification of sources of multiple disease resistance in mini-core collection of chickpea. Sci. Soc. 90 (9), 1214–1218.

Pande, S., Sharma, M., Gaur, P.M., Gowda, C.L.L., 2010. Host Plant Resistance to Ascochyta Blight of Chickpea. Information Bulletin No. 82. International Crops Research Institute for the Semi-Arid Tropics.

Parameshwarappa, S.G., Salimath, P.M., Upadhyaya, H.D., Patil, S.S., Kajjidoni, S.T., Patil, B.C., Narayana, Y.D., 2012. Variation in root characters of selected drought tolerant accessions of chickpea (*Cicer arietinum* L.) grown under terminal drought. Karnataka J. Agric. Sci. 25 (3), 389–391.

Parween, S., Nawaz, K., Roy, R., Pole, A.K., Suresh, B.V., Misra, G., Jain, M., Yadav, G., Parida, S.K., Tyagi, A.K., Bhatia, S., 2015. An advanced draft genome assembly of a desi type chickpea (*Cicer arietinum* L.). Sci. Rep. 5, 12806.

Petrova, S., Angelova, S., 2013. *In-situ* conservation of rare wild perennial chickpea (*Cicer montbretia*Jaub. &Spach) found in Strandja mountain, Bulgaria. Crop Wild Relative 44.

Pfaff, T., Kahl, G., 2003. Mapping of gene-specific markers on the genetic map of chickpea (*Cicer arietinum* L.). Mol. Genet. Genom. 269 (2), 243–251.

Roorkiwal, M., Von Wettberg, E.J., Upadhyaya, H.D., Warschefsky, E., Rathore, A., Varshney, R.K., 2014. Exploring germplasm diversity to understand the domestication process in *Cicer* s using SNP and DArT markers. PLoS One 9 (7), e102016.

Saeed, A., Darvishzadeh, R., 2017. Association analysis of biotic and abiotic stresses resistance in chickpea (*Cicer* s) using AFLP markers. Biotechnol. Biotechnol. Equip. 31 (4), 698–708.

Sandhu, J.S., 2004. Pers. Communication. Punjab Agricultural University, Ludhiana, Punjab, India.

Sandhu, J.S., Gupta, S.K., Singh, G., Sharma, Y.R., Bains, T.S., Kaur, L., Kaur, A., 2006. Interspecific hybridization between *Cicer arietinum* L. and *Cicer pinnatifidum* Jaub. et. Spach for improvement of yield and other traits. In: Fourth International Food Legumes Research Conference, October 18–22, 2005. Indian Society of Genetics and Plant Breeding, New Delhi, p. 192.

Sandhu, J.S., Singh, G., Bains, T.S., Sharma, Y.R., Singh, I., Sidhu, P.S., Singh, S., 2004. Pbg 5: a new multiple disease resistant desi chickpea variety for Punjab, India. Int. Chickpea Pigeonpea Newsl. 11, 18–20.

Serraj, R., Krishnamurthy, L., Upadhyaya, H.D., 2004. Screening chickpea mini-core germplasm for tolerence to soil salinity. Int. Chickpea Pigeonpea Newsl. 11, 29–32.

Sewak, S., Iquebal, M.A., Singh, N.P., Solanki, R.K., 2012. Genetic diversity studies in chickpea (*Cicer arietinum*) germplasm. J. Food Legumes 25 (1), 31–36.

Shah, T.M., Haq, M.A., Atta, B.M., Alam, S.S., Ali, H., 2005. Evaluation of *Cicer* species for resistance to ascochyta blight. Pak. J. Bot. 37 (2), 431.

Sharma, H.C., Bhagwat, M.P., Pampapathy, G., Sharma, J.P., Ridsdill-Smith, T.J., 2006. Perennial wild relatives of chickpea as potential sources of resistance to Helicoverpa armigera. Genet. Resour. Crop Evol. 53 (1), 131.

Sharma, H.C., Pampapathy, G., Lanka, S.K., Ridsdill-Smith, T.J., 2005. Exploitation of wild *Cicer reticulatum* germplasm for resistance to *Helicoverpa armigera*. J. Econ. Entomol. 98 (6), 2246–2253.

Sharma, K.D., Winter, P., Kahl, G., Muehlbauer, F.J., 2004. Molecular mapping of *Fusarium oxysporum* f. sp. *Cicer* is race 3 resistance gene in chickpea. Theor. Appl. Genet. 108 (7), 1243–1248.

Sharma, S., Upadhyaya, H.D., Roorkiwal, M., Varshney, R.K., Gowda, C.L.L., 2013. Chickpea. In: Singh, M., Upadhyaya, H.D., Bisht, I.S. (Eds.), Genetic and Genomic Resources of Grain Legume Improvement, first ed. Elsevier, London, pp. 81–111.

Siddique, K.H.M., Brinsmead, R.B., Knight, R., Knights, E.J., Paull, J.G., Rose, I.A., 2000. Adaptation of chickpea (*Cicer arietinum* L.) and faba bean (*Viciafaba* L.) to Australia. In: Linking Research and Marketing Opportunities for Pulses in the 21st Century. Springer, Dordrecht, pp. 289–303.

Simon, C.J., Muehlbauer, F.J., 1997. Construction of a chickpea linkage map and its comparison with maps of pea and lentil. J. Hered. 88 (2), 115–119.

Singh, K.B., Ocampo, B., Robertson, L.D., 1998. Diversity for abiotic and biotic stress resistance in the wild annual Cicer species. Genet. Resour. Crop Evol. 45 (1), 9–17.

Singh, G., Singh, K., Kapoor, M., 1982. Screening for sources of resistance to ascochyta blight of chickpea. Int. Chickpea Newsl. 7, 13–14.

Singh, K.B., Singh, O., 1997. Prospects of creating higher yield potential in chickpea. In: Recent Advances in Pulses Research. Indian Society of Pulses Research and Development, IIPR, Kanpur, 247(9).

Singh, K.B., Malhotra, R.S., Saxena, M.C., 1990. Sources for tolerance to cold in *cicer* species. Crop Sci. Soc. Am. 30 (5), 1136.

Singh, K.B., Pundir, R.P.S., Robertson, L.D., van Rheene, H.A., Singh, U., Kelley, T.J., Rao, P.P., Johansen, C., Saxena, N.P., 1997. Chickpea. In: Fuccillo, D., Sears, L., Stapleton, P. (Eds.), Biodiversity in Trust. Cambridge University Press, pp. 100–113.

Singh, M., Bhardwaj, C., Singh, S., Panatu, S., Chaturvedi, S.K., Rana, J.C., Rizvi, A.H., Kumar, N., Sarker, A., 2016. Chickpea genetic resources and its utilization in India: current status and future prospects. Indian J. Genet. Plant Breed. 76 (4), 515.

Singh, M., Bisht, I.S., Dutta, M., Kumar, K., Basandrai, A.K., Kaur, L., Sirari, A., Khan, Z., Rizvi, A.H., Sarker, A., Bansal, K.C., 2014. Characterization and evaluation of wild annual species for agro-morphological traits and major biotic stresses under Northwestern Indian conditions. Crop Sci. Soc. Am. 54 (1), 229.

Singh, M., Kumar, K., Bisht, I.S., Dutta, M., Rana, M.K., Rana, J.C., Bansal, K.C., Sarker, A., 2015. Exploitation of wild annual *Cicer* species for widening the gene pool of chickpea cultivars. Plant Breed. 134 (2), 186–192.

Singh, N.P., Pratap, A., 2016. Food legumes for nutritional security and health benefits. In: Biofortification of Food Crops. Springer, New Delhi, pp. 41–50.

Singh, R.P., Singh, I., Singh, S., Sandhu, J.S., 2012. Assessment of genetic diversity among interspecific derivatives in chickpea. J. Food. Legum. 25, 150–152.

Singh, S., Gumber, R.K., Joshi, N., Singh, K., 2005. Introgression from wild *Cicer reticulatum* to cultivated chickpea for productivity and disease resistance. Plant Breed. 124 (5), 477–480.

Singh, T.P., Deshmukh, P.S., Kumar, P., 2008. Relationship between physiological traits in chickpea (*Cicer arietinum* L.) under rainfed condition. Indian J. Plant Physiol. 13, 411–413.

Smartt, J., 1990. Grain Legumes: Evolution and Genetic Resources. Cambridge University Press.

Smithson, J.B., Thompson, J.A., Summerfield, R.J., 1985. Chickpea (*Cicer arietinum L.*). In: Summerfield, R.J., Roberts, E.H. (Eds.), Grain Legume Crops. Collins, London, UK, pp. 312–390.

Spillane, C., Gepts, P., 2001. Evolutionary and genetic perspectives on the dynamics of crop gene pools. In: Broadening the Genetic Base of Crop Production. CABI Publishing, pp. 25–70.

Srivastava, R., Bajaj, D., Malik, A., Singh, M., Parida, S.K., 2016. Transcriptome landscape of perennial wild *Cicer microphyllum* uncovers functionally relevant molecular tags regulating agronomic traits in chickpea. Sci. Rep. 6, 33616.

Stalker, H.T., 1980. Utilization of wild species for crop improvement. In: Advances in Agronomy, vol. 33. Academic Press, pp. 111–147.

Tayyar, R., Federici, C.V., Waines, G.J., 1996. Natural outcrossing in chickpea (*Cicer arietinum* L.). Crop Sci. 36 (1), 203–205.

Thompson, J.P., Reen, R.A., Clewett, T.G., Sheedy, J.G., Kelly, A.M., Gogel, B.J., Knights, E.J., 2011. Hybridisation of Australian chickpea cultivars with wild *Cicer* spp increases resistance to root-lesion nematodes (*Pratylenchus thornei* and *P. neglectus*). Australas. Plant Pathol. 40 (6), 601.

Thudi, M., Upadhyaya, H.D., Rathore, A., Gaur, P.M., Krishnamurthy, L., Roorkiwal, M., Nayak, S.N., Chaturvedi, S.K., Basu, P.S., Gangarao, N.V.P.R., Fikre, A., 2014. Genetic dissection of drought and heat tolerance in chickpea through genome-wide and candidate gene-based association mapping approaches. PLoS One 9 (5), e96758.

Thudi, M., Chitikineni, A., Liu, X., He, W., Roorkiwal, M., Yang, W., Jian, J., Doddamani, D., Gaur, P.M., Rathore, A., Samineni, S., Saxena, R.K., Xu, D., Singh, N.P., Chaturvedi, S.K., Zhang, G., Wang, J., Datta, S.K., Xu, X., Varshney, R.K., 2016. Recent breeding programs enhanced genetic diversity in both desi and kabuli varieties of chickpea (*Cicer arietinum* L.). Sci. Rep. 6 (1).

Toker, C., 2005. Preliminary screening and selection for cold tolerance in annual wild *Cicer* species. Genet. Resour. Crop Evol. 52 (1), 1–5.

Toker, C., Canci, H., Yildirim, T., 2007. Evaluation of perennial wild Cicer species for drought resistance. Genet. Resour. Crop Evol. 54 (8), 1781–1786.

Turner, N.C., Wright, G.C., Siddique, K.H.M., 2001. Adaptation of grain legumes (pulses) to water-limited environments. Adv. Agron. 193–231.

Tuwafe, S., Kahier, A.L., Boe, A., Ferguson, M., 1988. Inheritance and geographical distribution of allozyme polymorphisms in chickpea (*Cicer arietinum* L.). J. Hered. 79 (3), 170–174.

Upadhyaya, H.D., Ortiz, R., 2001. A mini core subset for capturing diversity and promoting utilization of chickpea genetic resources in crop improvement. Theor. Appl. Genet. 102 (8), 1292–1298.

Upadhyaya, H.D., Bajaj, D., Narnoliya, L., Das, S., Kumar, V., Gowda, C.L.L., Sharma, S., Tyagi, A.K., Parida, S.K., 2016. Genome-wide scans for delineation of candidate genes regulating seed-protein content in chickpea. Front. Plant Sci. 7, 302.

Upadhyaya, H.D., Bramel, P.J., Singh, S., 2001. Development of a chickpea core subset using geographic distribution and quantitative traits. Crop Sci. Soc. Am. 41 (1), 206.

Upadhyaya, H.D., Dronavalli, N., Dwivedi, S.L., Kashiwagi, J., Krishnamurthy, L., Pande, S., Sharma, H.C., Vadez, V., Singh, S., Varshney, R.K., Gowda, C.L.L., 2013. Mini core collection as a resource to identify new sources of variation. Crop Sci. 53 (6), 2506–2517.

Upadhyaya, H.D., Dwivedi, S.L., Baum, M., Varshney, R.K., Udupa, S.M., Gowda, C.L., Hoisington, D., Singh, S., 2008. Genetic structure, diversity, and allelic richness in composite collection and reference set in chickpea (*Cicer arietinum* L.). BMC Plant Biol. 8 (1), 106.

Upadhyaya, H.D., Furman, B.J., Dwivedi, S.L., Udupa, S.M., Gowda, C.L.L., Baum, M., Crouch, J.H., Buhariwalla, H.K., Singh, S., 2006b. Development of a composite collection for mining germplasm possessing allelic variation for beneficial traits in chickpea. Plant Genet. Res. 4 (1), 13–19.

Upadhyaya, H.D., Gowda, C.L.L., Buhariwalla, H.K., Crouch, J.H., 2006a. Efficient use of crop germplasm resources: identifying useful germplasm for crop improvement through core and mini-core collections and molecular marker approaches. Plant Genet. Res. 4 (1), 25–35.

Upadhyaya, H.D., Ortiz, R., Bramel, P.J., Singh, S., 2002. Phenotypic diversity for morphological and agronomic characteristics in chickpea core collection. Euphytica 123 (3), 333–342.

Usberti, R., Gomes, R.B.R., 1998. Seed viability constants for groundnut. Ann. Bot. 82 (5), 691–694.
Van der Maesen, L.J.G., 1972. *Cicer* L., a monograph of the genus, with special reference to the chickpea (*Cicer arietinum* L.), its ecology and distribution. Mendelingen Land bou who geschool Wageningen,Wageningen, The Netherlands, 1–341.
Van der Maesen, L., Pundir, R., 1984. Availability and use of wild Cicer germplasm. Plant Genetic Resources Newsletter (IBPGR/FAO) 57, 19–24.
Varshney, R.K., Song, C., Saxena, R.K., Azam, S., Yu, S., Sharpe, A.G., Cannon, S., Baek, J., Rosen, B.D., Tar'an, B., Millan, T., Zhang, X., Ramsay, L.D., Iwata, A., Wang, Y., Nelson, W., Farmer, A.D., Gaur, P.M., Soderlund, C., Penmetsa, R.V., Xu, C., Bharti, A.K., He, W., Winter, P., Zhao, S., Hane, J.K., Carrasquilla-Garcia, N., Condie, J.A., Upadhyaya, H.D., Luo, M.C., Thudi, M., Gowda, C.L.L., Singh, N.P., Lichtenzveig, J., Gali, K.K., Rubio, J., Nadarajan, N., Dolezel, J., Bansal, K.C., Xu, X., Edwards, D., Zhang, G., Kahl, G., Gil, J., Singh, K.B., Datta, S.K., Jackson, S.A., Wang, J., Cook, D.R., 2013. Draft genome sequence of chickpea (*Cicer arietinum*) provides a resource for trait improvement. Nat. Biotechnol. 31 (3), 240–246.
Vavilov, N.I., 1926. Studies on the origin of cultivated plants. Bull. Appl. Bot. Genet. Plant Breed. 16, 1–248.
Vavilov, N.I., 1951. Phytogeographic basis of plant breeding. The origin, variation, immunity and breeding of cultivated plants. Chron. Bot. 13, 1–366.
Williams, P.C., Singh, U., 1987. Nutritional quality and the evaluation of quality in breeding programmes. In: Saxena, M.C., Singh, K.B. (Eds.), The Chickpea. CAB International, Wallingford, pp. 329–356.
Zohary, D., Hopf, M., 1973. Domestication of pulses in the Old World: legumes were companions of wheat and barley when agriculture began in the Near East. Science 182 (4115), 887–894.

CHAPTER

Conventional cytogenetic manipulations

4

Gopal Katna[1], S.D. Nitesh[2], Kamal Dev Sharma[3]

[1]*Department of Organic Agriculture and Natural Farming, CSK Himachal Pradesh Agricultural University, Palampur, Himachal Pradesh, India;* [2]*Department of Crop Improvement, CSK Himachal Pradesh Agricultural University, Palampur, Himachal Pradesh, India;* [3]*Department of Agricultural Biotechnology, CSK Himachal Pradesh Agricultural University, Palampur, Himachal Pradesh, India*

4.1 Introduction

The primary aim of any breeding program is to evolve varieties with high yield, improved quality, early maturity, and resistance to biotic and abiotic stresses by genetic manipulation through conventional as well as modern biotechnological tools. The progenitors of a large variety of cultivated plants used by mankind today are wild relatives that evolved over time through selection or natural hybridization to produce high-yielding genotypes. In the words of Altman (1999), "the release of new improved genotypes of classical breeding is now too slow to cope up with the demands and is considerably limited by the lack of natural genes that can be introgressed through hybridization." Yield losses in legumes are due to various diseases (fungal, bacterial, viral) and parasitic weeds which affect the crops at different stages (Dita et al., 2006). Pulse seeds are also damaged by various storage insect-pests as they are rich source of protein (Clement et al., 1994, 1999). The sources for stress resistance are limited in available germplasm, so wild species can be sources of desirable genes for introgression to cultivated species. Among pulses, chickpea (*Cicer arietinum* L.) is one of the earliest grain crops cultivated by man and has been found in Middle Eastern archaeological sites in 7500–6800 BC (Zohary and Hopf, 2000). It is one of the most widely cultivated legume crop ranking second in area and third in production in the world. Due to its self-pollinating behavior (Gowda, 1981) and cultivation in selected environments in the Mediterranean and south Asia, chickpea did not face adequate selection pressure and consequently remained a crop with narrow genetic base. More recently, the breeding programs aimed at intervarietal/intraspecific hybridization further narrowed down its genetic base (Chaturvedi and Nadarajan, 2010). Chickpea germplasm lacks desirable agronomic traits for improvement of current day cultivars especially to abiotic stresses and ascochyta blight resistance (Millan et al., 2006). Because chickpea is a *Rabi* season crop, its productivity is hampered by low temperature, drought, and various abiotic and biotic stresses.

Chickpea: Crop Wild Relatives for Enhancing Genetic Gains. https://doi.org/10.1016/B978-0-12-818299-4.00004-X
Copyright © 2020 Elsevier Inc. All rights reserved.

The wild relatives of chickpea are a valuable source of genes for disease and pest resistance and stress tolerance. Transfer of these traits from wild *Cicer* species to cultigens is relatively easy if the trait is governed by the species of the primary gene pool such as *C. reticulatum* (the wild progenitor of chickpea) and *C. echinospermum*. It is difficult to transfer traits from *Cicer* species from the secondary and tertiary gene pools. Thus, there is scope for improvement of production as well as productivity of chickpea by transferring genes for various biotic (fungi, bacteria, viral, pod borer, etc.) and abiotic (cold, drought, salinity, etc.) stresses into the available germplasm from closely related species. Not only the specific characters but also wild *Cicer* collections can be exploited to broaden the genetic base in chickpea, i.e., increase in genetic variability through interspecific or wide hybridization (Hawkes, 1977; van der Maesen and Pundir, 1984; van Rheenen et al., 1993) and enhanced variability can be exploited to improve already existing cultivars by introgressing desirable traits such as yield and resistance to biotic and abiotic stresses. The crossability barriers between *C. arietinum* and wild relatives, however, are the major hindrance in exploitation of traits from wild species, especially from the species of secondary and tertiary gene pools. *C. echinospermum*, a species in the primary gene pool, also remained poorly exploited primarily due to poor crossability with *C. arietinum*. In addition to hybridization, induced mutagenesis is another means of generation of genetic variability, and several varieties from mutated chickpea material have been released worldwide.

4.2 Production-related problems

According to Millan et al. (2006), three major developments have taken place in the cultivated chickpea that has resulted into the narrow genetic base and reduced genetic diversity in the cultivars—shift in chickpea cultivation from autumn to spring-sown crop, selection of chickpea genotypes suitable for post–rainy season cultivation, and finally the replacement of landraces with elite cultivars developed by modern breeding tools.

Chickpea was traditionally grown in northern and central India due to its requirement of low temperature and prolonged winter for better growth. But due to the intensification of *Rabi* season wheat cultivation from green revolution, the area has been reduced significantly under winter cultivation, and cultivation has been shifted to comparative warmer and harsher environment of southern India (Dixit et al., 2019). Chickpea production and productivity is also affected by biotic (diseases like wilt, dry root rot, ascochyta blight, botrytis gray mold, etc.) and abiotic (heat, drought, salinity, cold, etc.) stresses during crop duration. The average crop loss is to the tune of 25%–30% to various biotic and abiotic stresses; it varies from 5% to 10% in temperate regions, 50%–100% in tropical regions; it is also hampered by various factors present in the field and storage conditions (van Emden et al., 1988; Clement et al., 2000; Dixit et al., 2019). Among the various biotic stresses, 10%–25% yield losses are caused by wild and dry root rot diseases, while the latter becoming severe in most of the chickpea growing regions in India (Mahendra, 1998; Masood and Shivkumar, 2001). Cultivation of the chickpea in marginal lands under a rainfed situation where soil fertility and soil moisture levels remain low is another abiotic stress causing low and unstable productivity of chickpea in South Asia (Kumar and van Rheenen, 2000). Major biotic and abiotic stresses in chickpea are discussed in the following sections.

4.2.1 Ascochyta blight

It was first described in the North-West Frontier Province of British India during 1911 by Butler. It occurs mainly in areas where cool, cloudy, and humid weather occurs during the crop season. It is the most important and destructive foliar disease that is caused by fungus *Ascochyta rabiei*. The disease occurs in chickpea growing areas throughout the world including India and causes yield losses ranging from 10% to 100% (Nene and Reddy, 1987; Singh, 1990). As per the report of Hawtin and Singh (1984), ascochyta blight, in the Mediterranean region, has reached the epidemic level due to the shift in the cropping season from spring to winter. The disease is more severe in India, Pakistan, the United States, Northwest Pacific, Australia, and Syria (Malhotra et al., 2003). Genetic resistance to the ascochyta blight has been reported to be governed by single dominant gene (Eser, 1976; Vir et al., 1975), single recessive gene (Singh and Reddy, 1983), or both recessive and dominant genes (Bhardwaj et al., 2010). In controlled environmental growth chamber, seedling screening and cut twig screening using excised trays are employed to screen chickpea genotype against ascochyta blight (Pande et al., 2011), while mini-dome technique of screening was developed by Chen and Muehlbauer (2003).

4.2.2 Fusarium wilt

Fusarium wilt is prevalent in almost all chickpea growing areas of the world irrespective of the climate. Its incidence has been reported from India, Pakistan, Bangladesh, Ethiopia, Malawi, Mexico, Peru, Syria, Spain, and the United States (Nene et al., 1984). The disease causes yield losses in the range of 10%–90% depending upon time and location of incidence (Singh and Dahiya, 1973; Jimenez-Diaz et al., 1989a,b; Singh and Reddy, 1991). Studies from India by Haware and Nene in 1982 found that the pathogen has four physiological races viz., 1, 2, 3, and 4, while Jimenez-Diaz et al. (1989a,b) reported races 0 and 5 from Spain, and Phillips (1988) reported race 6 from the United States. Host (chickpea) resistance to these races is available in *C. arietinum* and is governed by monogenes to polygenes (Upadhyaya et al., 1983; Sharma et al., 2004a,b, 2005a,b; Sharma et al., 2005; Sharma and Muehlbauer, 2005; Sharma and Muehlbauer, 2007). Because the pathogens are soil-borne, it is difficult to erode them from soil or exploit chemical management strategies, and host resistance is the best option available to manage the disease. Standardized field and laboratory screening techniques for evaluation of resistance in chickpea germplasm lines to *F. oxysporum* f. sp. *ciceris* are available (Nene et al., 1981; Sharma et al., 2004b).

4.2.3 Botrytis gray mold

The first report of this disease on chickpea was in 1915 (Shaw and Ajrekar, 1915). This is the second most important foliar disease of chickpea after ascochyta blight. The disease has been reported from Argentina, Australia, Bangladesh, Canada, Colombia, India, Nepal, Pakistan, Spain, and the United States (Nene et al., 1984). During 1979–82, the disease caused huge losses in Indo-Gangetic plains of India (Grewal and Laha, 1983). The presently available germplasm lacks stable sources of resistance to botrytis gray mold (Singh and Reddy, 1991); however, resistance to the disease exists in two wild species of chickpea viz., *C. pinnatifidum* and *C. judaicum* (Singh et al., 1982; van der Maesen and Pundir, 1984).

4.2.4 Dry root rot

The disease was reported for the first time in India by Mitra in 1931. Subsequently, the disease has been reported from Iran (Kaiser et al., 1968), the United States (Westerlund et al., 1974), Australia, Ethiopia, and Pakistan (Nene et al., 1984). The disease has gained the status of importance in several chickpea growing regions in India (Masood and Shivkumar, 2001). Mahendra (1998) reported 10%–25% losses in yields due to wilt and dry root rot disease.

4.2.5 Powdery mildew

Powdery mildew incidence was first reported by Nour in 1957 from Sudan. Later on, it was reported in Pakistan (Khan and Kamal, 1962) and India (Mishra and Kulkarni, 1973; Khune et al., 1979; Mahrshi and Mathur, 1981). During favorable conditions, the disease can reduce the yield significantly, though it is a minor disease in chickpea.

4.2.6 Pod borer

Cultivated chickpea is attacked by few insect-pests as compared with other tropical legumes. Among the insect-pests, the pod borer (*Helicoverpa armigera*) is the major insect-pest attacking the chickpea crop in Asia, Africa, Australia, and other regions. Chickpea is not the only host to this pest as *H. armigera* a polyphagous pest feeds on more than 182 plant species (Fitt, 1989; Sharma and Ortiz, 2002). Resistance against pod borer is not available in chickpea (Singh et al., 2008a,b).

4.2.7 Bruchids

Bruchids, a major storage pest, causes huge damage to stored seeds of chickpea (Southgate, 1978). It has a short life cycle of about 20 days; each female had high fecundity (>100 eggs), leading the production of huge numbers within a few months (Howe and Currie, 1964). Its infestation leads to postharvest losses of up to 13% in Mediterranean region and in India (Mookherjee et al., 1970; Dias and Yadav, 1988; Weigand and Tahhan, 1990).

4.2.8 Abiotic stresses

Drought, cold, and heat are the three major abiotic stresses in chickpea. All the three stresses affect chickpea at all developmental stages, but stress at the reproductive stage is more devastating as it induces pollen sterility and ovule sterility, inhibits pollen formation, and reduces stigma receptivity and pollen load on stigma (Fang et al., 2009; Kumar et al., 2013; Kiran et al., 2019). Like cold, it is well documented that flowers are the most sensitive part of the plant to temperature stress (Wery et al., 1993; Toker and Canci, 2006; Kiran et al., 2019). During flowering or reproductive period, if temperatures rise above the threshold level, these affect adversely the pod formation and seed set, resulting in reduced grain yield (Summerfield et al., 1984; Wang et al., 2006; Basu et al., 2009; Kumar et al., 2013). In comparison to other cool-season legume crops, chickpea is more tolerant to heat stress (Summerfield et al., 1984; Erskine et al., 1994; McDonald and Paulsen, 1997; Patrick and Stoddard, 2010).

These abiotic stresses play an important role in the reduction of chickpea productivity; e.g., during the mid-1970s, drought alone reduced chickpea cultivation to half in Ethiopia. Drought tolerance is a desirable characteristic for chickpea as the crop mainly grows on conserved soil moisture. The crop in most parts of the world faces terminal drought that is more devastating than stress at early stages (Khanna-Chopra and Sinha, 1987). In the Mediterranean region, winter-sown chickpea is more productive than that grown during spring season (Singh and Hawtin, 1979) due to longer growing season and better moisture availability; however, winter-sown chickpea is prone to cold stress and ascochyta blight when temperatures rise. During winter, the flowers drop and pods abort leading to major yield loss, as mean day temperature during winter falls below 15°C (Savithri et al., 1980; Srinivasan et al., 1999; Clarke and Siddique, 2004; Nayyar et al., 2005). Usually, heat and drought stresses occur in combination (Toker and Canci, 2009), thereby, further enhancing stress-induced crop yield losses. Little progress has been made on heat tolerance breeding (Malhotra and Saxena, 1993) as well as cold and drought stress breeding, though drought is the most studied stress in chickpea.

4.3 Narrow genetic bases: a bottleneck for enhancing productivity

During domestication of chickpea, a series of evolutionary bottlenecks resulted in a narrow genetic base among the cultivated chickpea. The domestication of chickpea made it an autumn-sown crop in India and East Africa, as well as the Mediterranean region. It is believed that cold stress during the vegetative stage and blight at the time of temperature increase forced a shift in cultivation from autumn to spring season in the Mediterranean basin to avoid both the cold and more probably the ascochyta blight (Abbo et al., 2003a,b). During this process, winter hardiness and vernalization requirement alleles were lost. The present day chickpea germplasm lines do not require vernalization, though its progenitor *C. reticulatum* require vernalization (Summerfield et al., 1989; Singh et al., 1997). According to seed morphology, there are two types of cultivated chickpea representing two gene pools namely, *desi* and *kabuli*. The seed of *desi* chickpea is relatively smaller with thick seed coats of dark-brown color and grown in semiarid tropical regions. Seeds of *kabuli* chickpeas have large seeds with thin seed coats of white-cream color and are usually grown in regions of temperate climate. In India, *desi* type of chickpea accounts for nearly 90% of total area under cultivation and remaining is occupied by *kabuli* type. Both the *desi* and *kabuli* types display limited variability and lack desirable agronomic traits needed for ready utilization in varietal improvement programs (Millan et al., 2006). As pointed out earlier, *C. arietinum* germplasm lacks resistance/tolerance to several biotic and abiotic stresses. Genetic advance for yield in chickpea is low because of limited genetic variation present in the germplasm which leads to classification of chickpea as a recalcitrant crop (van Rheenen et al., 1993). Furthermore, intermating between these accessions or elite lines of the same origin or intervarietal breeding programs have resulted in narrow genetic base of the existing cultivars (Chaturvedi and Nadarajan 2010).

4.4 Conventional chickpea breeding strategies

Chickpea is one of the oldest grain legume crop cultivated by man. During domestication, selection remained the method of choice for selecting superior lines for cultivation. In modern era, selection, hybridization, and mutation were used to develop new varieties and improve the existing ones. Hybridization included both *desi* × *desi*, *kabuli* × *kabuli*, and *desi* × *kabuli* crosses. *Desi* × *kabuli* crosses have been used extensively for exploiting genes present in both types and developed high-yielding progeny (Yadav et al., 2004).

Backcross breeding is commonly used to improve the popular varieties for disease resistance. Pedigree, bulk, or variants of these methods have been/are being used for varietal development in chickpea. Pedigree method of breeding was earlier used by many breeders, but later on, due to a cumbersome and limited number of crosses that can be handled, this method of breeding chickpea was soon replaced by other methods. The bulk method with its various modifications is now the most common selection method in cross-bred chickpea. During the early segregating generations (F_2 and F_3), the selection is done for simple traits, while single plant selection for yield usually starts from F_4 generation onward. Slinkard et al. (2000) suggested a recombinant-derived family method that uses early generation selection yield in F_2-derived F_4 or later generation families to eliminate inferior crosses and inferior F_2-derived families. Acceleration of breeding programs by advancing two or more generations per year is being practiced by using either off-season nurseries or greenhouse facilities. Single seed descent (SSD) method is commonly employed to advance generations in greenhouse due to space constraints. Sethi et al. (1981) induced early flowering by extending day length to 15–16 h through artificial lighting until floral initiation.

4.5 Cytogenetics and chickpea improvement

Genome-wide snapshots of chromosomes of some *Cicer* species including *C. arietinum* were also generated. The chromosome structure was also used to elucidate intervarietal differences. Position of primary and secondary constrictions and lengths of the individual chromosomes was used to find the intervarietal difference (Phadnis, 1970; Kutarekar and Wanjari, 1983; Sharma, 1983). In some cases, pachytene analysis karyotypes in *Cicer* species were prepared and compared with somatic karyotypes (Sharma and Gupta, 1986). Ahmad and Hymowitz (1993) prepared an idiogram of chickpea based on chromosome length, arm ratio, and distribution of heterochromatin and euchromatin. On the basis of descending order of chromosome length, they numbered the chickpea chromosomes from 1 to 8. At pachytene stage they observed 335.33 μ total lengths of the chromosome complement and chromosome size ranged from 58.05 to 30.53 μ. Ahmad et al. (1987) developed interspecific hybrid between *Cicer judaicum* Boiss. and *Cicer chorassanicum* (Bge.). The hybridity was confirmed using cytology and hybrid had $2n = 16$ chromosomes. The chiasmata frequency (5.4 ± 1.2 vs. 12.1 and 11.4 in the parental species) per cell of the hybrid was low, and it was highly sterile. The flowers were abnormal in that the stigma and style grew out of the keel, while the anthers remained inside. Venora et al. (1995) observed little difference among the four genotypes of *Triticum turgidum* var. *durum* Desf., *Pisum sativum* L., *Pisum fulvum* Sibth. and Sm., and *Cicer arietinum* L. for total genome length, chromosome length, and complement formula during the mitotic analysis. They carry a distinguishable satellite on the longest chromosome pair. *Kabuli* chickpea appears to have narrow adaptation as there was little variation in karyological parameters. Zatloukalova et al. (2011) established various genetic linkage maps and identified markers linked to resistance genes. They sorted mitotic chromosomes using flow cytometry. It was also established LG5 (linkage group 5) was linked to the largest chromosome A, LG4 to a medium-sized chromosome E, LG3 to the second-largest chromosome B, chromosomes F and G were anchored to LG1 and LG2 (Zatloukalova et al., 2011). Chromosomes C and D could not be flow-sorted separately and were jointly associated with LG6 and LG7 (Zatloukalova et al., 2011). They constructed a partial genomic bacterial artificial chromosome (BAC) library, and putative single/low-copy BAC clones were mapped cytogenetically.

4.6 Exploitation of marker-assisted selection

Markers linked to several useful traits have been identified/developed in chickpea; e.g., markers linked to fusarium wilt resistance genes in chickpea (Sharma et al., 2004b; Sharma and Muehlbauer, 2007), QTLs for vernalization response (Samineni et al., 2016), and QTLs for drought tolerance (Varshney et al., 2014) have been identified in chickpea. The markers associated with *goi* (gene of interest) or QTLs are used in marker-assisted selection (MAS) in monitoring the gene or QTL introgression. MAS is employed for increased precision and efficiency over the conventional plant breeding methods. MAS is specifically useful for those traits that are difficult to assess phenotypically, viz., root traits for drought avoidance, antinutritional factors, quality traits, a pyramiding of resistance genes from diverse sources, combo-resistance for two or more biotic or abiotic stresses. MAS will also be used for tracking introgression of resistance genes from transgenics to cultivars/elite breeding lines. MAS plays an important role for chickpea improvement for drought tolerance using "QTL hotspot" gene8, enhancing genetic diversity through multiparent advanced generation intercrossing (MAGIC) lines, introgressing resistance wilt and ascochyta blight diseases, etc. (Dixit et al., 2019). In addition to MAS, transgenic technology can also be exploited for chickpea improvement. The explants of chickpea variety ICCV-1, ICCV-6, and one local were transformed with *Agrobacterium tumefaciens* strain LBA4404. The explants of nptll in kanamycin showed resistance (Kar et al., 1996). Ignacimuthu and Prakash (2006) introduced α-amylase inhibitor gene isolated from *Phaseolus vulgaris* into chickpea cultivar K850 through *Agrobacterium*-mediated transformation. Similarly, Vigna Δ 1-pyrroline-5-carboxylate synthetase (P5CS) cDNA was transferred to chickpea (*C. arietinum* L.) cultivar Annigeri (Ghanti et al., 2011). The gene provided tolerance to salt stress and no yield penalties were found in transgenic plants (Ghanti et al., 2011). Mehrotra et al. (2011) introduced modified cry1Ab and cry1Ac from *Bacillus thuringiensis* into chickpea and reported that cry1Ac protein was more toxic to *H. armigera* than cry1Ab. Transgenic plants possessing resistance to pod borer were produced by incorporating fused Cry1Ab/Ac genes from *B. thuringiensis* to *C. arietinum* (Ganguly et al., 2014).

4.7 Generation of additional variability through hybridization

Early maturing varieties are suitable to the conditions of Peninsular India (ICRISAT, 1981) as it permits the crop to maintain favorable plant water potentials during the critical stages of crop growth (Sivakumar and Virmani, 1979). Early flowering is also vital for drought escape in chickpea crop i.e., the terminal drop. The line ICC 4958 is currently considered as a potential donor for drought tolerance and has been used to develop drought-tolerant cultivars in India and

Kenya (Gaur et al., 2012). Several other useful traits are present in some accessions of chickpea that can be transferred to agronomically superior variety through hybridization. For cold-tolerance, different cold-tolerant lines were identified by various workers e.g., from the cultivated (Singh et al., 1995; Kumar et al., 2011) and wild species (Robertson et al., 1995). The annual yield losses in chickpea due to terminal drought are estimated to be approximately 30%–40% (Yadav et al., 2004). Development of early maturing varieties is one of the major objectives in the chickpea breeding programs with the aim to avoid terminal drought. The first step in this direction was the development of an extra-short duration *kabuli* variety ICCV 2, which flowers in about 25 days and matures in about 85 days at Patancheru, India (altitude: 545 m above the mean sea level, latitude: 17027″ N, longitude: 78028″E). ICCV 2, derived from multiple crosses involving five parents (2 *kabuli* and 3*desi*), has been released for cultivation in India as "*Swetha,*" in Sudan as "*Wad Hamid,*" and in Myanmar as "*Yezin 3.*" It was instrumental in expanding the *kabuli* chickpea area from the cooler, long-season environments to the warmer, short-season environments (e.g., southern India and Myanmar). In addition to this, two super-early and cold-tolerant *desi* lines, ICCV 96029 and ICCV 96030, ere also developed by ICRISAT that mature in 75–80 days in southern India (Kumar and Rao, 1996).

4.8 Generation of variability through mutagenesis

The genetic variability in plants can also be created through mutagenesis (spontaneous and induced). Creation of genetic variability by induced mutagenesis provides a base for correcting one or two undesirable traits of already popular variety. Shah et al. (2008) reported that mutagens may cause genetic changes in an organism, breaking the linkages, and produce many new promising traits for the improvement of crop plants. Micke (1988) and Yildirim et al. (2013) revealed that genetic variability in chickpea can be generated effectively through induced mutations to broaden the genetic base of the chickpea. In the process of crop evolution and adaptation, legumes generally lose desirable alleles for quality, biotic stresses, and productivity (Dhumal and Bolbhat, 2012).

Mutation breeding remains an important tool for chickpea improvement as evident from generation of plants/varieties with improved plant architecture, yield, flowering and maturity duration, quality, and tolerance to biotic and abiotic stresses. Several improved mutants have been released as varieties or used in linkage studies (Dahiya et al., 1984; Pundir and Reddy, 1998; Gaur and Gour, 2002, 2003; McNeil et al., 2007; Rajesh et al., 2007; Salimath et al., 2007; Srinivasan et al., 2006; Wani and Anis, 2008; Ali et al., 2010; Kharkwal et al., 2010; Si et al., 2010; Wani, 2011). Globally different chickpea varieties are developed and released for commercial cultivation through mutation (Salimath et al., 2007). More than 27 varieties of chickpea are released through mutation by different countries till date (Anonymous, 2019). Shaikh et al. (1982) induced mutations using ^{60}Co gamma-rays source in dry chickpea cultivar Faridpur-l and isolated elite mutant line M-669 in the segregating generations (M8 and M9). This line was released as Hyprosola chickpea variety with the increase in yield and protein content of 20% and 4%, respectively, with some other traits, i.e., earliness of 10 days, more pods, high planting density, and harvest index over the mother cultivar. Patil et al. (1997) evaluated the mutants obtained by irradiation of Warangal and N31 with 10 kR and treatment with 0.05% EMS and observed that mutant ACM18 had the highest root respiration rate at the vegetative stage, Warangal had the highest root dry weight/plant at pod maturity, while ACM7 had highest seed yield/plant than Warangal, and its ratio of seed yield to root respiration was highest throughout the growth period; hence it is suggested that low root respiration during all stages of plant development is an index for productivity. Jagannath and Apparao (2012) isolated different mutants in locally popular chickpea (*C. arietinum* L.) cultivar 'Vijay' (Phule G 81-1-1). Five true mutant lines showed both positive and negative increases in quantitative traits in the M_4 generation. They found 10−11 days early mutant lines as compared to parent lines. *Gigas* mutant plants obtained after 400 Gy gamma radiation were tallest (44.2 cm), with a 2−3 fold increase in the pod and seed size over the control. The protein content in the M_4 generation ranged from 259.11 to 282.18 mg/g as compared to 255.59 mg/g in the control.

During the year 1984 first chickpea variety Kiran (RSG-2) was released through mutation with high pod number, early maturing, high yielding with salinity tolerance (Dua et al., 2001). Haq et al. (1984) identified a blight-resistant, high-yielding, brown-seeded mutant variety, CM72, which has helped greatly to stabilize chickpea production in Pakistan. Kharkwal et al. (1988) released three high-yielding and disease-resistant induced mutant varieties of chickpea namely, Ajay (Pusa-408), Atul (Pusa-413), and Gimar (Pusa-417) for commercial cultivation in India by the Ministry of Agriculture and Rural Development, Government of India, in 1985. Kharkwal et al. (2005) released another high-yielding chickpea mutant variety (BGM 547) having thin testa, attractive bold seeds, better cooking quality, and high yield performance under late sown conditions of North Western Plain Zone of India.

4.9 Wild *Cicer* species

The genus *Cicer* includes 9 annual and 33 perennial wild species and 1 species, *C. laeturn* Rass and Sharip. with unspecified status (van der Maesen, 1987; van der Maesen et al., 2007; Singh et al., 2008; Gaur et al., 2009). Out of the nine annuals, chickpea (*C. arietinum* L.) is the only cultivated one. Remaining eight annual species are wild: *C. reticulatum* Lad., *C. echinospermum* Davis, C. *pinnatifidum* J.&S., *C. judaicum* Boiss., *C. bijugum* Rich., *C. cuneatum* Rich., *C. chorassanicum* (Bge.) M.Pop., and *C. yamashitae* Kit. Various workers found that the wild species *C. reticulatum* and *C. echinospermum* are most closely related to the domesticated *C. arietinum* based on hybridization success, storage proteins, and DNA markers (Kazan and Muehlbauer, 1991; Ahmad and Slinkard, 1992; Singh and Ocampo, 1993). There seems to be a consensus that *C. reticulatum* is the primary progenitor, while *C. echinospermum* is the secondary wild progenitor of the cultivated chickpea (Ladizinsky and Adler, 1976a; Ahmad and Slinkard, 1992).

Based on morphological characteristics, life cycle, and geographical distribution, van der Maesen (1972, 1987) classified the *Cicer* species into four sections. First section: Monocicer comprises eight annual species (*C. arietinum, C. reticulatum, C. echinospermum, C. pinnatifidum, C. bijugum, C. judaicum, C. yamashitae*, and *C. cuneatum*), second section: Chamaecicer include two perennial species (*C. chorassanicum* and *C. incisum*), third section: Polycicer of 23 perennial species, and fourth section: Acanthocicer have seven woody perennial species.

Ladizinsky and Adler (1976a,b) had placed seven annual *Cicer* species (except *C. chorassanicum* and *C. yamashitae*) into three crossability groups based on their studies on crossability and fertility of hybrids in interspecific crosses:

Group I It is consisted of cultivated species, *C. arietinum,* and closely related wild relatives, *C. reticulatum* and *C. echinospermum.*

Group II It includes annual species *C. bijugum, C. judaicum*, and *C. pinnatifidum.*

Group III This includes only one annual species *C. cuneatum* which originated in Ethiopia. It is not cross compatible with any of the groups and thus, rated as distantly related to *C. arietinum.*

They had also placed cultivated species and landraces in the primary gene pool and the progenitor species, *C. reticulatum* and *C. echinospermum*, in secondary gene pool. The tertiary gene pool consists of all the annual and perennial *Cicer* that are not crossable with the cultivated species.

Based on the studies of seed protein electrophoresis pattern and crossability relationship, the wild relatives of chickpea has been arranged in three crossability groups and these findings were also supported by Croser et al. (2003).

Group 1 It consists of *C. arietinum*, C. *reticulatum*, and *C. echinospermum.*

Group 2 It consists of *C. judaicum, C. pinnatifidum*, and *C. bijugum.*

Group 3 It consists of *C. chorassanicum, C. cuneatum*, and *C. yamashitae.*

The crossing is possible within the groups but not between the groups due to some natural incompatibility barriers operating at pre- and postfertilization stages in interspecific hybridization evolved during the course of evolution.

van der Maesen et al. (2007) had revised the gene pool of genus *Cicer* following classification of Harlan and De Wet (1971). According to that the primary gene pool now contains *C. reticulatum* and *C. echinospermum* in addition to *C. arietinum* and secondary gene pool consists of *C. bijugum, C. judaicum*, and *C. pinnatifidum*. The tertiary gene pool consists of other *Cicer* species.

4.10 Utilization of wild species for crop improvement

Incorporation of traits from wild species into cultivars depends upon the biosystematics relationship among species. The cultivated species and their progenitors of same ploidy level constitute primary gene pool, they are cross-compatible, and their F_1 hybrids are fertile. Interspecific crosses between cultivars and species in the secondary gene pool are possible, but the hybrid is partially sterile due to ploidy level differences, genomic incompatibilities, cytoplasmic imbalances, or other factors which cause sterility. Crops and related species may vary greatly for ploidy levels, genomic groupings, and levels of incompatibilities. Thus, direct hybridization is possible for closely related wild species, whereas additional manipulations are needed to develop hybrids between two distantly related species. In general, following different methods are employed to transfer germplasm from wild to cultivated species.

4.10.1 Direct hybridization

It is the most common and easiest method for gene transfer. After hybridization, the filial or backcross progenies are generated and desirable progenies are advanced for agronomical evaluation. In cases where the cultivated and closely related species have different chromosome number, the resulted F_1 may be sterile due to chromosome imbalance. Sometimes the union of functional unreduced gametes from one parent and reduced gamete from the second parent leads to the production of hybrid seed; the fertility in F_1 is restored after colchicine treatment. Theoretically, the procedure looks simple, but practically undesirable linkage drag along with sterility makes the procedure difficult to accomplish while maintaining the genes of interest.

4.10.2 Chromosome doubling prior to hybridization

Doubling the chromosome number of the lower ploidy species to produce autoploids or amphiploid before crossing with the cultivated species offers an alternative to direct hybridization. This procedure avoids sterility associated with triploids, pentaploids, and aneuploids. Pundir et al. (1983) reported that colchicine 0.25% for 4 h was most effective in producing autotetraploid in chickpea out of three colchicine concentrations 0.25%, 0.05%, 0.025% and two treatment durations 4 and 6 h. The tetraploids showed increased plant vigor as evident from larger leaves, flowers, and seeds. Some of these showed normal meiosis and produced pods similar to the diploid parent, indicating the potential of induced autotetraploidy in chickpea improvement.

4.10.3 Other techniques

Pre- and postfertilization crossing barrier occur most frequently in interspecific hybridization (Hadley and Openshaw, 1980). Hence, growth regulators can be used to induce pollen tube growth, fertilization, seed maturation, and pod set.

Tissue culture techniques are often needed to recover ovules and embryos. Somatic hybridization and plant transformation are employed to overcome several obstacles to interspecific hybridization among divergent species.

4.10.3.1 The bridge crosses

Bridge crosses have been used in many genera as a mechanism to transfer genes among cross-incompatible species under the condition of traits being dominant or partial dominant and species should be readily crossable to other two cross-incompatible species. For example, no success was achieved in direct crosses, the F_1 of (*C. arietinum* × *C. reticulatum*) as female with the *C. judaicum;* thereby few viable bold seeds were obtained. The perennial wild species *C. anatolicum, C. reticulatum*, and *C. echinospermum* can be crossed with any one of the wild annual species; then it may be possible to transfer traits of interest to *C. arietinum* (Ahmad and Slinkard, 2003; Clarke et al., 2006).

4.10.3.2 Embryo culture

Embryo rescue can help to overcome the postfertilization barriers during wide hybridization using tertiary gene pool such as *C. bijugum* and *C. judaicum* (Ahmad et al., 1988; Clarke et al., 2006). With the help of modern biotechnology tools and tissue culture techniques, success has been found by various workers to incorporate the desirable traits from tertiary gene pool to the cultigens (Shiela et al., 1992; Mallikarjuna, 1999; Clarke et al., 2006) and, as a result, hybrids were obtained between *C. pinnatifidum* and *C. bijugum* (Mallikarjuna, 1999).

4.11 New germplasm/varieties developed through wide hybridization

Interspecific hybridization is known to improve productivity and resistance to disease in many crops. Since chickpea breeding program has initiated, it had been undoubtedly known that wild species are valuable gene pools, which offer a wide range of opportunities for enhancing genetic variability and introgressing desirable traits, viz., resistance to various biotic and abiotic stresses. Various workers used the wild species of *Cicer* for the transfer of desirable traits (resistance to biotic stresses, tolerance to abiotic stresses, high number of fruiting branches and pods per plant, etc.) to cultivated chickpea (Ladizinsky and Adler, 1976a,b; Jaiswal and Singh, 1989; Verma et al., 1990; Singh et al., 1994; Singh et al., 2005; Singh et al., 2015; Singh et al., 2016; Singh et al., 2018), although these wild species have been underutilized due to existence of crossability barriers. Thus far, only two annual wild species, *C. reticulatum* and *C. echinospermum* have been exploited in breeding programs, as the crossing of the cultigen with other species remained a challenge even with embryo rescue techniques. There is a need to continue efforts on exploitation of wild *Cicer* species of the tertiary gene pool, as these contain many useful genes, particularly resistance to some of the biotic and abiotic stresses (Ahmad et al., 2005). Through interspecific hybridization using *C. reticulatum* in the crosses, only one variety BG 1103 from Indian Agricultural Research Institute, New Delhi, has been released for commercial cultivation for North India in 2005. Mercy and Kakar (1975) attempted the first interspecific crosses between the cultivated annual *Cicer* species *C. arietinum* and a perennial species *C. soongaricum* but did not yield single hybrid seed.

Fertile hybrids with normal chromosome segregation were produced by Ladizinsky and Alder (1976a,b) by crossing *C. arietinum* and *C. reticulatum*. Early flowering segregants were isolated from *C. arietinum* × *C. reticulatum* by Singh et al. (1984), and these segregants escaped heat and drought stress. Similarly, Jaiswal et al. (1986, 1987) obtained variation in yield and yield-related traits in F_2, F_3, and F_4 progenies from the cross *C. arietinum* and *C. reticulatum*. Pundir and van der Maesen (1983) obtained seven successful crosses involving eight annual *Cicer* species (*C. echinospermum*). Similar results were also reported by Ladizinsky and Adler (1976a,b). Jaiswal et al. (1984) obtained successful hybridization between *C. arietinum* and *C. reticulatum*. Transgressive segregants from interspecific crosses had early maturity. Jaiswal et al. (1986) observed high variability for yield per plant in F_2 generation in *C. arietinum* × *C. reticulatum* crosses involving wild species in the primary gene pool.

Singh et al. (2005) introgressed the genes responsible for productivity and disease resistance from wild species into chickpea. To minimize the linkage drag of undesirable traits from *reticulatum* they attempted two backcrosses to the female parents. The segregating populations were advanced through the pedigree method for both straight crosses (without backcross) and BC_1/BC_2 generations to F_5–F_7. In the preliminary yield trial 16.9% increase was found in interspecific derivatives over the check and 25.2% increase over the female parents coupled with a high degree of resistance to wilt, foot rot, and root rot diseases.

Singh and Singh (2009) crossed five chickpea genotypes (K-850, PDG-84-10, Gaurav, Pusa 408, KPG 59 and *C. reticulatum* (ICCW 8)). They reported that 3.3% and 1.8% pod set following hybridization when wild species was used as a male parent and female parent, respectively.

Singh et al. (1984) crossed *C. arietinum* and *C. echinospermum*, and the progenies resembling phenotypically the cultigen had better agronomic attributes, viz., 100-seed weight, plant height, and biological yield. Although successful crosses between *C. arietinum* and *C. echinospermum* were produced by Pundir and Mengesha (1995), the hybrids were observed to have high pollen sterility with abnormal meiotic pairing. Singh and Ocampo (1997) made crosses among the cultivated chickpea (*Cicer arietinum*) ILC 482 with ILWC 179 (*C. echinospermum*) and ILWC 124 (*C. reticulatum*) for the introgression of desirable genes into cultivated chickpea and observed up to a 39% increase in seed yield following the pedigree method. Ahmad (1988) successfully accomplished crossing program involving all nine wild *Cicer* species including cultivated and also reported two new crosses, viz., *C. chorossanicum* × *C. pinnatifidum* and *C. judaicum* × *C. chorassanicum*.

Katna et al. (2017) attempted crosses between cultivated chickpea and two wild annual species *C. reticulatum* and *C. echinospermum*. They reported that varietal group, day length, temperature, and growing conditions influenced crossability and might be taken into the consideration while planning wide hybridization. Up to 148% and 71.70% hybrid vigor was observed in F_1 hybrids over mid parent and better parent, respectively, on per plant basis in initial studies. The results revealed that annual wild *Cicer* species (*C. reticulatum* and *C. echinospermum*) can successfully be exploited for chickpea improvement and broadening the genetic base of cultivated chickpea in respect of yield and stress tolerance.

Singh et al. (2018) studied segregating generations of seven crosses of *C. reticulatum* Ladiz. and *C. echinospermum* Davis with cultivated varieties of chickpea along with parents under natural field condition. They confirmed the fertility of F_1 by their normal meiotic chromosomal configuration including high pollen stainability percentage and normal seed set. They observed a wide range of variation for cold stress in F_3. They found higher vigor in inbreds for pod numbers and earliness than that of better parent in F_2 and F_3 generations.

Verma et al. (1990) reported that successful interspecific hybridization in genus *Cicer* was largely dependent on a number of pollinations attempted and application of growth regulator (100 ppm gibberellic acid + 25 ppm NAA + 10 ppm kinetin). Naik (1993) recommended application of a mixture of growth regulators containing 75 mg/L gibberellic acid (GA_3), 10 mg/L naphthalene acetic acid (NAA), and 10 mg/L kinetin (Kn) to a cotton pad at the base of the pedicel of the flowers about half an hour after pollination following interspecific pollinations in genus *Cicer*. Mallikarjuna and Jadhav (2008) observed successful interspecific hybridization between *C. arietinum* and *C. pinnatifidum* by hormone-aided pollinations, embryo rescue, and tissue culture techniques to save hybrid embryo abortion and grafted hybrid shoots to cultivated chickpea stocks. Singh et al. (2008) suggested the use of *C. arietinum* as female parent and application of a combination of growth regulators to the pollinated pistils for successful hybridization with wild *Cicer* species. Mallikarjuna and Muehlbauer (2011) also applied hormone solution containing gibberellic acid, naphthalene acetic acid, and kinetin (GA_3 50 mg/L + NAA 10 mg/L + Kn 10 mg/L) to prevent flower abscission in interspecific crosses with *Cicer* species.

4.12 Wide hybridization for transfer of resistance to biotic stresses

Different diseases, e.g., wilt, blight, gray mold, viruses, influence the overall performance of chickpea and cause variable yield reduction in the crop. Singh et al. (1994) and Ahmad et al. (2005) observed the multiple resistance in wild species, e.g., *C. bijugum* (ILWC 7-1) for ascochyta blight, fusarium wilt, leaf miner, cyst nematode, and cold; and *C. pinnatifidum* (ILWC 33/S-4) for ascochyta blight, fusarium wilt, seed beetle, and cyst nematode. A higher level of resistance for gray mold (*Botrytis cinerea* Pars.) has been identified in some wild *Cicer* species, including closely related *C. echinospermum*; this resistance has yet to be incorporated in the cultivated species (Pande et al., 2002). Significantly better resistance was subsequently identified in accessions of *C. echinospermum* (Knights et al., 2003).

Due to the absence of genetic resistance sources for ascochyta blight in the cultivated genotypes, different gene pools of *Cicer* sp. such as *C. bijugum, C. echinospermum, C. pinnatifidum, C. judaicum*, and *C. montbretii* have been exploited for blight resistance. Two accessions of *C. echinospermum* were classified as resistant to blight and both the lines were cross-compatible with *C arietinum*, thereby are valuable source of resistance to blight in chickpea (Collard et al., 2001; Ahmad and Slinkard, 2004). Similarly the wild accessions of *C. judaicum*, viz., ATC 46994, ICC 17211, IG 69986, IC 70030, IG 70037, and IG 70038, were blight resistant (Nguyen et al., 2004; Pande et al., 2006a, 2006b). Accession of *C. bijugum,* ILWC 7-1, has valuable source of genes for multiple disease resistance, viz., ascochyta blight, fusarium wilt, leaf miner, cyst nematode along with abiotic stress resistance like cold Singh et al. (1994) and Ahmad et al. (2005); while the wild species accession of *C. pinnatifidum,* ILWC 33/S-4, also found to harbor multiple resistance genes for ascochyta blight, fusarium wilt, seed beetle, and cyst nematodes. Feasibility of introgression of novel sources of resistance from the tertiary to the domestic gene pool is the prime breeding importance for chickpea breeder (Mallikarjuna, 1999). Using marker-assisted backcrossing (MABC) program, two quantitative trait loci (QTL) regions—ABQTL-I and ABQTL-II—were transferred from the ILC 3279 resistant line to the C214 cultivar using eight SSR markers linked with the ABQTL-I and ABQTL-II for the introgression of the ascochyta blight resistance (Varshney et al., 2013).

Genetic studies for wilt resistance have shown that vertical type resistance is governed by single major gene against race 3 and 5 (Sharma et al., 2005a,b); two independent genes govern resistance against race 0 (Rubio et al., 2003), whereas resistance to races 1A, 2, and 4 are governed by digenes or trigenes. Source of valuable genes for resistance have been reported by many workers to be found in indigenous and wild chickpea germplasm (Singh et al., 2013, 2014); else among the wild species like *C. bijugum, C. judaicum*, and *C. reticulatum* (van der Maesen and Pundir, 1984; Kaiser et al., 1994; Infantino et al., 1996). Winter et al. (2000) developed recombinant inbred lines (RILs) from a wide cross between a cultivated chickpea *C. arietinum* resistant to Fusarium wilt and *C. reticulatum* accession PI 489777.

Madrid et al. (2008) identified a gene that controls resistance to chickpea rust in an RIL population derived from an interspecific cross between *C. arietinum* (ILC72) × *C. reticulatum* (Cr5-10), susceptible and resistant to rust, respectively. For botrytis gray mold, wide hybridization can play an important role in developing stable resistant lines of cultivated chickpea. Wide hybridization breeding method is employed for incorporating the resistant genes from wild relative into the cultivar background. Using wide hybridization between *C. pinnatifidum* and cultivated chickpea developed resistance lines for this disease (Kaur et al., 2013). Clement et al. (2000) reported that chickpeas are damaged by a large number of insect species, both under field and in storage conditions. Among the many insect pests damaging food legumes, the pod borers, *H. armigera* (Hubner), sap-sucking pests, especially *Aphis craccivora*, and nematodes are the most devastating pests of chickpea in Asia, Africa, and Australia (van Emden et al., 1988). High level of resistance for leaf miner (*Liriomyza cicerina* Rondani) has been identified in C. *echinospermum* (ILWC 35), *C. bijugum* (ILWC 227), *C. judaicum* (ILWC 206, 207, 210, 211, 223), *C. pinnatifidum* (ILWC 9, 97, 226, 252), *C. cuneatum* (ILWC 40, 185, 187, 232), and *C. chorassanicum* (ILWC 147). Some of the accessions of wild species *C. reticulatum* (ILWC 104), *C. echinospermum* (ILWC 39, 179, 181), *C. bijugum* (ILWC 65, 67, 68, 70, 73, 74, 75, 83, 177), *C. judaicum* (ILWC 46, 54, 173, 174 176 189), and *C. cuneatum* (ILWC 187) showed no damage to seed beetle or bruchid (*Callosobruchus* spp). However, there are no reports of successful transfer of bruchids resistance from wild to the cultivated species (Robertson et al., 1995). In the case of pod borer resistance or tolerance has been reported in wild *Cicer* species (Rai et al., 2004). Wild species of chickpea, viz., *C. bijugum, C. echinospermum, C. judaicum, C. pinnatifidum, C. reticulatum*, and *C. cuneatum*, are reported showing tolerance against the *Helicoverpa* pod borer (Kaur et al., 1999; Sharma, 2004). Hence, for the reduction in yield losses due to pod borer, introgression of resistance from wild *Cicer* species into elite chickpea cultivar(s) is an important breeding objective. Many public-funded institutions are in the process of developing transgenic lines against pod borer using different sources of *Cry* genes/*Bt* genes.

Di Vito et al. (1996) introgressed resistance against cyst nematode into the cultivated chickpea from its wild species *C. reticulatum*. Wild *Cicer* species viz., *C. echinospermum, C. bijugum* and *C. judaicum* process some valuable genes for resistance against bruchids (Singh et al., 1994, 1998; Singh and Weigand, 1994) and play an important role in resistance breeding. Losses can also be minimized by use of good and clean grain storage facilities since there are no resistance sources available among the cultivated chickpea species.

Toker et al. (2007) reported that wild species such as *C. anatolicum, C. microphyllum, C. montbretii, C. oxydon* and *C. songaricum* have the tolerance to drought. High levels of resistance or tolerance to important abiotic stresses such as drought (Croser et al., 2003; Kashiwagi et al., 2005; Toker et al., 2007) and cold (Croser et al., 2003; Toker, 2005; Berger et al., 2012) have been reported in wild *Cicer* species. Similarly, few accessions of wild species such as *C. reticulatum* and *C. pinnatifidum* possess resistance to heat stress up to 41.8°C (Canci and Toker, 2009).

4.13 Future strategy and conclusions

Two major goals of chickpea breeding i.e., widening the genetic base and development of agronomically superior cultivars can be achieved by transfer of genes from the chickpea germplasm accessions or wild species of primary gene pool using conventional as well as molecular breeding approaches. The exploitation of *Cicer* species of secondary and tertiary gene pools remains remote due to unsuccessful crossability. Marker-assisted breeding can further provide speed and precision in breeding programs if markers linked closely to the genes/QTLs of the interest are identified. Marker-assisted backcross breeding is another tool in reducing linkage drag in wide crosses. Molecular markers identified as being close to the gene(s) of interest can be used to break undesirable linkages and eliminate nonadvantageous genes (Muehlbauer et al., 1994). MASs, though in infancy in chickpea, have been exploited to improve drought tolerance by transferring "QTL hotspot" (Varshney et al., 2014). Among the biotic stresses, pod borer (*H. armigera*) is the most challenging insect-pest of chickpea due to nonavailability of sources with high levels of resistance in germplasm. To overcome this problem, *Cry* genes (*Bt* genes) can play a major role and cry gene transgenic lines display insect resistance.

Since crossing barriers exist between *C. arietinum* and wild species of the secondary and tertiary pool, the protoplast fusion may be a technique of choice for combining genomes of different species without pollination (Sagare and -Krishnamurthy, 1991).The technique is, however, currently limited by the absence of protocols for the regeneration of whole plants from single cells in any of the *Cicer* species. Progress toward the development of a microspore culture system in *Cicer* (Lulsdorf et al., 2001; Croser, 2002) resulting in embryogenesis from single-celled microspores in liquid culture may assist in the future development of this technique wherein haploid plants from F_1 plants may be generated and evaluated for agronomic superiority after chromosome doubling. International collaborative efforts should be made to identify chickpea genes governing different traits and exploitation of those in chickpea improvement including the use of transgenic technology. The challenge is to identify the novel genes in the wild relatives of chickpea and find the best strategy to utilize them. Enhancement of minimum support price will also encourage the farmers to adopt the chickpea cultivation as a profitable venture as productivity per unit area of chickpea is far less than cereals such as wheat and rice. To sustain production, several steps, e.g., assured procurement and inclusion, lucrative minimum support price, buffer stock, insurance schemes, safe storage and postharvest, are still required.

Furthermore, improvement in yield potential, introgression of resistance/tolerance to biotic/abiotic stresses, wider adaptation, development of suitable plant ideotype, and grain characteristics will be the focus of chickpea plant breeders in near future. Search for novel genes governing different traits including those for resistance/tolerance to biotic and abiotic stresses in the available germplasm and in the wild species of chickpea is another potential area of interest to breeders. Attempts to cross the wild species with chickpea have been made worldwide with mixed results. The majority of these attempts go unpublished since only two annual species, *C. reticulatum* and *C. echinospermum*, are compatible and produce fertile hybrids. Recent advances in biotechnology techniques, including embryo rescue for interspecific hybrids, linkage mapping, marker-assisted breeding, and transformation, hold much promise for successful introgression of genes from wild annual species into cultivated chickpea. An efficient embryo rescue protocol to enable crosses between chickpea and the more distantly related *Cicer* species will be crucial to the utilization of traits from these wild relatives. Through wide hybridization, genetic base of chickpea can be broadened which will be beneficial for chickpea improvement. The combination approach by application of biotechnological and conventional strategies will lead to rapid progress in the chickpea improvement. In the future, upcoming breeding programs are likely to take into account the industry's emerging quality requirements and also giving emphasis on the enhancement of nutritional quality. MAS for hidden traits which are not visible for direct selection in the field should be initiated on a priority basis.

References

Abbo, S., Berger, J., Turner, N.C., 2003a. Evolution of cultivated chickpea: four bottlenecks limit diversity and constrain adaptation. Funct. Plant Biol. 30, 1081–1087.

Abbo, S., Shtienberg, D., Lichtenzveig, J., Lev-Yadun, S., Gopher, A., 2003b. The chickpea, summer cropping, and a new model for pulse domestication in the ancient near East. Quant. Rev. Biol. 78, 435–448.

Ahmad, F., Gaur, P.M., Croser, J., 2005. Chickpea (*Cicer arietinum* L.). In: Singh, R.J., Jauhar, P.P. (Eds.), Genetic Resources, Chromosome Engineering, and Crop Improvement. CRC Press, Boca Raton, FL, pp. 187–217.

Ahmad, F., Hymowitz, T., 1993. The fine structure of chickpea (*Cicer arietinum* L.) chromosomes as revealed by pachytene analysis. Theor. Appl. Genet. 86 (5), 637–641.

Ahmad, F., Slinkard, A.E., 1992. Genetic relationships in the genus *Cicer* L. as revealed by polyacryl amide gel electrophoresis of seed storage proteins. Theor. Appl. Genet. 84, 688–692.

Ahmad, F., Slinkard, A.E., 2003. Limitations to bridge-species facilitated alien gene transfers in chickpea: pre-fertilization events (*Cicer arietinum* L.). J. Genet. Breed. 57 (1), 69–74.

Ahmad, F., Slinkard, A.E., 2004. The extent of embryo and endosperm growth following interspecific hybridization between *Cicer arietinum* L. and related annual wild species. Genet. Resour. Crop Evol. 51, 765–772.

Ahmad, F., Slinkard, A.E., Scoles, G.J., 1988. Investigation into the barrier(s) to interspecific hybridization between *Cicer arietinum* L. and eight other annual *Cicer* species. Plant Breed. 100, 193–198.

Ahmad, F., 1988. Interspecific Hybridization and Genetic Relationships Among the Annual *Cicer* L. Species (Ph.D. thesis). University of Saskatchewan, Saskatoon, Canada, p. 5.

Ahmad, F., Slinkard, A.E., Scoles, G.J., 1987. The cytogenetic relationship between Cicer judaicum Boiss. and Cicer chorassanicum (Bge.) M. Pop. Genome 29 (6), 883–886.

Ali, H., Shah, T.M., Iqbal, N., Atta, B.M., Haq, M.A., 2010. Mutagenic induction of double-podding trait in different genotypes of chickpea and their characterization by STMS marker. Plant Breed. 129, 116–119.

Altman, A., 1999. Plant biotechnology in the 21st century: the challenges ahead. Electron. J. Biotechnol. 2, 51–55.

Anonymous, 2019. FAO/International Atomic Energy Agency (IAEA), Mutant Variety Database. International Centre, Vienna, Austria.

Basu, P.S., Ali, M., Chaturvedi, S.K., 2009. Terminal heat stress adversely affects chickpea productivity in Northern India – strategies to improve thermo tolerance in the crop under climate change. In: Panigrahy, S., Shankar, S.R., Parihar, J.S. (Eds.), ISPRS Archives XXXVIII-8/W3 Workshop Proceedings: Impact of Climate Change on Agriculture, 23–25 February. ISPRS Archives, Ahmedabad, pp. 189–193.

Berger, J.D., Kumar, S., Nayyar, H., Street, K.A., Sandhu, J.S., Henzell, J.M., 2012. Temperature-stratified screening of chickpea (*Cicer arietinum* L.) genetic resource collections reveals very limited reproductive chilling tolerance compared to its annual wild relatives. Field Crop. Res. 126, 119–129.

Bhardwaj, R., Sandhu, J.S., Kaur, L., Gupta, S.K., Gaur, P.M., Varshney, R., 2010. Genetics of resistance to Ascochyta blight in chickpea. Euphytica 171, 337–343.

Butler, E.J., 1911. Fungi and Disease in Plants. Bishen Singh Mahendra Pal Singh, New Connaught Place, Dehradun Periodical Experts, 42-D, VivekVihar, Delhi 32, 547 (reprinted 1973).

Canci, H., Toker, C., 2009. Evaluation of annual wild Cicer species for drought and heat resistance under field conditions. Genet. Resour. Crop Evol. 56 (1), 1–6.

Chaturvedi, S.K., Nadarajan, N., 2010. Genetic enhancement for grain yield in chickpea-accomplishments and resetting research agenda. Electron. J. Plant Breed. 1 (4), 611–615.

Chen, W., Muehlbauer, F.J., 2003. An improved technique for virulence assay of *Ascochyta rabie* on chickpea. Int. Chickpea Pigeonpea Newsl. 10, 31–33.

Clarke, H.J., Siddique, K.H.M., 2004. Response of chickpea genotypes to low-temperature stress during reproductive development. Field Crop Res. 90, 323–334.

Clarke, H.J., Wilson, J.M., Kuo, I., Lulsdorf, M.M., Mallikarjuna, N., Kuo, J., Siddique, K.H.M., 2006. Embryo rescue and plant regeneration in vitro of selfed chickpea (*Cicer arietinum* L.) and its wild annual relatives. Plant Cell Tiss. Org. Cult. 85 (2), 197–204.

Clement, S.L., Cristofaro, M., Cowgill, S.E., Weigand, S., 1999. Germplasm resources, insect resistance and grain legume improvement. In: Clement, S.L., Quisenberry, S.S. (Eds.), Global Plant Genetic Resources for Insect Resistant Crops. CRC Press, Boca Raton, FL, pp. 131–148.

Clement, S.L., Sharaf El-Din, N., Weigand, S., Lateef, S.S., 1994. Research achievements in plant resistance to insect pests of cool season food legumes. Euphytica 73, 41–50.

Clement, S.L., Wightman, J.A., Hardie, D.C., Bailey, P., Baker, G., McDonald, G., 2000. Opportunities for integrated management of insect pests of grain legumes. In: Knight, R. (Ed.), Linking Research and Marketing Opportunities for Pulses in the 21st Century. Kluwer Academic, Dordrecht, The Netherlands, ISBN 0-7923-5565-2, pp. 467–480.

Collard, B., Ades, P., Pang, E., Brouwer, J., Taylor, P., 2001. Prospecting for sources of resistance to ascochyta blight in wild *Cicer* species. Australas. Plant Pathol. 30, 271–276.

Croser, 2002. Haploid and Zygotic Embryo Culture of *Cicer* (Ph.D. thesis). University of Melbourne, Australia.

Croser, J.S., Ahmad, F., Clarke, H.J., Siddique, K.H.M., 2003. Utilisation of wild *Cicer*in chickpea improvement – progress, constraints, and prospects. Aust. J. Agric. Res. 54, 429–444.

Dahiya, B.S., Lather, V.S., Solanki, I.S., Kumar, R., 1984. Useful spontaneous mutants in chickpea (*Cicer arietinum* L.). Int. Chickpea Newsl. 11, 4–7.

Dhumal, K.N., Bolbhat, S.N., 2012. Induction of genetic variability with gamma radiation and its applications in improvement of horsegram. In: Adrovic, F. (Ed.), Gamma Radiation. InTech, New York, USA, ISBN 9789535103165, pp. 207–228.

Di Vito, M., Singh, K.B., Greco, N., Saxena, M.C., 1996. Sources of resistance to cyst nematode in cultivated and wild *Cicer* sps. Genet. Resour. Crop Evol. 43, 103–107.

Dias, C.A.R., Yadav, T.D., 1988. Incidence of pulse beetles in different legume seeds. Indian J. Entomol. 50, 457–461.

Dita, M.A., Rispail, N., Prats, E., Rubiales, D., Singh, K.B., 2006. Biotechnology approaches to overcome biotic and abiotic stress constraints in legumes. Euphytica 147, 1–24.

Dixit, G.P., Srivastava, A.K., Singh, N.P., 2019. Marching towards self-sufficiency in chickpea. Curr. Sci. 116 (2), 239–242.

Dua, R.P., Chaturvedi, S.K., Sewak, S., 2001. Reference Varieties of Chickpea for IPR Regime. Indian Institute of Pulses Research, Kanpur.

Erskine, W., Hussain, A., Tahir, M., Bahksh, A., Ellis, R.H., Summerfield, R.J., Roberts, E.H., 1994. Field evaluation of a model of photothermal flowering responses in a world lentil collection. Theor. Appl. Genet. 88, 423–428.

Eser, D., 1976. Heritability of some important plant characters, their relationships with plant yield and inheritance of Ascochyta blight resistance in chickpeas (*Cicer arietinum* L.). Ankara Universitesi Ziraat Facultesi Yayinlari 620, 40.

Fang, X., Turner, N.C., Yan, G., Li, F., Siddique, K.H., 2009. Flower numbers, pod production, pollen viability, and pistil function are reduced and flower and pod abortion increased in chickpea (*Cicer arietinum* L.) under terminal drought. J. Exp. Bot. 61 (2), 335–345.

Fitt, G.P., 1989. The ecology of *Heliothis*species in relation to agro-ecosystems. Annu. Rev. Entomol. 34, 17–52.

Ganguly, M., Ali Molla, K., Karmakar, S., Datta, K., Datta, S.K., 2014. Development of pod borer-resistant transgenic chickpea using a pod-specific and a constitutive promoter-driven fused cry1Ab/Ac gene. Theor. Appl. Genet. 127 (12), 2555–2565.

Gaur, P.M., Gour, V.K., 2002. A gene producing one to nine flowers per flowering node in chickpea. Euphytica 128, 231–235.

Gaur, P.M., Gour, V.K., 2003. Broad-few-leaflets and outwardly curved wings: two new mutants of chickpea. Plant Breed. 122, 192–194.

Gaur, P.M., Jukanti, A.K., Srinivasan, S., Gowda, C.L.L., 2012. Chickpea (*Cicer arietinum* L.). In: Bharadwaj, D.N. (Ed.), Breeding of Field Crops. Agrobios (India), Jodhpur, pp. 165–194.

Gaur, P.M., Mallikarjuna, N., Knights, T., Beebe, S., Debouck, D., Mejia, A., Malhotra, R.S., Muhammad, I., Sarker, A., Tripathi, S., Gowda, C.L.L., 2009. Gene introgression in grain legumes. In: International Conference on Grain Legumes: Quality Improvement, Value Addition and Trade. Indian Society of Pulses Research and Development, Inc., Kanpur, India, pp. 1–17.

Ghanti, S.K.K., Sujata, K.G., Kumar, B.V., Karba, N.N., Janardhan Reddy, K., Rao, M.S., Kishor, P.K., 2011. Heterologous expression of P5CS gene in chickpea enhances salt tolerance without affecting yield. Biol. Plant. 55, 634–640.

Gowda, C.L.L., 1981. Natural outcrossing in chickpea. Int. Chickpea Newsl. 5, 6.

Grewal, J.S., Laha, S.K., 1983. Chemical control of Botrytis blight of chickpea. Indian Phytopathol. 27, 643–645.

Hadley, H.H., Openshaw, S.J., 1980. Interspecific and intergeneric hybridization. In: Gehr, W.R., Hadley, H.H. (Eds.), Hybridization of Crop Plants. American Society of Agronomy, Madison, WI, pp. 133–159.

Haq, M.A., Sadiq, M., Hassan, M., 1984. Induction of ascochyta blight resistance in chickpea through induced mutations. In: Proceedings of Res. Co-ordination Mtg Faisalabad, Pakistan.

Harlan, J.R., De Wet, J.M.J., 1971. Toward a rational classification of cultivated plants. Taxon 20, 509–517.

Haware, M.P., Nene, Y.L., 1982. Races of *Fusarium oxysporum* f. sp. *ciceri*. Plant Dis. 66, 809–810.

Hawkes, J.G., 1977. The importance of wild germplasm in Plant Breeding. Euphytica 26, 615–621.

Hawtin, G.C., Singh, K.B., 1984. Prospects and potential of winter sowing of chickpea in Mediterranean region. In: Saxena, M.C., Singh, K.B. (Eds.), Ascochyta Blight and Winter Sowing of Chickpeas. Martinus Nijhoff/Dr. W. Junk. Publishers, Hague, The Netherlands, pp. 11–16.

Howe, R.E., Currie, J.E., 1964. Some laboratory observations on the rates of development, mortality and oviposition of several species of Bruchidae breeding in stored pulses. Bull. Entomol. Res. 55 (3), 437–477.

ICRISAT, 1981. Annual Report, 1979–80, ICRISAT Patancheru, India, pp. 81–82.

Ignacimuthu, S., Prakash, S., 2006. Agrobacterium-mediated transformation of chickpea with α-amylase inhibitor gene for insect resistance. J. Biosci. 31 (3), 339–345.

Infantino, A., Porta-Puglia, A., Singh, K.B., 1996. Screening of wild *Cicer* species for resistance to *Fusarium* wilt. Plant Dis. 42–44.

Jagannath, B., Apparao, B., 2012. Genetic improvement of chickpea (*Cicer arietinum L.*) cultivar 'Vijay' (Phule G 81-1-1) through induced mutagenesis. Biorem. Biodivers. Bioavailab. 6 (1), 103–106.

Jaiswal, H.K., Singh, B.D., 1989. Analysis of gene effects for yield and certain yield traits in crosses between *Cicer arietinum* L. and *Cicer reticulatum* Ladz. Indian. J. Genet. 49 (1), 9–17.

Jaiswal, H.K., Singh, B.D., Singh, A.K., Singh, R.M., 1986. Introgression of genes for yield and yield traits from *C. reticulatum* to *C. arietinum*. Int. Chickpea Newsl. 14, 5–8.

Jaiswal, H.K., Singh, B.D., Singh, R.M., 1987. Improvement of chickpea (*Cicer arietinum*) through introgression of genes form *C. reticulatum*. Indian J. Agric. Sci. 57, 880–883.

Jaiswal, H.K., Singh, B.D., Singh, R.M., 1984. Possible origin of *desi* chickpeas through introgression of *C. reticulatum* genes. Int. Chickpea Newsl. 11, 18–20.

Jimenez-Diaz, R.M., Basallote-Ureba, M.J., Rapoport, H., 1989a. Colonization and pathogenesis in chickpea infected by races of *Fusarium oxysporum* f. sp. *ciceri*. In: Tjamos, E.C., Beckman, C. (Eds.), Vascular Wilt Disease of Plants. NATO ASI Series, Vol H28. Springer-Verlag, Berlin, Germany, pp. 113–121.

Jimenez-Diaz, R.M., Trapero-Casas, A., Cabrera de la Colina, J., 1989b. Races of *Fusarium oxysporum* f. sp. *Ciceris* infecting chickpea in southern Spain. In: Tjamos, E.C., Beckman, C.H. (Eds.), Vascular Wilt Disease of Plants. NATO ASI Series, Vol. H28. Springer, Berlin, pp. 515–520.

Kaiser, W.J., Alcala-Jimenez, A.R., Hervas-Vargas, A., Trapero-Casas, J.L., Jimenez-Diaz, R.M., 1994. Screening of wild *Cicer* species for resistance to races 0 to 5 of *Fusarium oxysporum* f.sp. *ciceris*. Plant Dis. 78, 962–967.

Kaiser, W.J., Danesh, D., Okhovat, M., Mossahebi, H., 1968. Disease of pulse crops (edible legumes) in Iran. Plant Dis. Rep. 52, 687–691.

Kar, S., Johnson, T.M., Nayak, P., Sen, S.K., 1996. Efficient transgenic plant regeneration through Agrobacterium-mediated transformation of Chickpea (*Cicer arietinum* L.). Plant Cell Rep. 16 (1−2), 32−37.

Kashiwagi, J., Krishnamurthy, L., Upadhyaya, H.D., Krishna, H., Chandra, S., Vadez, V., Serraj, R., 2005. Genetic variability of drought-avoidance root traits in the mini-core germplasm collection of chickpea (*Cicer arietinum* L.). Euphytica 146, 213−222.

Katna, G., Panwar, A., Sharma, K.D., Chaudhary, H.K., 2017. Broadening the genetic of chickpea (*Cicer arietinum* L.) through wide hybridization. In: National Symposium on 'Pulses for Nutritional Security and Agricultural Sustainability' Held on December 2−4, 2017. IIPR, Kanpur, pp. 82−83.

Kaur, L., Sirari, A., Kumar, D., Sandhu, J.S., Singh, S., Singh, I., Kapoor, K., Gowda, C.L.L., Pande, S., Gaur, P., Sharma, M., Imtiaz, M., Siddique, K.H.M., 2013. Harnessing Ascochyta blight and Botrytis grey mould resistance in chickpea through interspecific hybridization. Phytopathol. Mediterr. 52 (1), 157−165.

Kaur, S., Chhabra, K.S., Arora, B.S., 1999. Incidence of *Helicoverpa armigera* (Hubner) on wild and cultivated species of chickpea. Int. Chickpea Pigeonpea Newsl. 6, 18−19.

Kazan, K., Muehlbauer, F.J., 1991. Allozyme variation and phylogeny in annual species of *Cicer* (Leguminosae). Plant Syst. Evol. 175, 1−21.

Khan, S.A., Kamal, M., 1962. Some additional hosts of *Leveillula taurica* (lev.) arnaud from Pakistan. Pak. J. Sci. Res. 5, 41.

Khanna-Chopra, R., Sinha, S.K., 1987. Chickpea: physiological aspects on growth and yield. In: Saxena, M.C., Singh, K.B. (Eds.), The Chickpea. CAB International, Wallingford, pp. 163−189.

Kharkwal, M.C., Cagirgan, C., Toker, T., Shah, M.M., Islam, H., 2010. Legume mutant varieties for food, feed and environmental benefits. In: Proceedings of the 5th International Food Legumes Research Conference and 7th European Conference on Grain Legumes, April 26−30, 2010, Antalya, Turkey, p. 196.

Kharkwal, M.C., Jain, H.K., Sharma, B., 1988. Induced mutations for improvement of chickpea, lentil, pea and cowpea. In: Proceedings of the FAO/IAEA Workshop on Improvement of Grain Legume Production Using Induced Mutations, 1−5 July, 1986, Pullman, Washington (USA). IAEA, Vienna, pp. 89−109.

Kharkwal, M.C., Nagar, J.P., Kala, Y.K., 2005. BGM 547-A high yielding chickpea (*Cicer arietinum* L.) mutant variety for late sown conditions of North Western Plain Zone of India. Indian J. Genet. 65, 229−230.

Khune, N.N., Charde, J.D., Raut, J.G., 1979. Chickpea: a new record of *Leveillula taurica*. Indian Phytopathol. 32, 655.

Kiran, A., Kumar, S., Nayyar, H., Sharma, K.D., 2019. Low temperature induced aberrations in male and female reproductive organ development cause flower abortion in Chickpea. Plant Cell Environ. 42, 2075−2089.

Knights, E.J., Southwell, R.J., Schwinghamer, M.W., 2003. Evaluation of wild *Cicer* species for resistance to Phytophthora root rot. In: Sharma, R.N., Shrivastava, G.K., Rathore, A.L., Sharma, M.L., Khan, M.A. (Eds.), Chickpea Research for the Millennium: Proceedings of International Chickpea Conference, 20−22 January 2003. Indira Gandhi Agricultural University, Raipur, India, pp. 54−57.

Kumar, J., Rao, B.V., 1996. Super early chickpea developed at ICRISAT Asia Center. Int. Chickpea Pigeonpea Newsl. 3, 17−18.

Kumar, J., van Rheenen, H.A., 2000. A major gene for time of flowering in chickpea. J. Hered. 91, 67–68. https://doi.org/10.1093/jhered/91.1.67.

Kumar, S., Malik, J., Thakur, P., Kaistha, S., Sharma, K.D., Upadhyaya, H.D., Berger, J.D., Nayyar, H., 2011. Growth and metabolic responses of contrasting chickpea (*Cicer arietinum* L.) genotypes to chilling stress at reproductive phase. Acta Physiol. Plant. 33, 779–787.

Kumar, S., Thakur, P., Kaushal, N., Malik, J.A., Gaur, P., Nayyar, H., 2013. Effect of varying high temperatures during reproductive growth on reproductive function, oxidative stress and seed yield in chickpea genotypes differing in heat sensitivity. Arch. Agron Soil Sci. 9, 823–843.

Kutarekar, D.R., Wanjari, K.B., 1983. Karyomorphological studies in some of the varieties of Bengal gram (*Cicer arietinum* L.). Cytologia 48, 699–705.

Ladizinsky, G., Adler, A., 1976a. Genetic relationships among the annual species of *Cicer* L. Theor. Appl. Genet. 48, 197–203.

Ladizinsky, G., Adler, A., 1976b. The origin of chickpea, *Cicer arietinum* L. Euphytica 25, 211–217.

Lulsdorf, M., Dament, T., Croser, J., Warkentin, T., Vandenberg, A., 2001. Development of microspore culture technology for the production of doubled-haploid lines of chickpea (*Cicer arietinum* L.). In: Proceeding of the North American Pulse Improvement Association Meeting. Fargo, USA, pp. 28–30.

Madrid, E., Rubiales, D., Moral, A., Moreno, M.T., Millan, T., Gil, J., Rubio, J., 2008. Mechanism and molecular markers associated with rust resistance in a chickpea interspecific cross (*Cicer arietinum* × *Cicer reticulatum*). Eur. J. Plant Pathol. 121, 43–53.

Mahendra, P., 1998. Diseases of pulse crops, their relative importance and management. J. Mycol. Plant Pathol. 28, 114–122.

Mahrshi, R.P., Mathur, A.K., 1981. Outbreak of powdery mildew of chickpea (*Cicer arietinum* L.) in Rajasthan, India. Indian J. Plant Prot. 9, 118–119.

Malhotra, R.S., Baum, M., Udupa, S.M., Bayaa, B., Kababbeh, S., Khalaf, G., 2003. Ascochyta blight research in chickpea- present status and future prospects. In: Sharma, R.N., Srivastava, G.K., Rahore, A.L., Sharma, M.L., Khan, M.A. (Eds.), Proceedings of the International Chickpea Conference Chickpea Research for the Millennium, 20–22 January 2003. Indira Gandhi Krishi Vishwavidyalaya, Raipur, Chhattisgarh, India, pp. 108–117.

Malhotra, R.S., Saxena, M.C., 1993. Screening for cold and heat tolerance in cool season food legumes. In: Singh, K.B., Saxena, M.C. (Eds.), Breeding for Tolerance in Cool Season Food Legumes. Wiley, Chichester, pp. 227–244.

Mallikarjuna, N., 1999. Ovule and embryo culture to obtain hybrids from interspecific incompatible pollinations in chickpea. Euphytica 110, 1–6.

Mallikarjuna, N., Jadhav, D., 2008. Techniques to produce hybrid between *Cicer arietinum* L. × *C. pinnatifidum* Jaub. Indian J. Genet. 68, 398–405.

Mallikarjuna, N., Muehlbauer, F.J., 2011. Chickpea hybridization using in vitro techniques. In: Trevor, A., et al. (Eds.), Plant Embryo Culture: Methods and Protocols, Methods in Molecular Biology. Springer, New York, NY, p. 93.

Masood, A., Shivkumar, 2001. An overview of chickpea research in India. Indian J. Pulse Res. 14, 81–89.

McDonald, G.K., Paulsen, G.M., 1997. High temperature effects on photosynthesis and water relations of grain legumes. Plant Soil 196, 47–58.

McNeil, D., Ahmad, F., Abbo, S., Bahl, P.N., 2007. Genetics and cytogenetics. In: Yadav, S.S., Redden, R., Chen, W., Sharma, B. (Eds.), Chickpea Breeding and Management. CAB International, London, UK, pp. 321–337.

Mehrotra, M., Singh, A.K., Sanyal, I., Altosaar, I., Amla, D.V., 2011. Pyramiding of modified cry1Ab and cry1Ac genes of *Bacillus thuringiensis* in transgenic chickpea (*Cicer arietinum* L.) for improved resistance to pod borer insect *Helicoverpa armigera*. Euphytica 182 (1), 87–102.

Mercy, S.C., Kakar, S.K., 1975. Barrier to interspecific crosses in *Cicer*. Proc. Indian Natl. Sci. Acad. 41, 78–82.

Micke, A., 1988. Genetic improvement of food legumes in developing countries by mutation induction. In: Summerfield, R.J. (Ed.), World Crops: Cool Season Food Legumes. Kluwer Academic Publishers, Dordrecht, The Netherlands, ISBN 9789400927643, pp. 1031–1047.

Millan, T., Clarke, H.J., Siddique, K.H.M., Buhariwala, H.K., Gaur, P.M., Kumar, J.K., Gil, J., Khal, G., Winter, P., 2006. Chickpea molecular breeding: new tools and concepts. Euphytica 147, 81–103.

Mishra, R.P., Kulkarni, S.N., 1973. Two new records of powdery mildews from Madhya Pradesh. JNKVV Res. J. 7, 287.

Mitra, M., 1931. Report of the imperial mycologist. Science Report of the Agricultural Research Institute, pp. 58–71.

Mookherjee, P.B., Jotwani, M.G., Yadav, T.D., Sircar, P., 1970. Studies on incidence and extent of damage due to insect pests in stored seeds. II. Leguminous and vegetable seeds. Indian J. Entomol. 32, 350–355.

Muehlbauer, F.J., Kaiser, W.J., Simon, C.J., 1994. Potential for wild species in cool season food legume breeding. Euphytica 73, 109–114.

Naik, S.V., 1993. Interspecific Hybridization and Characterization of Hybrids in Genus *Cicer* L. Masters thesis). Department of Plant Breeding and Genetics, Indira Gandhi Krishi Vishwa Vidyalaya, Raipur, India.

Nayyar, H., Bains, T.S., Kumar, S., 2005. Chilling stressed chickpea seedlings: effect of cold acclimation, calcium and abscisic acid on cryoprotective solutes and oxidative damage. Environ. Exp. Bot. 54, 275–285.

Nene, Y.L., Reddy, M.V., 1987. Chickpea diseases and their control. In: Saxena, M.C., Singh, K.B. (Eds.), The Chickpea. CAB International, Wallingford, pp. 233–270.

Nene, Y.L., Sheila, V.K., Sharma, S.B., 1984. A World List of Chickpea (Cicer *arietinum* l.) and Pigeonpea (*Cajanuscajan* l. millsp.) Pathogens. ICRISAT Pulse Pathology Progress Report, 32, 19.

Nene, Y.L., Haware, M.P., Reddy, M.V., 1981. Chickpea diseases, resistance-screening techniques. ICRISAT Inf. Bull. 10.

Nguyen, T.T., Taylor, P.W.J., Redden, R.J., Ford, R., 2004. Genetic diversity in *Cicer* using AFLP analysis. Plant Breed. 123, 173–179.

Nour, M.A., 1957. Control of powdery mildew disease in the Sudan with special reference to broadbean. Imp. J. Exp. Agric. 25, 119–131.

Pande, S., Kishore, G.K., Upadhyaya, H.D., Narayana Rao, J., 2006a. Identification of sources of multiple disease resistance in the mini-core collection of chickpea. Plant Dis. 90, 1214–1218.

Pande, S., Ramgopal, D., Kishore, G.K., Mallikarjuna, N., Sharma, M., Pathak, M., Narayana Rao, J., 2006b. Evaluation of wild *Cicer* species for resistance to Ascochyta blight and Botrytis grey mould in a controlled environment at ICRISAT, Patancheru, India. J. SAT Agric. Res. 2, 1–3.

Pande, S., Sharma, M., Gaur, P.M., Tripathi, S., Kaur, L., Basandarai, A., Khan, T., Gowda, C.L.L., Siddique, K.H.M., 2011. Development of screening techniques and identification of new sources of resistance to Ascochyta blight disease of chickpea. Australas. Plant Pathol. 40, 149–156.

Pande, S., Singh, G., Narayana Rao, J., Bakr, M.A., Chaurasia, P.C.P., Joshi, S., Johansen, C., Singh, S.D., Kumar, J., Rahman, M.M., Gowda, C.L.L., 2002. Integrated Management of Botrytis Grey Mould of Chickpea. Information Bulletin No. 61. ICRISAT, Patancheru, India, p. 26.

Patil, B.N., Dhopte, A.M., Meshram, S.M., 1997. Yield potential of chickpea mutants in relation to growth and biological oxygen demand of roots. J. Maharashtra Agric. Univ. 2, 145–146.

Patrick, J.W., Stoddard, F.L., 2010. Physiology of flowering and grain filling in faba bean. Field Crop Res. 115, 234–242.

Phadnis, B.A., 1970. Karyotypic studies in varieties of *Cicer arietinum* L. Cytol. Genet. 5, 80–88.

Phillips, J.C., 1988. A distinct race of chickpea wilt in California. Int. Chickpea Pigeonpea Newsl. 18, 19–21.

Pundir, R.P.S., Rao, N.K., van der Maesem, L.J.G., 1983. Induced autotetraploidy in chickpea (*Cicer arietinum* L.). Theor. Appl. Genet. 65 (2), 119–122.

Pundir, R.P.S., Mengesha, M.H., 1995. Cross compatibility between chickpea and its wild relative *C. echinispermum* Davis. Euphytica 83, 241–245.

Pundir, R.P.S., Reddy, G.V., 1998. Two new traits-open flower and small leaf in chickpea (*Cicer arietinum* L.). Euphytica 102, 357–361.

Pundir, R.P.S., van der Maesen, L.J.G., 1983. Interspecific hybridization in *cicer*. Int. Chickpea Newsl. 8, 4–5.

Rai, K.N., Gaur, P.M., Hash, C.T., Sharma, K.K., Gowda, C.L.L., Serraj, R., 2004. Development of crop cultivars for increased and stable production in dry lands of the semi-arid tropics. J. Arid Land Stud. 14, 69–72.

Rajesh, P.N., McPhee, K.E., Ford, R., Pittock, C., Kumar, J., Muehlbauer, F.J., 2007. Ciceromics: advancement in genomics and recent molecular techniques. In: Yadav, S.S., Redden, R., Chen, W., Sharma, B. (Eds.), Chickpea Breeding and Management. CAB International, London, UK, pp. 445–457.

Robertson, L.D., Singh, K.B., Ocampo, B., 1995. A Catalog of Annual Wild *Cicer* Species. ICARDA, Aleppo, Syria, p. 171.

Rubio, J., Hajj Moussa, E., Kharrat, M., Moreno, M.T., Millan, T., Gill, J., 2003. Two genes and linked RAPD markers involved in resistance to *Fusarium oxysporum* f.sp. *ciceris* race 0 in chickpea. Plant Breed. 122, 188–191.

Sagare, A.P., Krishnamurthy, K.V., 1991. Protoplast regeneration in chickpea, *Cicer arietinum* L. Indian J. Exp. Biol. 29, 930–932.

Salimath, P.M., Toker, C., Sandhu, J.S., Kumar, J., Suma, B., Yadav, S.S., Bahl, P.N., 2007. Conventional breeding methods. In: Yadav, S.S., Redden, R.J., Chen, W., Sharma, B. (Eds.), Chickpea Breeding and Management. CABI Publishing, pp. 369–390.

Samineni, S., Kamatam, S., Thudi, M., Varshney, R.K., Gaur, P.M., 2016. Vernalization response in chickpea is controlled by a major QTL. Euphytica 207 (2), 453–461.

Savithri, K.S., Ganapathy, P.S., Sinha, S.K., 1980. Sensitivity to low tolerance in pollen germination and fruit set in *Cicer arietinum* L. J. Exp. Bot. 31, 475–481.

Sethi, S.C., Byth, D.E., Gowda, C.L.L., Green, J.M., 1981. Photoperiodic response and accelerated generation turnover in chickpea. Field Crop Res. 4, 215–225.

Shah, T.M., Mirza, J.I., Haq, M.A., Atta, B.M., 2008. Radio sensitivity of various chickpea genotypes in M_1 generation I. Laboratory studies. Pak. J. Bot. 40, 649–665.

Shaikh, M.A.Q., Ahmed, Z.U., Majid, M.A., Wadud, M.A., 1982. A high-yielding and high-protein mutant of chickpea (*Cicer arietinum* L.) derived through mutation breeding. Environ. Exp. Bot. 22 (4), 483–489.

Sharma, H.C., 2004. A Little Help from Wild: Exploiting Wild Relatives of Chickpea for Resistance to *Helicoverpa armigera*. ICRISAT, Patancheru, India.

Sharma, H.C., Ortiz, R., 2002. Host plant resistance to insects: an eco-friendly approach for pest management and environmental conservation. J. Environ. Biol. 23, 11–35.

Sharma, H.C., Pampapathy, G., Lanka, S.K., Ridsdill-Smith, T.J., 2005a. Antibiosis mechanism of resistance to legume pod borer, *Helicoverpa armigera*in wild relatives of chickpea. Euphytica 142, 107–117.

Sharma, K.D., Chen, W., Muehlbauer, F.J., 2004a. A consensus set of differential lines for identifying races of *Fusarium oxysporum* f. sp. *ciceris*. Int. Pigeonpea Chickpea Newsl. 11, 34–36.

Sharma, K.D., Chen, W., Muehlbauer, F.J., 2005b. Genetics of chickpea resistance to five races of Fusarium wilt and a concise set of race differentials for *Fusarium oxysporum* f. sp. *ciceris*. Plant Dis. 89 (4), 385–390.

Sharma, K.D., Muehlbauer, F.J., 2005. Genetic mapping of *Fusarium oxysporum* f. sp. *ciceris* race-specific resistance genes in chickpea (*Cicer arietinum* L.). In: Abstract of the International Food Legume Research Conference—IV. Indian Agricultural Research Institute, New Delhi, India, pp. 18–22.

Sharma, K.D., Muehlbauer, F.J., 2007. Fusarium wilt of chickpea: physiological specialization, genetics of resistance and resistance gene tagging. Euphytica 157 (1–2), 1–14.

Sharma, K.D., Winter, P., Kahl, G., Muehlbauer, F.J., 2004b. Molecular mapping of *Fusarium oxysporum* f. sp. ciceris race 3 resistance gene in chickpea. Theor. Appl. Genet. 108 (7), 1243–1248.

Sharma, P.C., 1983. Cytogenetic Studies in the Genus *Cicer* L. and Some Other Pulse Crops (Ph.D. thesis). Meerut University, Meerut.

Sharma, P.C., Gupta, P.K., 1986. Cytogenetics of legume genera *cicer* L. And *lens* L. In: Gupta, P.K., Bahl, J.R. (Eds.), Genetics and Crop Improvement. Rastogi and Company, Meerut, India, pp. 321–339.

Shaw, F.J.F., Arjikar, L., 1915. The genus Rhizoctonia in India. In: Memoires of the Department of Agriculture India Botany Series, vol. 7, p. 177.

Shiela, V.K., Moss, J.P., Gowda, C.L.L., Rheenen, H.A., 1992. Interspecific hybridization between *Cicer arietinum* and wild *Cicer* species. Int. Chickpea Newsl. 27, 11–13.

Si, P., Chen, Y., Weerakoon, S., Quealy, J., Powles, S., Erskine, W., 2010. Chickpea (*Cicer arietinum* L.) breeding lines tolerant to metribuzin applied post-emergence. In: Proceedings of the 5th International Food Legumes Research Conference and 7th European Conference on Grain Legumes, April 26–30, 2010, Antalya, Turkey, p. 216.

Singh, M., Bhardwaj, C., Singh, S., Panatu, S., Chaturvedi, S.K., Rana, J.C., Rizvi, A.H., Kumar, N., Sarker, A., 2016. Chickpea genetics resources and its utilization in India: current and future prospects. Indian J. Genet. 76, 515–529.

Singh, A., Singh, N.P., 2009. Estimation of genetic parameters in recombinant inbred population derived from interspecific cross of chickpea (*Cicer arietinum* L.). Indian J. Genet. 69, 122–126.

Singh, B.D., Jaiswal, H.K., Singh, R.N., Singh, A., 1984. Isolation of early flowering recombinants from interspecific cross between *Cicer arietinum* and *Cicer reticulatum*. Int. Chickpea Newsl. 11, 14–16.

Singh, G., Kapoor, S., Singh, K., 1982. Screening of chickpea for grey mould resistance. Int. Chickpea Newsl. 7, 13–14.

Singh, K.B., 1990. Patterns of resistance and susceptibility to races of *Ascochytarabiei*among germplasm accessions and breeding lines of chickpea. Plant Dis. 74, 127–129.

Singh, K.B., Dahiya, B.S., 1973. Breeding for wilt resistance in chickpea. In: Symposium on Wilt Problems and Breeding for Wilt Resistance in Bengal Gram, September 1973 at IARI, New Delhi, India, pp. 13–14.

Singh, K.B., Hawtin, G.C., 1979. Winter planting. Int. Chickpea Newsl. 1, 4.

Singh, K.B., Malhotra, R.S., Halila, M.H., Knights, E.J., Verma, M.M., 1994. Current status and future strategy in breeding chickpea for resistance to biotic and abiotic stresses. Euphytica 73, 137–149.

Singh, K.B., Malhotra, R.S., Saxena, M.C., 1995. Additional sources of tolerance to cold in cultivated and wild *Cicer* species. Crop Sci. 35, 1491–1497.

Singh, K.B., Ocampo, B., 1993. Interspecific hybridization in annual *Cicer* species. J. Genet. Breed. 47, 199–204.

Singh, K.B., Ocampo, B., 1997. Exploitation of wild *Cicer*species for yield improvement in chickpea. Theor. Appl. Genet. 95, 418–423.

Singh, K.B., Ocampo, B., Robertsonm, L.D., 1998. Diversity for abiotic and biotic stress resistance in the wild annual *Cicer* species. Genet. Resour. Crop Evol. 45, 9–17.

Singh, K.B., Reddy, M.V., 1983. Inheritance of resistance to ascochyta blight in chickpea. Crop Sci. 23, 9–10.

Singh, K.B., Reddy, M.V., 1991. Advances in disease resistance breeding in chickpea. Adv. Agron. 45, 191–222.

Singh, K.B., Weigand, S., 1994. Identification of resistant sources in *Cicer* species to *Liriomyza cicerina*. Genet. Resour. Crop Evol. 41, 75–79.

Singh, K.B., Malhotra, R.S., Saxena, M.C., Bejiga, G., 1997. Superiority of winter sowing over traditional spring sowing of chickpea in the Mediterranean region. Agron. J. 89, 112–118.

Singh, M., Bisht, I.S., Kumar, K., Dutta, M., Kaur, L., Kumar, A., Sarker, A., Bansal, K.C., 2014. Characterization and evaluation of wild annual *Cicer* species for agro-morphological traits and major biotic stresses under North-Western Indian conditions. Crop Sci. 54, 229–239.

Singh, M., Khan, Z., Kumar, K., Dutta, M., Pathania, A., 2013. Sources of resistance to *Fusarium* wilt and root knot nematode in indigenous chickpea germplasm. Plant Genet. Resour. 10, 258–260.

Singh, M., Kumar, K., Bisht, I.S., Dutta, M., Rana, M.K., Rana, J.C., Bansal, K.C., Sarker, A., 2015. Exploitation of wild annual *Cicer* species for widening the gene pool of chickpea cultivars. Plant Breed. 134, 186–192.

Singh, M., Rani, S., Malhotra, N., Katna, G., Sarker, A., 2018. Transgressive segregations for agronomic improvement using interspecific crosses between *C. arietinum* L. × *C. reticulatum* Ladiz. and *C. arietinum* L. × *C echinospermum* Davis species. PLoS One 13 (9), e0203082.

Singh, R., Sharma, P., Kumar, R., Varshney, R.K., Sharma, S.K., Singh, N.K., 2008a. Chickpea improvement: role of wild species and genetic markers. Biotechnol. Genet. Eng. Rev. 25, 267–314.

Singh, S., Gumbler, R.K., Joshi, N., Singh, K., 2005. Introgression from wild *Cicer reticulatum* to cultivated chickpea for productivity and disease resistance. Plant Breed. 124 (5), 477–480.

Singh, T.P., Deshmukh, P.S., Kumar, P., 2008b. Relationship between physiological traits in chickpea (*Cicer arietinum* L.) under rainfed condition. Indian J. Plant Physiol. 13, 411–413.

Sivakumar, M.V.K., Virmani, S.M., 1979. Measuring shoot water potential in chickpea with a pressure chamber. Exp. Agric. 15, 377–383.

Slinkard, A.E., Soth, M.B., Vandenberg, A., 2000. Breeding for yield: the direct approach. In: Knight, R. (Ed.), Linking Research and Marketing Opportunities for Pulses in the 21st Century. Kuwar Academic Publishers, Dordrecht, The Netherlands, pp. 1183–1190.

Southgate, B.J., 1978. The importance of Bruchidae as pests of grain legumes, their distribution and control. In: Singh, S.R., van Emden, H.F., Taylor, T.A. (Eds.), Pests of Grain Legumes: Ecology and Control. Academic Press, London, pp. 219–229.

Srinivasan, A., Saxena, N.P., Johansen, C., 1999. Cold tolerance during early reproductive growth of chickpea (*Cicer arietinum* L.): genetic variation in gamete development and function. Field Crop Res. 60, 209–222.

Srinivasan, S., Gaur, P.M., Chaturvedi, S.K., Rao, B.V., 2006. Allelic relationships of genes controlling number of flowers per axis in chickpea. Euphytica 152, 331–337.

Summerfield, R.J., Hadley, P., Roberts, E.H., Minchin, F.R., Rawsthrone, S., 1984. Sensitivity of chickpea (*Cicer arietinum* L.) to hot temperatures during the reproductive period. Exp. Agric. 20, 77–93.

Summerfield, R.J., Ellis, R.H., Roberts, E.H., 1989. Vernalization in chickpea (*Cicer arietinum*): fact or artefact? Ann. Bot. 64, 599–604.

Toker, C., 2005. Preliminary screening and selection for cold tolerance in annual wild *Cicer*species. Genet. Resour. Crop Evol. 52, 1–5.

Toker, C., Canci, H., 2006. Selection for drought and heat resistance in chickpea under terminal drought conditions. In: Kharkwal, M.C. (Ed.), Food Legumes for Nutritional Security and Sustainable Agriculture, 4th International Food Legumes Research Conference, Indian Agricultural Research Institute, New Delhi, October 2006. Indian Society of Genetics and Plant Breeding, New Delhi, pp. 18–22.

Toker, C., Canci, H., 2009. Evaluation of yield criteria for drought and heat resistance in chickpea (*Cicer arietinum* L.). J. Agron. Crop Sci. 195, 47–54.

Toker, C., Canci, H., Yildirim, T., 2007. Evaluation of perennial wild *Cicer* species for drought resistance. Genet. Resour. Crop Evol. 54, 1781–1786.

Upadhyaya, H.D., Smithson, J.B., Haware, M.P., Kumar, J., 1983. Resistance to wilt in chickpea. II. Further evidence for two genes for resistance to race 1. Euphytica 32 (3), 749–755.

van der Maesen, L.J.G., 1972. *Cicer* L., A monograph of the genus with special reference to the chickpea (*Cicer arietinum* L.), its ecology and cultivation. Meded Land Bouwhooge School Wageningen 72 (10), 9–136.

van der Maesen, L.J.G., 1987. Origin, history and taxonomy of the chickpea. In: Saxena, M.C., Singh, K.B. (Eds.), The Chickpea. CAB International, Wallingford, pp. 11–34.

van der Maesen, L.J.G., Maxted, N., Javadi, F., Coles, S., Davies, A.M., 2007. Taxonomy of *cicer* revisited. In: Yadav, S.S., et al. (Eds.), Chickpea Breeding and Management. CABI International, pp. 14–46.

van der Maesen, L.L.G., Pundir, R.P.S., 1984. Availability and use of wild *Cicer*germplasm. Plant Genet. Resour. Newsl. 57, 19–24.

van Emden, H.F., Ball, S.L., Rao, M.R., 1988. Pest disease and weed problems in pea lentil and faba bean and chickpea. In: Summerfield, R.J. (Ed.), World Crops: Cool Season Food Legumes. Kluwer Academic Publishers, Dordrecht, the Netherlands, pp. 519–534.

van Rheenen, H.A., Pundir, R.P.S., Miranda, J.H., 1993. How to accelerate the genetic improvement of a recalcitrant crop species such as chickpea. Curr. Sci. 65, 414–417.

Varshney, R.K., Mohan, S.M., Gaur, P.M., Chamarthi, S.K., Singh, V.K., Srinivasan, S., Swapna, N., Sharma, M., Pande, S., Varshney, R.K., Singh, S., Kaur, L., 2013. Marker-Assisted backcrossing to introgress resistance to fusarium wilt race 1 and Ascochyta blight in C 214, an Elite cultivar of Chickpea. Plant Genome 7, 1–11.

Varshney, R.K., Thudi, M., Nayak, S.N., Gaur, P.M., Kashiwagi, J., 2014. Genetic dissection of drought tolerance in chickpea (*Cicer arietinum* L.). Theor. Appl. Genet. 127, 445–462.

Venora, G., Ocampo, B., Singh, K.B., Saccardo, F., 1995. Karyotype of the kabuli-type chickpea (*Cicer arietinum* L.) by image analysis system. Caryologia 48 (2), 147–155.

Verma, M.M., Sandhu, J.S., Brar, H.S., Brar, J.S., 1990. Crossability studies in different species of *Cicer* (L.). Crop Improv. 17, 179–181.

Vir, S., Grewal, J.S., Gupta, V.P., 1975. Inheritance of resistance to Ascochyta blight in chickpea. Euphytica 24, 209–211.

Wang, J., Gan, Y.T., Clarke, F., McDonald, C.L., 2006. Response of chickpea yield to high-temperature stress during reproductive development. Crop Sci. 46, 2171–2178.

Wani, A.A., 2011. Spectrum and frequency of macromutations induced in chickpea (*Cicer arietinum* L.). Turk. J. Biol. 35, 221–231.

Wani, A.A., Anis, M., 2008. Gamma ray- and EMS-induced bold-seeded high-yielding mutants in chickpea (*Cicer arietinum* L.). Turk. J. Biol. 32, 161–166.

Weigand, S., Tahhan, O., 1990. Chickpea insect pests in the Mediterranean zones and new approaches to their management. In: Chickpea in the Nineties: Proceedings of the Second International Workshop on Chickpea Improvement, 4–8 December 1989. ICRISAT, Hyderabad, pp. 169–175.

Wery, J., Turc, O., Lecoeur, J., 1993. Mechanism of resistance to cold, heat and drought in cool-season legumes, with special reference to chickpea and pea. In: Singh, K.B., Saxena, M.C. (Eds.), Breeding for Tolerance in Cool Season Food Legumes. Wiley, Chichester, pp. 271–291.

Westerlund Jr., F.V., Cambell, R.N., Kimble, K.A., 1974. Fungal root-rots and wilt of chickpea in California. Phytopathology 64, 432–436.

Winter, P., Benko-Iseppon, A.M., Huttel, B., Ratnaparkhe, M., Tullu, A., Sonnante, G., Pfaff, T., Tekeoglu, M., Santra, D., Sant, V.J., Rajesh, P.N., Khal, G., Muehlbauer, F.J., 2000. A linkage map of the chickpea (*C. arietinum* L.) genome based on recombinant inbred lines from a *C. arietinum* × *C. reticulatum* cross: localization of resistance genes for *Fusarium* wilt races 4 and 5. Theor. Appl. Genet. 101 (7), 1155–1163.

Yadav, S.S., Kumar, J., Turner, N.C., Berger, J., Redden, R., McNeil, D., Materne, M., Knights, E.J., Bahl, P.N., 2004. Breeding for improved productivity, multiple resistance and wide adaptation in chickpea (*Cicer arietinum* L.). Plant Genet. Resour. 2, 181–187.

Yildirim, T., Canci, H., Inci, N.E., Baloglu, F.O.C., Ikten, C., Toker, C., 2013. Inheritance of female sterility in induced *Cicer* species. Turk. J. Field Crops 18, 78–81.

Zatloukalova, P., Hribova, E., Kubalakova, M., Suchankova, P., Simkova, H., Adoracion, C., Kahl, G., Millan, T., Dolezel, J., 2011. Integration of genetic and physical maps of the chickpea (*Cicer arietinum* L.) genome using flow-sorted chromosomes. Chromosome Res. 19 (6), 729.

Zohary, D., Hopf, M., 2000. Pulses. In: Domestication of Plants in the Old World. The Origin and Spread of Cultivated Plants in West Asia, Europe, and the Nile Valley, third ed. Oxford University Press, New York, pp. 108–111.

CHAPTER

Embryo rescue and chromosomal manipulations

5

Biswajit Mondal[1], S.K. Chaturvedi[2], Alok Das[3], Yogesh Kumar[1], Aneeta Yadav[4], Shiv Sewak[1], N.P. Singh[5]

[1]*Division of Crop Improvement, ICAR-Indian Institute of Pulses Research, Kanpur, Uttar Pradesh, India;* [2]*College of Agriculture, RLBCAU, Jhansi, Uttar Pradesh, India;* [3]*Division of Plant Biotechnology, ICAR-Indian Institute of Pulses Research, Kanpur, Uttar Pradesh, India;* [4]*Faculty of Agriculture and Allied Industries, RAMA University, Kanpur, Uttar Pradesh, India;* [5]*ICAR-Indian Institute of Pulses Research, Kanpur, Uttar Pradesh, India*

5.1 Introduction

Chickpea (*Cicer arietinum* L.) is an important food legume globally after dry beans and peas. It is an Old World pulse crop consumed in different forms and serves as an important source of dietary protein for the vegetarian population of South Asia (especially India) and other parts of the world. Food legumes with high protein content like soybean and groundnut are majorly consumed as vegetarian fat (oils, butter, etc.), but chickpea is a commodity which provides considerable diversity in the consumption pattern (green immature seeds, soaked seeds, germinated seeds, salad combos, roasted seeds, flour from raw and roasted seeds, split cotyledons or pulse, other different processed products). Chickpea seeds have a rich nutrition profile comprising protein, fiber, calcium, potassium, iron, phosphorus, magnesium, zinc, selenium, sodium, and copper (USDA, 2019), which can address malnutrition to a great extent. Chickpea is the second most important grain legume after soybean and contributes 86.73% of global pulse production from 89.89% area under pulses (Singh et al., 2016). Global production of chickpea reached 14.77 million tonnes in 2017 (FAOSTAT), with India contributing a giant share of 9.075 million tonnes, followed by Australia (2.004 million tonnes). The production of chickpea in India has recorded a highest of 11.23 million tonnes in 2018, with a record productivity level of 1063 kg/ha from an area of 10.56 million hectares (MoAF and W, 2018). This achievement is being considered as a significant step toward attaining self-sufficiency in chickpea in India (Dixit et al., 2019), which is the largest consumer of the commodity. Conventional approaches of plant breeding, namely, pedigree breeding, mass-pedigree, convergent breeding approaches, shuttle breeding, etc. have led to this breakthrough in chickpea. Sustaining the genetic advancement

Chickpea: Crop Wild Relatives for Enhancing Genetic Gains. https://doi.org/10.1016/B978-0-12-818299-4.00005-1
Copyright © 2020 Elsevier Inc. All rights reserved.

and enhancing the yield under different stresses by development of superior breeding materials is a tough challenge. Alleles existing in the varieties released have been utilized to a great extent through crossing programs, and broadening of the genetic base with novel alleles from landraces, wild accessions through prebreeding, and wide hybridization is a focus of the chickpea breeding programs across the globe. With the emerging need of new plant type amenable to mechanical harvesting, resistance to biotic stresses like ascochyta blight, dry root rot, collar rot, botrytis gray mold, etc. along with the abiotic stresses which limit horizontal spread of the crop to new ecologies; distant hybridization and crossing program with the landraces, wild relatives/accessions is the research approach. Until now, accomplishment in the form of varieties through wide hybridization in chickpea is limited; however, genomic resources in the form of linkage maps, polymorphic markers, mapping populations, breeding populations, and introgressed lines have been developed (Winter et al., 2000; Singh et al., 2008; Millan et al., 2010; Kaur et al., 2013). After adoption of pureline selection and pedigree breeding in genetic enhancement of chickpea during 1970–80s, wide hybridization and mutation breeding were the novel approaches which followed next. Embryo rescue technique facilitated the wide hybridization. Chromosomal manipulation techniques flourished along with basic cytogenetic studies in the mutants developed by mutation breeding, where structural aberrations and chromosomal disjunction leading to numeral aberrations were the focus of research. Later cytogenetic characterization of F_1s and segregating generations derived out of wide hybridization became a major area of discourse. Embryo rescue and chromosomal manipulation were two discrete fields which grew together and contributed in the genetic enhancement of commercial crops. In chickpea, embryo rescue has been applied in wide hybridization and chromosomal characterization is majorly limited to karyotyping the different species of *Cicer*. Until now, limited results could be achieved in the area of embryo rescue and chromosomal manipulations in chickpea and their application in varietal development is yet awaited. However, lack of suitable donor lines for various emerging biotic and abiotic stresses (for example, dry root rot, collar rot, botrytis gray mold, wilt in *kabuli* chickpea, saline/alkaline soil conditions) and broadening of the genetic base of breeding program has focused of prebreeding and wide hybridization in contemporary research. Under this purview, embryo rescue and chromosomal manipulation are relevant research approaches which are to be duly reviewed and discoursed. In this chapter, an account of crossability and wide hybridization in *Cicer* species, and their potential to be utilized in breeding will be discussed. Accomplishments in embryo rescue and cytogenetic studies will be dealt as well.

5.2 Origin and taxonomy

The genus Cicer L. (Leguminosae, Cicereae) comprises 9 annual and 35 perennial species. The genus is the member of the monogeneric tribe Cicereae Alef., subfamily Papilionoideae, family Leguminosae. Only one of the 44 species currently recognized is cultivated on a large scale, the chickpea (*Cicer arietinum* L.). Today it is cultivated in 49 countries. The monogeneric tribe *Cicereae* comprises genus *Cicer* belonging to the family Fabaceae (Kupicha, 1981). The genus *Cicer* has 49 taxa including nine annual species (van der Maesen et al., 2007). Genus *Cicer* has two subgenera, namely, Pseudononis and Viciastrum and four sections—Cicer, Chamaecicer, Polycicer, and Acanthocicer. The four sections vary based on morphological characteristics and geographical distribution (Popov, 1929; van der Maesen, 1972, 1987). Davies et al. (2007) carried out a conclusive characterization of 152 herbarium specimens representing 34 taxa (out of 44) reported in chickpea considering 104 traits and classified the accessions by hierarchical clustering following Euclidean coefficient. The authors also reviewed the taxonomy of *Cicer* in timeline and concluded a new subgeneric classification with three subgenera, five sections, and two series. This is one of the most recent taxonomical classification study based on morphological descriptors in chickpea. Insights into this classification also provide a basis of gene pool structure.

5.3 Genus *cicer*

1. Subgenus Cicer
   - 1.1. Section Cicer
     - ➢ Series Cicer

   C. arietinum L., C. echinospermum P.H. Davies, C. reticulatum Ladiz.
     - ➢ Series Pinnatifida

   C. bijugum Rech.f, C. judaicum Boiss, C. pinnatifidum Jaub. Sapach
   - 1.2. Section Chamaecicer

   C. atlanticum Coss. Ex Maire, *C. incisum* (Willd.) K. Maly
2. Subgenus Viciastrum
   - 2.1. Section Annua

   C. chorassanicum (Bunge) Popov, C. yamashitae Kitam
   - 2.2. Section Polycicer

   C. floribundum Fenzl., *C. graceum* Orph., *C. heterophyllum* Contandr. Pamuk C and Quezel, *C. isauricum* P.H. Davies, *C. montbretii* Jaub. and Spach.
   - 2.3. Section Vicioides

   C. acanthophyllum Borris, *C. anatolicum* Alef., *C. balcaricum* Galushko, *C. baldshuanicum* (Popov) Lincz., *C. fedtschenkoi* Lincz, *C. flexuosum* Lipsky, *C. grande* (Popov) Korotkova, *C. incanum* Korotkova, *C. korshinsky* Lincz., *C. laetum* Rassulova and Sharipova, *C. luteum* Rassulova and Sharipova, *C. macracanthum* Popov, *C. microphyllum* Benth., *C. multijugum* Maesen, *C. nuristanicum* Kitam, *C. paucijugum* (Popov) Nevski, *C. pungens* Boiss, *C. rassulovie* Lincz., *C. rechingeri* Podlech, *C. songaricum* Steph. Ex DC, *C. stapfianum* Rech.f, *C. subaphyllum* Boiss, *C. tragacanthoides* Jaub and Spach.

3. Subgenus Stenophylloma

C. canariense A.G. Guerra and G.P. Lewis, *C. cuneatum* Hochst. Ex A. Rich

Van der Maesen at al. (2007) reviewed all the previously reported classifications and validated only the classification reported by Popov (1929) among the previous reports. The authors also attempted classification based on sequence data of amplicons based on primers designed on nuclear ITS (internal transcribed spacer regions in nucleus) and trnK/matK and trns-trnG regions of chloroplast (Van der Maesen et al., 2007).

5.4 Gene pools and wide hybridization

Gene pools of chickpea have been proposed by various authors based on crossability between different species. On the basis of the results derived in wide hybridization, different authors had proposed the gene pool structure of chickpea which are broadly similar. Gene pool structure classifies the annual *Cicer* species and forms the basis on which further wide hybridization breeding program is decided. Harlan and de Wet (1971) first proposed the gene pool classes of chickpea which has been validated and contrasted by various authors. Smartt (1984) reviewed the gene pools of legumes and referring to the studies of Ladinsky and Adler (1976a,b) proposed division of the primary gene pool into GP 1a and GP 1b, housing *C. arietinum* and *C. reticulatum*, respectively. Secondary gene pool (GP 2) was proposed to be comprising *C. echinospermum*. The tertiary gene pool (GP3) was proposed to be comprising *C. bijugum, C. pinnatifidum, C. judaicum, C. cuneatum, C. yamashitae*, and other annual and perennial species. The gene pool structure proposed by Harlan and de Wet (1971) has been validated by Van der Maesen et al. (2007). Here *C. echinospermum* was put in primary gene pool. Six annual species have been grouped within the crossable limit. The gene pool structure comprises *C. arietinum* in the primary gene pool (GP 1a); *C. reticulatum* and *C. echinospermum* in GP 1b; *C. bijugum, C. pinnatifidum*, and *C. judaicum* in GP 2; whereas rest of the *Cicer* species belong to GP 3.

Gene pool structure concluded in the report of by Van der Maesen et al. (2007) is as follows:

- GP1a: *C. arietinum*
- GP1b: *C. reticulatum* and *C. echinospermum*
- GP2: *C. bijugum, C. pinnatifidum, C. judaicum*
- GP3: other *Cicer* species (annual and perennial)

An equivalent system was proposed by Redden and Berger (2007), who included *C. arietinum* and *C. reticulatum* in the primary gene pool, *C. echinospermum* in the secondary gene pool, and rest of the annual and perennial species in the tertiary gene pool. Mallikarjuna et al. (2011) modified the above gene pool structure placing *C. arietinum* in the primary gene pool (GP 1), *C. reticulatum* and *C. echinospermum* in the secondary gene pool (GP 2), and rest of the annual and perennial species in the tertiary gene pool (GP 3).

Gene pools of *Cicer* proposed by Mallikarjuna et al. (2011) are as following:

- Primary gene pool: *C. arietinum* (landraces and cultivars)
- Secondary gene pool: *C. reticulatum, C. echinospermum*
- Tertiary gene pool: *C. bijugum, C. judaicum, C. pinnatifidum, C. chorassanicum, C. yamashitae, C. cuneatum, C. atlanticum, C. incisum, C. incisum* ssp. serpentinica, *C. floribundum, C. floribundum* var. amanicola, *C. graecum, C. heterophyllum, C. heterophyllum* var. kassianum, *C. uludereensis, C. isauricum, C. montbretii, C. acanthophyllum, C. anatolicum, C. balcaricum, C. baldshuanicum, C. fedtschenkoi, C. flexuosum, C. grande, C. incanum, C. korshinskyi, C. laetum, C. luteum, C. macracanthum, C. microphyllum, C. multijugum, C. nuristanicum, C. paucijugum, C. pungens, C. rassuloviae, C. rechingeri, C. songaricum, C. stapfianum, C. subaphyllum, C. tragacanthoides, C. kermanense, C. mogoltavicum, C. oxyodon, C. spiroceras, C. canariense.*

Out of all the GP structures proposed in chickpea, it can be concluded that three annual species, namely, *C. arietinum, C. reticulatum*, and *C. echinospermum* are readily crossable and trait transfer can be attempted through conventional hybridization. In case of wide hybridization involving any other annual or perennial species, interventions like hormonal treatment of the peduncle to prevent abscission, embryo rescue, etc. are essential. Strategic locations often help wide hybridization activities, particularly locations with prolonged congenial temperature window (<25°C) with normal relative humidity and less wind. It has been observed in the wide hybridization garden facility at the ICAR-Indian Institute of Pulses Research, Kanpur, that such congenial conditions prevailing after pollination enhance the chances for fertilization and further development. Zygotic embryogenesis through embryo culture is a cornerstone of wide hybridization activities, but chickpea per se has been observed to be responding better to somatic embryogenesis from embryonal axis or other explants. Pod and seed setting in wide hybridization have been an issue. Besides, development of the chlorotic mutants like green, partially green, and albino plants from the rescued embryo limits the embryo rescue technique to a considerable extent in chickpea. Limited literature and reports are available in embryo rescue in this crop. Accomplishments in wide hybridization in chickpea are summarized in Table 5.1 based on the available literatures and reports.

Out of the different wide hybridizations attempted, conventional hybridization leads to viable F_1 in *C. arietinum/C. reticulatum* and *C. arietinum/ C. echinospermum* crosses only. In these crosses, F_1 sterility (based on pollen staining data) has not been observed and F_2 populations show segregation for the contrasting characters (Singh et al., 2015). Considerable success has been achieved in these crosses and traits related to plant architecture and yield attributes have been successfully transferred to *C. arietinum* from *C. reticulatum* and *C. echinospermum*, respectively. In ICAR-IIPR, segregants with three to four seeds per pod (in >40% of the total pods set in a single plant) have been reported from

Table 5.1 Accomplishments in wide hybridizations in chickpea. Attempted crosses between species and the consequent results are summarized.

S.No.	Wide hybridization attempted between species of chickpea	Observations and results
1	*C. arietinum/C. reticulatum*	Fertile embryo, F_1 and F_2[1,4]; 8%–15% pod setting in F_1, seed setting up to 14%, heterosis observed for yield attributes, >90% F_1 pollen staining[15]
2	*C. arietinum/C. echinospermum*	Partially fertile F_1[1,3,4]; viable embryo and seed setting[8]; 4%–8% pod setting in F_1, seed setting up to 10%, heterosis observed for yield attributes, >90% F_1 pollen staining[15]
3	*C. arietinum/C. pinnatifidum*	Albino plants derived after embryo rescue[5]; sterile F_1 after embryo rescue[6]; embryo abortion[7,8]; green, partially green, and albino plants derived after embryo rescue[11,12]; viable embryo and F_1 derived through hormonal treatment in peduncle of crossed flower, desirable advanced breeding lines derived[17]
4	*C. arietinum/C. bijugum*	Embryo abortion[7,8,9]; green F_1 plants derived through embryo rescue[10]
5	*C. arietinum/C. cuneatum*	Failed[1]; embryo abortion[8]; F_1 embryos rescued through ovule culture, hybrid plants did not flower[16]
6	*C. arietinum/C. songaricum*	Failure in pollen tube development and fertilization[2]
7	*C. arietinum/C. judaicum*	Embryo abortion[7,8]; albino, partially green and green plants derived after embryo rescue[11,12]
8	*C. arietinum/C. chorassanicum*	Embryo abortion[8]
9	*C. arietinum/C. yamashitae*	Embryo abortion[8]
10	*C. reticulatum/ C. echinospermum*	Sterile F_1[1]
11	*C. judaicum/C. pinnatifidum*	Partially fertile F_1 and F_2[1]; F_1 with protruding pistils and 30% pollen staining, poor F_2 population growth[13]
12	*C. judaicum/C. bijugum*	Partially fertile F_1 and F_2[1]; F_1 partly sterile (50% pollen staining) and poor F_2 population and growth[13]
13	*C. judaicum/C. cuneatum*	Failed[1]
14	*C. judaicum/C. arietinum*	Embryo abortion[8]; F_1 partly sterile (54% pollen staining)[14]
15	*C. pinnatifidum/C. bijugum*	Partially sterile F_1 and F_2[1]
16	*C. pinnatifidum/C. cuneatum*	Failed[1]
17	*C. pinnatifidum/C. arietinum*	Embryo aborted[8]
18	*C. echinospermum/C. arietinum*	Embryo aborted[8]

Table 5.1 Accomplishments in wide hybridizations in chickpea. Attempted crosses between species and the consequent results are summarized.—*cont'd*

S.No.	Wide hybridization attempted between species of chickpea	Observations and results
19	*C. chorassicum/C. arietinum*	Embryo aborted[8]
20	*C. yamashitae/C. arietinum*	Embryo aborted[8]
21	*C. cuneatum/C. canariense*	50% pollen sterility in F_1, poor F_2 population and growth[13]

Corresponding references are provided with numbers in superscript against the results, [1]Ladizinsky and Adler (1976a,b), [2]Mercy and Kakkar (1975), [3]Pundir and Mangesha (1995), [4]Singh and Ocampo (1997), [5]Badami et al. (1997), [6]Mallikarjuna (1999), [7]Stamigna et al. (2000), [8]Ahmad and Slinkard (2004), [9]Clarke et al. (2006), [10]Mallikarjuna et al. (2007), [11]Kumari et al. (2011), [12]Clarke et al. (2011), [13]Abbo et al. (2011), [14]Singh and Singh (2012), [15]Singh et al. (2015), [16]Singh and Singh (1989); [17]Kaur et al. (2013).

C. arietinum/C. echinospermum cross (Chaturvedi et al., 2018). Growth hormone treatment of the peduncle of the emasculated female flower to retain it after pollination by avoiding abscission has also been found to be effective; it provides due time to the slowly growing pollen tube to fertilize the ovule. To combine resistance to ascochyta blight and botrytis gray mold from *C. pinnatifidum* (ICC 17152 or *C. pinnatifidum* accession no. 188) in *C. arietinum* (ICC 96030) wide hybridization was attempted through growth hormone treatment of the flower stalk to avoid abscission and promote the floral retention. Growth hormone–containing solution with 75 ppm GA + 1 ppm NAA +10 ppm kinetin was applied on flower peduncle soon after pollination (Kaur et al., 2013). Zygotic embryo culture methods have helped to derive F_1s in many wide hybridization crosses (Table 5.1) e.g., embryo rescue in *C. arietinum/C. pinnatifidum, C. arietinum/C. bijugum, C. arietinum/C. judaicum*. However, chlorotic mutant F_1 plants are a major concern in this approach which prevents further generation advancement and deriving the segregants. Mallikarjuna et al. (2007) derived green F_1 plants in *C. arietinum/C. bijugum*; in majority of the other reports, chlorotic mutants had been an impediment. A target cross for enhancing number of seeds per pod, *C. arietinum/C. cuneatum*, has remained unsuccessful in majority of wide hybridization attempts reported. Singh and Singh (1989) rescued the F_1s through ovule culture in *C. arietinum/C. cuneatum* cross; the hybrid plants expressed the phenotype close to *C. cuneatum*, but they did not flower.

Based on the wide hybridization studies, strategies to derive fertile F_1s can be worked out before initiating any wide cross addressing a particular target trait(s) introgression. From the studies done until now, chance of getting success by using bridge species in distant hybridization is not so promising in chickpea (Singh and Singh, 1989). However, in the study carried out by Kaur et al. (2013), targeting a large number of emasculations and application of hormonal mix in the peduncle to retain the flower appears promising. Use of strategic locations with long congenial temperature window during grain filling/pod setting increases pod and seed setting percentages to some extent and has been observed in the wide hybridization program of ICAR-IIPR and NBPGR (Singh et al., 2015). Alleles of various genes/traits possessed by different species of chickpea have been studied to a considerable extent, reported through screening studies in literature, and reviewed well (Singh et al., 2008, 2016). In Table 5.2, an account of sources of donor alleles for different morphological parameters and biotic and abiotic stresses present in different species of chickpea is presented. In *C. arietinum*, alleles of resistance to ascochyta blight and fusarium wilt are present along with alleles for some morphological parameters (tallness, yield attributes, etc.). To address all the other breeding objectives, wide hybridization is the only approach in chickpea, and importance of zygotic embryo rescue techniques is worth a mention. Restructuring plant architecture, breeding for nutritional quality, and addressing abiotic stresses (heat and drought, especially heat through enhanced podding and seed setting under high temperature) bank on alleles from *C. reticulatum* and *C. pinnatifidum*. From those wide hybridizations, promising advanced breeding lines have been developed at ICAR-IIPR through the prebreeding program, which are under field trial for evaluation of their yield performance. To scale up the utilization of the wild alleles, however, embryo rescue technique need to be further standardized between different species of *Cicer* (Fig. 5.1).

5.4.1 Embryo rescue

Embryo culture deals with sterile isolation and in vitro culturing of a mature or immature embryo to obtain a viable and fertile plant. This tissue culture intervention is adopted to overcome the postfertilization developmental barriers in the growing embryo originating from the differences in the pattern/rate of embryogenesis of two distant parental lines. Embryo culture can be broadly classified in two ways, namely on the basis of stage and histological origin of the isolated embryo explants. On the basis of stage of the isolated embryo explants, embryo culture is classified into mature embryo culture and immature embryo culture (embryo rescue). Whereas, considering the histological origin of the isolated embryo explants, embryo culture has been classified into zygotic and somatic embryo culture. Those four broad classes of embryo culture rather overlap when the approaches adopted in different research initiatives are studied.

In chickpea, wide hybridization leads to all possible undesirable fates, namely, flower drop, early embryo abortion along with pod development up to limited extent, late embryo abortion after considerable pod development, nonviable F_1 seeds with poor germination, weak F_1 seedlings unable to sustain the sees-seed cycle, chlorotic F_1 mutant plants, F_1 plants with pollen sterility up to different extents, F_2 sterility and distorted segregation, etc. To address the abovementioned fates and derive the viable hybrid F_1 plants, embryo culture provides a wide array of optional protocols whose applicability and related standardization need to be worked out.

Embryo culture from mature embryos is advantageous with respect to immature embryo explants, because of the advanced growth stage of the mature embryos and

Table 5.2 Genes for a wide range of traits present in different chickpea species revealed by the large-scale screening studies at the national and Consultative Group of International Agricultural Research (CGIAR) institutes reviewed in recent literatures.

Chickpea species	**Resistance/tolerance response of chickpea species to different stresses and desirable morphological traits in them as reported in literatures and reviewed.**				
	Pathogenic diseases (fungal)	**Insect-pests**	**Nematodes**	**Abiotic stresses**	**Morphological traits**
Cicer arietinum	Ascochyta blight, fusarium wilt				Plant architecture and yield attributes
C. reticulatum	Ascochyta blight, fusarium wilt, botrytis gray mold, rust, phytophthora root rot, stem rot	Bruchids, *Helicoverpa*, leaf miner, seed beetle	Cyst, root knot, root lesion nematode	Cold, drought	Yield attributes
C. echinospermum	Ascochyta blight, fusarium wilt, rust, phytophthora root rot	*Helicoverpa*, seed beetle	Root knot, root lesion	Cold, drought	
C. pinnatifidum	Ascochyta blight, fusarium wilt, botrytis gray mold, phytophthora root rot, stem rot	*Helicoverpa*	Cyst, root knot	Cold, drought	Yield attributes
C. judaicum	Ascochyta blight, fusarium wilt, botrytis gray mold, stem rot	Leaf miner, seed beetle	Root knot	Cold	
C. bijugum	Ascochyta blight, fusarium wilt, botrytis gray mold, rust, phytophthora root rot	*Helicoverpa*, leaf miner	Cyst, root knot	Cold	
C. montbretii	Ascochyta blight			Drought	High number of seeds per plant
C. cuneatum	Ascochyta blight, fusarium wilt	*Helicoverpa*, seed beetle, leaf miner			
C. yamashitae	Stem rot				
C. microphyllum		*Helicoverpa*		Drought	
C. songaricum				Drought	

Singh et al. (2008) and (2016)

FIGURE 5.1

Pods with three to four seeds identified in the advanced breeding lines derived from the *Cicer arietinum* × *Cicer echinospermum* wide cross IPC 2006–88/ILWC 179.

Chaturvedi, S.K., Mondal, B., Srivastava, A.K., Tewari ,Y., 2018. Chickpea breeding lines with stable green grains developed. Pulses Newsl. 29 (1), 2.

their autotrophic mode of survival. Culturing media is also an important deciding factor depending on the explants, but invariably, liquid nutrient media with explants embedded in suitable filter paper bridge have been observed to provide better culturing condition with respect to semisolid media on which direct explant is put for further growth. Mature embryos are isolated from developed seeds prior to drying and cultured in vitro. Mature embryo culture can be resorted to under following conditions:

- **i.** Prolonged seed dormancy of the hybrid embryos
- **ii.** Low in vivo survival of embryos
- **iii.** To bypass germination inhibitions in the hybrid seeds
- **iv.** Deriving viable seedlings from sterile seeds

Embryo rescue involves the interventions to derive fertile hybrid plant through sterile isolation of the immature embryo and culturing in suitable nutrient media. Embryo rescue is achieved through embryo culture, ovule culture, and ovary or pod culture approaches (Reed, 2005). Embryo rescue intervention can be done under following situations:

- **i.** Failure of development of the hybrid endosperm
- **ii.** Production of toxin-like metabolites from the hybrid endosperm which suppress the embryo growth
- **iii.** Slow differentiation in the hybrid embryo due to wide variation in the developmental pattern of the embryos in the parents leading to flower drop
- **iv.** To avoid chlorotic mutants in F_1
- **v.** Avoiding germination problem in F_1 arising due to dormancy or other issues and F_1 sterility

Categories of embryo cultures, based on the stage of zygotic/somatic explants used, with their objectives, purposes, and corresponding explants used are provided in Table 5.3. Not only wide hybridization, but also embryo culture has a much wider purview and applicability. Broad-scale applications of embryo culture can be summarized into following heads:

Table 5.3 Different categories of embryo cultures depending on type of explants used and corresponding objectives.

S.No.	Explants for embryo culture	Objective/Purpose
1	Mature and intact seed embryo	Studying metabolic and biochemical aspects of dormancy and germination; analysis of the various parameters of embryonic growth.
2	Surgically dissected embryo	Dissected embryo parts can be studied for their relative capacity of dedifferentiation and redifferentiation leading to development of the shoot/plantlet.
3	Immature embryos or pro-embryos	Differentiation pattern and efficacy of the globular or heart-shaped developing embryos to plantlets studied. Those explants may be grown in artificial media directly or cultured after grafting in mature cotyledons with excised embryos. Nutrient requirements in the growth media can be standardized.
4.	Intact seed with undifferentiated embryo	Orchids and several epiphytes produce seeds with undifferentiated or premature embryos. Culturing such seeds improve seed-plant turnover ratio.
5	Adventitious embryos from polyembryonic seeds	Culturing of adventitious embryos developed from nucellar or chalazal tissues in polyembryonic seeds facilitates clonal propagation.
6	Inviable or abortive embryos	Abortive embryos produced in distant hybridization due to contrast in the rate and pattern of differentiation of the parental lines; their growth rate is often very slow leading to their abortion or flower abscission. Culturing such embryos in suitable nutrient media helps to derive the F_1 plants.

i. Prevention of embryo abortion (mature/immature)
ii. Overcoming seed dormancy (F_1)
iii. Shortening of breeding cycle
iv. Handling wild accessions of perennial nature
v. Production of haploids
vi. Overcoming seed sterility (F_1)
vii. Clonal propagation
viii. Development of transgenic plants through somatic embryogenesis from genetically transformed explants (nodal zone in chickpea)

Out of all the fields of application of embryo culture technology mentioned, in chickpea, somatic embryogenesis—enabled development of transgenic plants from the transformed cells in the explants (embryonal axis) has been achieved with perfection in ICAR-IIPR and in other transgenic labs working on BT chickpea. However, zygotic embryogenesis and embryo rescue technology on chickpea are still evolving.

5.4.2 Chickpea embryology

Under the present context, where embryo rescue is going to be important as an intervention to broaden the genetic base and utilization of wild alleles in the breeding program, embryology of chickpea is a major point of consideration to enable the wide hybridization crosses. Out the five chief types of embryogenies discussed by Maheshwari (1950), i.e., *Crucifer* or *Onagrad* type, *Asterad* type, *Solanad* type, *Caryophyllad* type, and *Chenopodiad* type, embryogeny of chickpea cannot be classified into any of the classes mentioned above (Mercy et al., 1974). Mercy et al. (1974) studied the embryogenesis in *C. arietinum* and *C. soongaricum* and reported the similarity in stages of embryogenesis in these distant species of genus *Cicer*. The authors discussed about the difference in the rate of embryogenesis in these distant species (considering the resting phase between the cell divisions and differentiations) which might be a major cause for failures in development of hybrid embryo in wide hybridizations. Details of microsporogenesis and macrosporogenesis and corresponding gametogenesis stages of *C. arietinum* and *C. soongaricum* have been described by Mercy et al. (1974). Depending on the formation of "suspensor" of "haustoria" from the primordial basal cell, chickpea embryogenesis has some link with the *Solanad* type only; however, discourse in this field is needed for adoption of strategies to facilitate the hybrid embryo survive through wide hybridizations. Embryo sac classification provided by Maheshwari (1937), i.e., monosporic, bisporic, tetrasporic, and their modifications will provide a framework to deal with the cases of hybrid embryo sacs of different combinations. The report of Mosjidis et al. (1986) regarding the prediction of developmental stages of a developing chickpea embryo is a major consideration while troubleshooting the hybrid embryo development in wide hybridizations. While working out the embryo rescue strategy in a wide hybridization, the stage of harvest of the immature hybrid embryo is most important and needed to be worked out for each pair of species crossed. In the report of Mercy et al. (1974) the difference in rate of development of embryos of *C. arietinum* and *C. soongaricum* was mentioned and to address such cases the proper stage of harvest of the developing immature embryo is a key information. Work of Mosjidis et al. (1986) provides a way out and provides us with some mathematical equations framed on the basis of variables which were found to be closely related to the embryo growth stage (ES). The authors standardized the equations in chickpea cultivar *Hermosillo*. The half open flowers (pollination is over) were tagged and developing pods were harvested in the sixth, seventh, and eighth days after pollination. Six parameters were scored in the sampled pods, namely, length, width, fresh weight, dry weight, total number of ovules, and number of growing ovules. The embryos were excised to chronologically study the developmental stages. Between 6–7 and 7–8 days after pollination the globular and heart stages were identified in majority of the sampled embryos. After 8 days after pollination, torpedo stage initiated in the embryos.

Mosjidis et al. (1986) coded embryo stages as:

(1) Earlier than globular stage
(2) Globular stage
  (2.3) Early-heart stage, with incipient lobules formed
  (2.7) Early-heart stage, with larger lobules, heart shape not attained
(3) Heart-shaped stage
  (3.3) Late-heart stage, with lobules large enough to make the embryo look like an elongated heart
  (3.7) Late-heart stage, with enlarged lobules, tending to attain torpedo shape
(4) Torpedo stage

Broadly chickpea embryo is classified into globular, early-heart, heart, late-heart, and torpedo stages. Primary and secondary traits like days after pollination, pod fresh weight, pod dry weight, pod length, pod width, and length × width were observed to be significantly correlated with the embryo stage (ES). Regression analysis revealed that the best estimate of ES (as a dependent variable) using one to three independent variables can be obtained by framing the regression equations with pod length ($R^2 = 0.726$); pod fresh weight ($R^2 = 0.726$); days after pollination and fresh weight ($R^2 = 0.797$); days after pollination, fresh weight, and number of growing ovules ($R^2 = 0.836$), respectively. Path coefficient analysis showed that pod fresh weight is the most important variable affecting the embryo stage. The equations with single correlated variables (pod length and pod fresh weight) and corresponding coefficients of determinations are following:

$$\text{ES} = -0.134 + 0.187 \text{ pod length (in mm) })R^2 = 0.726)$$

$$\text{ES} = 1.35 + 8.24 \text{ pod fresh weight (in g) } (R^2 = 0.726)$$

When another correlated independent variable, days after pollination is added and regression equations are reframed with equivalent coefficients of determination:

$$\begin{aligned}\text{ES} = & -1.37 + 0.448 \text{ days after pollination} \\ & + 5.54 \text{ pod fresh weight (in g) } (R^2 = 0.797)\end{aligned}$$

$$\begin{aligned}\text{ES} = & -2.08 + 0.403 \text{ days after pollination} \\ & + 0.128 \text{ pod length (in mm) } (R^2 = 0.774)\end{aligned}$$

Adding of the additional significantly correlated variables in a regression equation always increases the robustness and applicability of a regression equation for predicting the dependent variable (ES in this case). However, all the above equations comprise the independent variables which can be scored nondestructively (without damaging the developing pod). If pod weight parameter may be a cumbersome parameter to score when the pods are on plant, preliminary decisions can be taken on the basis of pod dimensions (length, breadth, and thickness). Mosjidis et al. (1986) found number of growing ovules to be correlated with embryo stage. Number of developing ovules is a parameter with limited variability between the chickpea accessions, varying from 1 to 3. However the authors framed a regression equation considering this trait as well.

$$\begin{aligned} \text{ES} = {} & 0.475 + 0.181 \text{ pod length (in mm)} \\ & - 0.346 \text{ no. of growing ovules } (R^2 = 0.771). \end{aligned}$$

Mosjidis et al. (1986) included all the four correlated independent variables (days after pollination, pod fresh weight, pod width, and number of growing ovules) to form two more regression equations with a high coefficient of determination to predict the embryo stage.

$$\begin{aligned} \text{ES} = {} & -0.788 + 0.439 \text{ days after pollination} + 5.32 \text{ pod fresh weight (in g)} \\ & - 0.326 \text{ number of growing ovules } (R^2 = 0.836) \end{aligned}$$

$$\begin{aligned} \text{ES} = {} & -1.65 + 0.449 \text{ days after pollination} + 0.239 \text{ pod width (in mm)} \\ & -0.423 \text{ number of growing ovules } (R^2 = 0.818) \end{aligned}$$

If the heart stage is targeted for prediction and isolation of the immature developing embryo for culturing, the above two equations predict that out of the sampled pods, heavier pods to be selected on eighth and lighter pods to be selected on ninth day after pollination, respectively. For example, a pod with two growing ovules on eighth day after pollination weighed 0.252 g and had the embryo at heart stage; whereas a pod with two growing ovules weighed 0.174 g on ninth day after pollination had embryo at the heart stage. Mosjidis et al. (1986) also proposed similar equations can be framed considering the pod width trait for embryo stage prediction. All the above equations impart flexibility and provide option to the plant breeders working in wide hybridization to select for different measurable parameters correlated with the embryo development. Based on such parameters, equations can be framed in different pairs of wide crosses involving *Cicer* species to predict the stage of harvesting the viable pro-embryos which can be rescued through suitable embryo culture technique. Studies on variability in the embryological parameters (especially phenological parameters related to embryo development) in chickpea accessions used in wide hybridization need to be done. Contrasting lines for those traits are to be identified as a new strategy to address postfertilization barriers. Tentative cross compatible lines can be identified depending on the equivalent rate of growth of the immature embryo up to 15 days after pollination.

Another embryological consideration in the wide hybridization of chickpea is the development of chlorotic mutants from the hybrid embryos (Table 5.1).

Clarke et al. (2006) studied wide hybridization using *C. pinnatifidum* and *C. bijugum* as donors with *C. reticulatum* and did not find any particular growth stage of embryo in which abortion was frequent. Thereby variability in the embryo growth was present in different hybrid embryos which provide us scope for further research on standardization of the timing of harvest of the immature embryo explant. Hybrid embryos of *C. bijugum* aborted at early stages of embryo development (4–8 days after pollination), whereas in *C. pinnatifidum* hybrids, the embryo reached the heart-shaped stage in 10–12 days after pollination and abortion was observed in later stages. Some of the wide crosses involving *C. pinnatifidum* developed on the plant up to torpedo stage (22 days after pollination). Studies on embryo rescue in selfed ovules showed positive correlation ($P < .001$) between embryo germination percentage and age of the rescued ovule. Embryo germination percentage ranged from 0 in 5-day-old ovules to 43% in 12-day-old ovules developed by selfing. The authors also observed positive correlation ($P < .001$) between germination percentage and actual size of the rescued ovule (selfed). Only 10% germination was observed in selfed ovules measuring 3 mm and 40% germination observed in ovules with 8 mm length. Positive correlation ($P < .001$) was also observed between the selfed ovule size and age of the embryo (number of days after pollination); however, this relation did not hold well in hybrid embryos. These reports of Clarke et al. (2006) validated some of the findings of Mosjidis et al. (1986) discussed above.

Clarke et al. (2006) also reported variation between chickpea genotypes for ability of immature selfed embryos to germinate and respond to the in vitro embryo rescue protocol. With respect to germination percentage under embryo rescue conditions, significant difference was observed between *C. arietinum* and *C. bijugum*. The advanced breeding lines derived from an interspecific cross involving *C. echinospermum* did not differ significantly for in vitro germination percentage under embryo rescue conditions. No statistically significant contrast was observed within *C. arietinum* genotypes for this trait, whereas the difference was significant between the genotypes of *C. bijugum*. Within the wide hybridization crosses studied by Clarke et al. (2006), germination of hybrid embryos was lowest in crosses involving *C. bijugum*. The interspecific variability in the embryogenesis traits in chickpea reduces the success rate of wide hybridization.

Microscopic studies by Clarke et al. (2006) revealed some histological abnormalities in the hybrid ovules from wide crosses involving *C. bijugum*, particularly difference in the shape of the cells of suspensor and integument wall with respect to the selfed ovules. Healthy endosperm was observed in the aborting ovules indicating proper double fertilization in contrast with previous reports where improperly formed, deficient or degenerating endosperm was identified as the major cause for abortion of the hybrid embryos (Geerts et al., 2002; Ahmad and Slinkard, 2004; Stamigna et al., 2000). Singh at al. (1999) in IIPR, Kanpur, observed endosperm breakdown in hybrid ovules of *C. bijugum*, but Clarke et al. (2006) recorded viable and actively diving endosperm in some of their hybrid embryos up to later stages of embryo development. Clarke et al. (2006) opined that the variability in the responses of the hybrid ovules following culture depended on their age and size, which are important points of consideration in embryo rescue. The authors observed 12-day-old selfed ovules (heart-shaped stage of embryogenesis) had

40% probability of germination and gave rise to a plantlet, whereas 5-day-old immature ovules (proembryo or globular embryo) did not germinate in best culture media with all type of supplements. In wide crosses involving *C. bijugum*, the hybrid embryo seldom reached to the heart-shaped stage on the mother plant. Under in vitro culture conditions, significant genotypic variation in the germination response of immature ovules was a major finding of Clarke et al. (2006). This finding paves the avenue for further research in the embryological compatibility between the parents of distant hybridizations.

Plant tissue culture division of the National Chemical Laboratory, Pune, India, had carried out some pioneering research in the field of chickpea embryology. Suhasini et al. (1997) studied the zygotic and somatic embryogenesis in chickpea and reported the similarities and contrasts. The group also studied somatic embryogenesis from the embryonal axis explants (Suhasini et al., 1994) and histological aspects of somatic embryogenesis in chickpea from different explants (Sagare at al. 1995). A brief account of the chickpea embryo (zygotic) reported in Suhasini et al. (1997) is provided herewith.

The zygotic embryo of chickpea is an elongated structure composed of two parts, an apically located embryo proper and a basal suspensor composed of two layers of columnar cells (biseriate). Six tiers of columnar biseriate cells comprise the suspensor with two basal single cells. Expansion of the cells of the suspensor increases the embryo length; however, the cell count of the suspensor remains unchanged during the embryo growth. Developmental stages of the chickpea embryo growth have been divided into eight stages (Suhasini et al., 1997) starting from 3 to 12 days after pollination, in which duration the ovule size increases from 1 to 7 mm and length of the embryo increases from 500 to 1700 μm. Chronologically those stages are early-globular stage (3 days after pollination), mid-globular stage (5 days after pollination), globular stage (7 days after pollination), late-globular stage (7 days after pollination), early-heart stage (8 days after pollination), mid-heart stage (9 days after pollination), late-heart stage (9 days after pollination) and early-cotyledonary stage (12 days after pollination). In the globular stage, the size (of cells) of the embryo proper and suspensor further increased. In the bright-field microscope or in electron microscope, the cytoplasmic density of apical cells of embryo proper appears to be high with respect to the basal suspensor cells. Maheshwari (1950) while describing the *Solanad* embryos indicated the importance of suspensor in the thriving of the embryo proper and mentioned it as the first differentiated structure produced by the developing embryo. In the embryogeny of chickpea, suspensor plays a major role for the establishment, development, and differentiation of the embryo proper. Suspensor provides anchorage to the embryo in the developing ovule. It is a filamentous structure expected to be made of fixed number of biseriate columnar cells and it is reported to be supporting the early development of the embryo proper by transport of nutrients and metabolites (growth regulators) essential for determination of the polarity (Yeung, 1980; Yeung and Meinke, 1993). Along with the growth and maturation of the embryo, the suspensor degenerates (Yeung, 1980; Yeung and Meinke, 1993; Warren Wilson and Warren Wilson, 1993).

At the early-heart-shaped stage (8 days after pollination), homeotic differentiation of the embryo proper is initiated and the radial symmetry is gradually converted to bilateral symmetry with the formation of the apical cleavage or furrow. Later it leads to the formation of the heart-shaped embryo where the differentiation for cotyledons and embryonal axis starts. With the formation of the shoot meristem at the apex of the embryonal axis, bilateral symmetry is established in the developing embryo which is an important checkpoint in the embryogenesis (Berger and Brownlee, 1995). Raghavan (1986) reported the differentiation of the heart-shaped embryo is attained due to viable cell division behavior of the cell clusters at the apex of the embryo proper, which promotes centrifugal cell division and suppresses the centripetal cell division thereby giving rise to the heart-shaped embryo. The physiological basis of this transition of symmetry from radial to bilateral is achieved due to polar transport of auxin as reported by Liu et al. (1993). Wide variation in homeosis of suspensor is observed in Leguminosae, extensively reviewed by Lersten (1983). Variation in number and arrangement of cells in suspensors in genus *Cicer* has also been observed (Suhasini et al., 1997). Thereby enhancement of the suspensor-specific traits may be achieved in different cross combinations of wide hybridizations. Basic differentiation pattern of zygotic and somatic embryogenesis is quite similar; however, the structure of suspensor differs. Failure of development of the hybrid embryo in wide hybridization is largely accounted to the early abortion within 18–20 days of pollination, when the formation of the suspensor and polarity delineation of the embryo proper takes place. There is scope for future research in this area of embryology to overcome the postfertilization barriers and promote the development of the embryo in the initial stages of cell division and differentiation. Suspensor structure being the major difference in the developing zygotic (elongated suspensor) and somatic embryos (suppressed suspensor), the undesirable fate of the hybrid embryo (mentioned in Table 5.1) may be correlated with it along with many other factors and further studies can be continued.

The seedlings developed through the two approaches have some contrasts, summarized in Table 5.4. Suhasini et al. (1997) reported major difference in the size of cotyledon of the seedlings derived from somatic and zygotic embryogenesis, where seedlings from zygotic embryos had a larger cotyledon (up to 5.4 mm diameter, *desi* chickpea cultivar Phule G 12). The authors also reported initial smaller size of the zygotic embryo makes it difficult to isolate in early stages and study, thereby the somatic embryos which are comparatively larger in size can be isolated and studied vividly.

5.4.3 Somatic embryogenesis in chickpea

Somatic embryogenesis in chickpea has been achieved from different explants (embryonal axis, immature cotyledon, leaf, etc.). Somatic embryogenesis provides a better feasibility to work in distant hybridization because of routing the process through differentiation of the explants into callus, which helps to derive several embryos or embryogenic cell clusters from a single callus developing from a single explant. Under suitable culture conditions, these embryogenic cell clusters developing on the callus in form of protuberances re-differentiate into seedling after shooting and rooting supplemented by proper growth media. In contrast, except the case of polyembryony, a single zygotic embryo always develops into a single seedling only. Suhasini et al. (1994) at NCL, Pune, reported development of globular stage somatic embryos directly in the form of protuberances from the shoot tip surface of the embryonal axis explants when cultured under proper conditions. Embryonal axis is the most potent explant for somatic embryogenesis, and the authors reported 65% of the mature embryonal axis explants gave rise to globular stage somatic embryos. Each of the mature embryonal axis explant gave rise to 12–15 somatic globular embryos, average size of each embryo being larger (300–1200 mm) with respect to the zygotic globular embryo (70–120 mm). The globular somatic embryo differs in histological structure from the globular embryo with respect to the suspensor. In zygotic globular embryos, the suspensor comprises biseriate cells whereas in somatic globular embryos, the suspensor is a stalk-like structure comprising multiseriate cells developing from the callus or shoot tip of

Table 5.4 Comparing and contrasting the seedlings developed from somatic and zygotic embryos. Structurally the seedlings are similar and somatic embryogenesis follows largely the zygotic pattern, but considerable differences can be observed in their development.

S.No.	Criteria	Seedling from zygotic embryo	Seedling from somatic embryo
1	Globular embryo	Developed from differentiation of a smaller globular embryo (70 ± 120 mm)	Developed from differentiation of a larger size globular embryo (300 −1200 mm)
2	Embryonal axis	Prominent	Weak/suppressed
3	Cotyledons	Prominent, devoid of testa actively supporting seedling growth	Rudimentary
4	Radicle and developing root system	Prominent radical and developing tap root system	Adventitious root development
5	Plumule	Prominent plumule, with slowly differentiating apical bud	Fast differentiating slender plumule, with early dichotomous branching and leaflets
6	Branching	Prominent apical bud, branching initiates comparatively later	Often branching initiates early on the slender shoot

Source: Suhasini, K., Sagare, A.P., Sainkar, S.R., Krishnamurthy, K.V., 1997. Comparative study of the development of zygotic and somatic embryos of chickpea (Cicer arietinum *L.). Plant Sci. 128, 207–216.*

embryonal axis explant. The electron microscopic images of the suspensors of the zygotic and somatic embryos have been published in the report of Suhasini et al. (1997) which are for reference of the readers. The cell division and differentiation of the apical embryo proper portion of zygotic embryo and apical part of somatic globular embryo are similar, where centripetal cell division leads to formation of depression at the apex of the globular embryo further differentiating it into the heart-shaped stage. Cotyledons elongate in the torpedo stage and a mature cotyledonary stage embryo is observed with a distinct shoot apical meristem between the cotyledons. Thus the somatic embryogeny recapitulates the zygotic embryogeny in chickpea (Suhasini et al., 1997).

Somatic embryogenesis in chickpea has been broadly utilized in transgenic event development across different laboratories. Somatic embryogenesis from the transformed explants (embryonal axis of chickpea seeds) also shows similar developmental pattern with respect to the hybrid embryos developed from wide hybridization. It also starts through formation of embryogenic cell clusters and development of protuberance, followed by formation of globular embryo which led to further differentiation. Embryologically hemizygous transformed explant or callus and distant allelic hybrid embryos behave similarly; any strategy worked out to enable proper embryogenesis will be a breakthrough intervention in wide hybridization and transgenic event development as well (Fig. 5.2).

5.4.4 Considerations regarding somatic embryogenesis

Somatic embryogenesis has certain considerations based on the morphological variations observed in the developing somatic embryos and their different efficiency to be converted to healthy seedlings. Standardization of the embryo culture conditions for maximization of frequency of normal type (zygotic type) embryos is a prior prerequisite for the parental lines taken for wide hybridization. Suhasini et al. (1996) from National Chemical Laboratory, Pune, India, presented a classification of the somatic embryos based on the shape of the developing embryo and number of cotyledons which is necessary consideration while upscaling embryo rescue technique to the product delivery mode. The authors used electron microscopy and classified the developing somatic embryos into six types (in *desi* chickpea genotype Phule G 12), as mentioned below:

a. Type 1 (horn-shaped embryo): Type 1 embryos (horn shaped) are the most prevalent type of embryos in somatic embryogenesis with 63.61% (±6.43 standard deviation) occurrence and 0% conversion to plantlets. In such embryos poorly developed fused cotyledons form a ring structure in the apex and are devoid of shoot apical meristem. Defective polar auxin transport leads to formation of such embryos. Four morphological variants were observed in horn-shaped embryos, namely, small fused cotyledon type, broad hypocotyl type, fused broad and flattened cotyledon type, and embryo with cotyledonary ring with an apical pore. Horn-shaped embryos profuse rooting and absence of shoot apical meristem prevents shoot development.

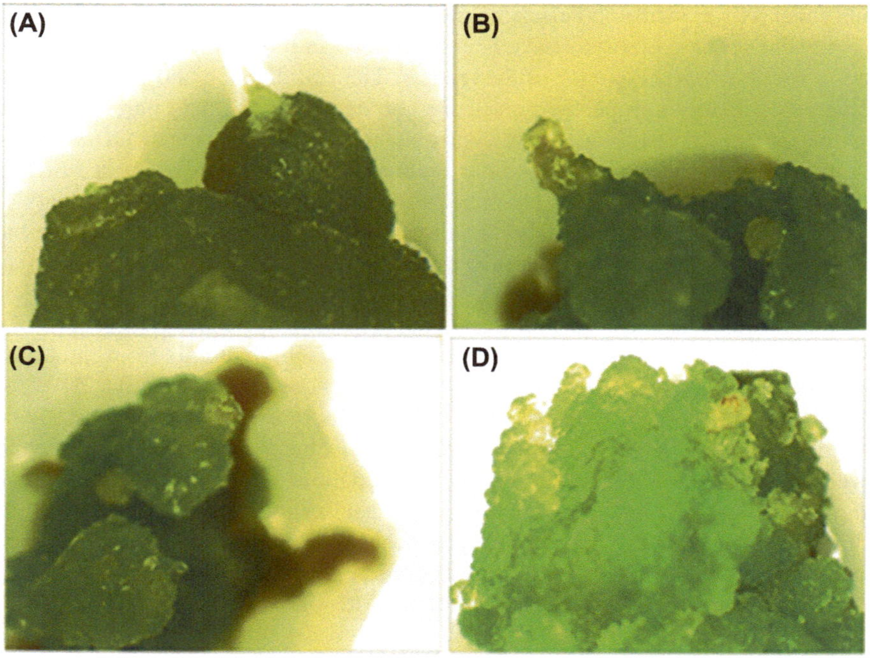

FIGURE 5.2

Early stages of differentiation of the somatic embryo from the genetically transformed explants of chickpea (embryonal axis) observed at Indian Institute of Pulses Research, Kanpur. (A) Initiation of protuberance from the embryogenic cell cluster, (B) Early globular embryo developed from the protuberance, (C) Multiple globular embryos from the transformed callus at later stage, and (D) Three months old callus with differentiating embryos embedded in it, just before transferring into shooting media.

Photo: Courtesy Dr. Alok Das, IIPR, Kanpur.

b. Type 2 (monocotyledonary embryo): Somatic embryos with a single cotyledon and a cleft at the apex. Occurrence of such embryos is 7.38% (±2.23 standard deviation) and 46.67% (±20.82 standard deviation) conversion to plantlets.

c. Type 3 (normal embryo): Normal somatic embryos have morphology similar to zygotic embryos with an occurrence of 9.06% (±7.33 standard deviation) and conversion to plantlets up to 73.33% (±15.28 standard deviation). Such embryos have a trident type appearance at the apex with the two elongated cotyledons on both sides and a prominent shoot meristem centrally placed between them. Sometimes, in the dicotyledonous embryos with well-differentiated cotyledons and shoot meristem, an additional rudimentary cotyledon is also observed.

d. Type 4 (tricotyledonary embryo): Such somatic embryos have trifid apex and no prominent shoot apical meristem. Their occurrence is 2.81% (±1.91 standard deviation) and conversion to plantlets is insignificant.

e. Type 5 (multicotyledonary embryo): Such somatic embryos have a diffused structure with numerous primordial cotyledons. Occurrence of such somatic embryos is quite low, i.e., 0.51% (±0.89 standard deviation) and conversion to plantlets is insignificant.
f. Type 6 (fasciated embryo): In such somatic embryos, diaxial fusion of two distinct embryos takes place. Their occurrence is 16.63% (±5.27 standard deviation) and conversion to plantlets is insignificant.

Suhasini et al. (1996) also reported about secondary embryogeny from hypocotyls or cotyledonary regions of developing somatic embryos, but such embryos seldom differentiate. Out of all the type of somatic embryos described, only Type 2 and Type 3 have fair conversion rate into seedling. From the above classification, it is clear that shoot apical meristem is an indispensable part of the embryo for conversion into seedling and in vitro shooting is a more important phenomenon than rooting (rooting is ample in the Type 1 embryos, which has insignificant conversion percentage). Thereby, the differentiation pattern from the somatic embryo is a major determining factor while maximizing the conversion to a healthy seedling. There is a research gap between the embryo culture protocols and corresponding conversion percentage. Even external provision of auxin has not enhanced the conversion percentage of somatic embryos significantly.

5.4.5 Considerations in the seedlings derived from the hybrid embryos

The second major consideration regarding embryo culture or embryo rescue is the recovery of the chlorotic mutant seedlings from the hybrid and somatic embryos. Expression of chlorotic mutant seedlings could not be associated to any embryological factor, whereas appearance of defective or morphological variant of somatic embryos has been attributed to changes in suspensor, cotyledon count, and apical meristem.

Albinism and chlorotic variations of the recovered seedlings from the hybrid embryos have been observed to be a major hindrance in wide hybridization. Clarke et al. (2011) revealed the physiological basis of this chlorotic variation in seedlings derived from wide hybridization. The authors accomplished wide hybridization in *C. arietinum* with *C. judaicum* and *C. pinnatifidum* as donors and rescued the hybrid embryos through in vitro culturing. Finally transfer to soil could be achieved for few seedlings only. Under in vitro conditions, Clarke et al. (2011) classified the developing primordial seedlings from hybrid embryos into three classes, namely, green, pale green, and albino. Parameters like regeneration, pod setting, and albinism (Chlorotic mutations) are affected by the plant genotype, and such mutations were observed to be appearing in certain wide cross combinations. Clarke et al. (2011) studied the platid ultrastructure through electron microscopy (TEM) and identified the structural modification in the thylakoid membrane responsible for the chlorotic mutants (albino). Thylakoid membranes were properly formed in normal seedlings, whereas in albino and other chlorotic mutants those were not properly formed. Spectrophotometric data also confirmed three- to fourfold presence of chlorophyll a, b and total carotenoids in normal seedlings with respect to albinos. However, contrast in carotenoids could not be confirmed in healthy seedlings and chlorotic mutants but, concerned to chlorophyll, the contrast was distinct. Reports of Clarke et al. (2011) revealed genetic variation for this trait and frequency of recovery of green healthy seedlings was higher in certain wide crosses which opens an avenue of research for identification strategy of parents for wide crossing where recovery of healthy seedlings can be maximized from hybrid embryos.

In the wide crosses using *C. arietinum* (Australian genotypes) as recipient and *C. pinnatifidum*, *C. judaicum,* and *C. bijugum* as donor parents, Clarke et al. (2006) recovered albino to pale green and green seedlings through embryo rescue. Badami et al. (1997) also recovered healthy seedlings from wide cross between *C. arietinum* (Indian genotypes) taken as recipient and *C. pinnatifidum* accession taken as male. Mallikarjuna et al. (2007) derived green healthy seedlings from the hybrid embryos recovered through embryo rescue technique in wide hybridization of *C. arietinum*/*C. bijugum* (accession ILWC 73) in ICRISAT, Hyderabad. Clarke et al. (2011) concluded that during cell division and maturity in albino seedlings, the proplastids underwent improper differentiation giving rise to abnormal chloroplasts with poorly formed thylakoid membranes. Such disorganized thylakoids are unable to maintain the structural integrity of the photosystem and support photosynthesis. Genetic regulation of postembryogenesis pigment disintegration is a field that needs discourse and further research.

5.4.6 Embryo rescue protocols in chickpea

The embryo rescue protocols in chickpea have few cornerstone factors like other embryo culture protocols; different laboratories worked on those criteria and developed their own modified protocols developing on the cumulative knowledge of the previous researchers. Here we will discuss few model protocols in terms of basal media, growth regulators, embryogenesis triggering chemical, culturing conditions, and illumination used. The protocols are discussed based on the basal culture medium taken for culturing the hybrid ovules. A common observation across several protocols is, for initiating embryogenesis, a liquid culture medium with a setup in which the hybrid ovules are placed on filter paper bridges placed in it was congenial. For standardization of the embryo culture media for the hybrid embryo developed from a pair of parents involved in distant hybridization, gradient range of the growth regulators is valuable information. While discussing the protocols, gradient setting of growth regulators (2,4,5-trichlorophenoxy acetic acid, indole acetic acid, zeatin, benzyl amino purine, naphthalene acetic acid, etc.) in the basal media has been focused.

5.4.6.1 Basic culture media: L6 media (Kumar et al., 1988)

Clarke et al. (2006) standardized a protocol for rescue of the selfed embryos of chickpea and attempted rescue of *C. arietinum*/*C. bijugum* and *C. arietinum*/*C. pinnatifidum* hybrid embryos. The authors reported durations of retention of the hybrid ovules on the mother plant, which is a crucial factor for recovery of healthy plantlets. Microscopy studies revealed deleterious histological changes in certain hybrid ovules as early as 2–4 days after pollination. The authors reported rescue duration of *C. arietinum*/*C. bijugum* hybrid ovules ranging from 2 to 7 days (globular embryo stage) and for *C. arietinum*/*C. pinnatifidum* hybrid ovules, the rescue duration ranged from 15 to 20 days (torpedo stage of embryo). The salient features of the embryo culture protocol adopted by the authors are briefly discussed.

Pods from the wide crosses were retained on the mother plant until they were healthy. The pods were harvested and embryo rescued with the appearance of yellowing or any symptom of weathering. The harvested pods were surface sterilized and hybrid ovules exercised along with funicle. The ovules were dissected and cultured on L6 media (Kumar et al., 1988) with 1 mg/L zeatin and 0.25 mg/L indole-3 acetic acid (Mallikarjuna, 1999) under 16:8 h photoperiod at 25°C.

In case of culture of selfed embryos, excision was done after germination of the seed and transferred on filter paper bridges in tubes with liquid media containing L6 media without any plant growth regulator. After differentiation of plantlets, those were micropropagated in MS media (Murashige and Skoog, 1962), supplemented with vitamin B5 (Gamborg et al., 1968), 30 g/L sucrose, 0.1 mg/L benzyl amino purine, 0.01 mg/L naphthalene acetic acid, and 9 g/L agar (modified from Pigeaire et al., 1997) for 28 days. After shooting, subculturing for root induction was done in L6 media with 3 mg/L indole butyric acid for 14 days.

Finally the plantlets were cultured in L6 media (devoid of any growth regulator) with 4.5 g/L agar for 7 days prior to transfer to pots in glasshouse.

Mallikarjuna (1999) from ICRISAT, India, reported a detailed study of wide hybridization of chickpea with *C. judaicum* and *C. pinnatifidum* as donors and taking *desi* and *kabuli* chickpea lines (*C. arietinum*) as recipients. The flowers on the female parent were tagged on the day of opening and harvested at 6, 8, 10, and 12 days after cross pollination depending on the appearance of the weathering symptoms. Ovules with funicles were dissected out and cultured. Mallikarjuna (1999) took L6 medium (Kumar et al., 1988) as the basal embryo culture medium. Hybrid ovules were cultured in L6 medium with 3% sucrose, indole-3-acetic acid, and zeatin at pH 5.6. Gradient medium was prepared for the growth regulators for standardization. Indole-3-acetic acid was used at 0.0, 0.1, 0.25, 0.5, 1.0, and 2.0 mg/L and zeatin at 0.0, 0.1, 0.5, 1.0, 2.0, and 5.0 mg/L concentrations. Ovules were cultured on filter paper bridges placed in the liquid L6 media with growth promoters and incubated at 26°C under 18 h daily illumination. Mallikarjuna (1999) studied the embryogenesis in wide hybridization involving three *kabuli* (ICC 92318, ICC 95311, ICCV 2) and four *desi* (ICCV 6, ICCV 10, ICCC 37, GL 769) chickpea genotypes (*C. arietinum*) with donor lines of *C. pinnatifidum* (ICCW 37, ICCW 38). The authoress adopted a unique strategy for retention of the hybrid ovules on the female plant, a factor observed to be enhancing in vitro embryogenesis in previous reports. A mixture of growth regulators (gibberellin 75 mg/L + naphthalene acetic acid 10 mg/L + kinetin 10 mg/L) was applied to the base of the pollinated flowers to prevent the premature pod abscission. Kaur et al. (2013) at Punjab Agriculture

University, Ludhiana, India, combined resistance to ascochyta blight and botrytis gray mold through wide hybridization involving *C. judaicum* and *C. pinnatifidum* as donors. The wide crossings were achieved in vivo by applying growth regulator mixture in the stalk of flowers. The selfed pods were removed from the female parent to facilitate the source: sink relationship for grain filling in the hybrid pods and reducing competition. The hybrid pods recovered out of crossing between *C. arietinum/C. pinnatifidum* accessions were harvested between 18 and 25 days with appearance of weathering symptom (yellowing at the apex or base of immature developing pod). Ovules were cultured on filter paper bridges in liquid L6 media (Kumar et al., 1988) with 3% sucrose, 1 mg/L zeatin and 0.25 mg/L indole acetic acid. Thirty days after incubation, the developing embryos were exercised and subcultured in the fresh L6 growth media with growth regulators mentioned above (embryogenesis media). Along with the emergence of the plantlets subculturing was done as per need; finally the healthy plants were transferred to soil. For the plants reaching flowering stage, pollen fertility study was done. Mallikarjuna (1999) used 36 combinations of growth regulators for optimization of embryogenesis, out of which L6 media with growth regulator supplements indole acetic acid (0.1−2 mg/L) and zeatin (0.5−2 mg/L) were observed to be most congenial for the hybrid embryos for embryogenesis. Maximum number of embryos developed from the ovules when the growth regulator combination was indole acetic acid (0.25 mg/L) and zeatin (1 mg/L); plantlets derived from those embryos were healthy as well. Within the *kabuli* chickpea genotypes used in wide hybridization program, ICC 92318 and *desi* cultivar ICCV 10 set maximum number of pods and also had maximum conversion to seedlings when pollinated with *C. pinnatifidum* accession ICCW 37. Genotypic variation in response to wide crossing was evident from this study since distinct *desi* and *kabuli* chickpea genotypes were included in the study. Pod setting percentage was >50% in both the abovementioned wide crosses. The authoress concluded that stimulus created by the distant hybridization per se might be responsible for the high pod setting percentage, and by more extensive wide hybridization program, such parental combinations need to be identified. The plantlets were cultured for 75 days; during this phase, conversion of chlorotic mutant seedlings was observed. Few of the albino seedlings got converted to pale green ones and some of the pale green seedlings got converted to normal green seedlings. This conversion of chlorotic mutants is a significant finding which also indicates that there is need of further research for standardization of the illumination regimes to facilitate the conversion of albino or pale green seedlings to green ones, by the conversion of leucoplast to chloroplast.

5.4.6.2 Basic culture media: MS media (Murashige and Skoog, 1962)

Suhasini et al. (1994) standardized an embryo rescue protocol in selfed chickpea embryos and carried out a comparative study of zygotic and somatic embryogenesis (Suhasini et al., 1996). The seeds were surface sterilized and soaked in distilled water in darkness for germination. The embryonal axes of the germinating seeds were taken as explants and somatic embryogenesis was induced. Mature embryonal axes with shoot tip were cultured in MS media (Murashige and Skoog, 1962) 3% sucrose and 0.6% agar. The authors took 2,4,5-trichlorophenoxyacetic acid as the somatic embryogenesis triggering agent and set up a gradient of 0.1, 1.0, 3.0, 5.0, 10.0, 20.0, and 40.0 mg/L to optimization. pH of the media was maintained acidic (5.8) prior to autoclaving at 121°C for 20 min. The explants were incubated in the induction media at 25 ± 2°C in darkness for 10–12 weeks for embryogenesis. In an alternate setup, the authors also cultured the explants on filter paper bridges (Whatman no. 1) in liquid MS media supplemented with 3.0 mg/L 2,4,5-trichlorophenoxyacetic acid (T3 media) which proved to be better for the embryogenesis induction. Globular embryos were formed (12–15 on each explant) after 10–12 weeks incubation in the T3 media and conversion percentage from explants was up to 65. 2,4,5-Trichlorophenoxyacetic acid induced somatic embryogenesis at 0.1mg/L–20 and 40 mg/L concentrations. 2,4,5-Trichlorophenoxyacetic acid was observed to be supporting embryogenesis at higher concentrations (5 and 10 mg/L) as well, but the number of embryos per explants reduced.

The globular embryos were exercised from the explants and subcultured in solid T3 media with 3 mg/L 2,4,5-trichlorophenoxyacetic acid with incubation at 25 ± 2°C for 6 weeks with 16 h illumination per day. Illumination was provided with cool white fluorescent tubes at 23 $\mu mol/m^2/s$ until the somatic embryos attain cotyledonary stage. On an average, duration of 14–18 weeks were taken by the developing embryos to reach the cotyledonary stage from the explants in soild T3 culture medium. However, culturing the developing embryos on filter paper bridges placed in liquid T3 media reduced the duration and hastened embryogenesis by 3–4 weeks. Cotyledonary embryos were isolated and subcultured in maturation media comprising MS medium with 3% sucrose, 0.6% agar, and 0.1 mg/L abscisic acid and incubated for 10 days under similar illumination conditions mentioned above.

The differentiating cotyledonary embryo structures were transferred to conversion medium of half-strength MS medium supplemented with 3% sucrose, 0.6% agar, and 1.0 mg/L zeatin for 2–3 weeks under similar illumination conditions, which led to the formation of roots and shoots. Regenerated plantlets were subcultured in half-strength MS medium with 3% sucrose and 0.16% Phytagel (Sigma) in 250-mL Erlenmeyer flasks and incubated for 2–3 weeks under similar illumination conditions. 6–7 cm height plantlets were transferred to pots with peat, vermiculite, and soil (1:1:1 by volume) and hardened for 2 weeks at 25 ± 2°C under similar illumination conditions and then transferred to green house.

From the above-discussed research findings, it is evident that stage of the hybrid embryo is a major determining factor for the embryo rescue. It is important to retain the cross-pollinated flower on the female plant as long as possible, avoiding weathering and abscission, and equally important is the culture media composition to support the mature and immature embryo in vitro. Chauhan et al. (2003) from the Indian Institute of Pulses Research, Kanpur, optimized the media requirement of the mature and immature chickpea embryo for in vitro regeneration. The authors took MS media (Murashige and Skoog, 1962) as the basal media. Flowers were tagged at anthesis and pods were harvested at 5, 10, 12, 15, and 18 days. Immature ovules were excised from surface sterilized pods. After surface sterilization under aseptic conditions, the ovules were cultured until development of embryos for 2 weeks. The MS media was supplemented with benzyl amino purine (2.2–8.8 μM), kinetin (2.3–4.6 μM), 2,4,5-trichlorophenoxyacetic acid (11.38 μM), indole butyric acid (0.6–14.7 μM), indole acetic acid (2.8–17.1 μM), and 3%–6% sucrose. Before adding the solidifying agent agar (0.8%), the pH of the media was maintained to 5.8 ± 2. The embryos were incubated for 2 weeks at 25 ± 2°C and daily 16 h illumination (3000 Lux. intensity). For rooting, the shoots were exercised and subcultured in one-fourth-strength MS medium supplemented with naphthalene acetic acid (10.7 μM) and indole butyric acid (9.8 μM) with different sucrose concentrations (1%–3%). After rooting, the plantlets were transferred to pots with sand, soil, vermiculite, and farmyard manure in different ratios and hardened in growth chambers at 28°C with 16 h of illumination. The plantlets were supported with one-fourth-strength Hoagland solution.

Maximum regeneration frequency of the mature embryos was observed in the solid MS media with vitamin B5 and 4% sucrose without any growth regulator. High shoot proliferation was observed in the MS media supplemented with 0.6 μM indole butyric acid, 8.8 μM benzyl amino purine, and 40% sucrose. The authors observed difference in response to in vitro embryogenesis within the chickpea genotypes used for study; *desi* cultivar K 850 showed maximum regeneration frequency of mature embryos followed by C235, BG 256, and PDG 84–10.

The authors observed better regeneration of immature embryos in liquid media with a filter paper bridge setup. Ten- to twelve-day-old embryos showed high regeneration frequency (52%) in liquid MS media supplemented with 2.8 μM indole acetic acid and 2.3 μM kinetin. Liquid MS media supplemented with 2.8 μM indole acetic acid and 2.2 μM benzyl amino purine also supported embryogenesis up to 44%. Authors observed a better regeneration response in below 15-day-old embryos in MS media supplemented with kinetin. Multiple shoot was induced in immature embryos (18 days old) cultured in liquid MS media supplemented with vitamin B5, 2.5 μM indole acetic acid, 2.2 μM benzyl amino purine, and 4% sucrose. The authors reported importance of sucrose in supporting embryogenesis of the immature embryos (10–12 days old) where its concentration may be taken up to 6% for positive regeneration response. After initiation of the shoot buds, subculturing was carried out in the same media with reduced sucrose content (3%).

Thus embryo rescue is indeed an intriguing field of discourse. A holistic approach considering the classical embryology, tissue culture, developmental genetics, and compatibility in wide crossing is needed to address the intervention for widening the genetic base of chickpea and utilizing the wild alleles in breeding program.

5.5 Chromosomal manipulation

Research accomplishments in chickpea cytogenetics are comparatively limited with respect to advances in other areas of genetics and breeding like classical genetics, conventional breeding (yield, biotic and abiotic stresses), molecular marker–assisted interventions and molecular breeding, genetic mapping and sequencing, etc. In the first phase of cytogenetics research (1970 and 1980s) meiosis studies were carried out in wide hybridizations, on the basis of which the gene pool structure in chickpea was proposed. Diploid number of chromosomes in different species of *Cicer* was gradually reported. Mutation breeding or mutagenesis studies being a major research trend in the 1970s and the 1980s, colchicines-induced polyploidy and chromosome behavior in the progenies were studied upon. Onset of the 1990s witnessed reports regarding karyotypes and idiograms of different *Cicer* species, *desi* and *kabuli* type chickpea accessions; DNA quantification; pachytene chromosome studies; and banding patterns of different chromosomes (linkage groups), etc. From 2000 onward after the first draft of human genome project published, sequencing studies gradually established as the major research trend, and genome sequence projects in chickpea were initiated under the flagship of ICRISAT which has revealed many valuable information regarding the genome size, linkage group sizes, etc. Molecular cytogenetics, flow cytometry, and hybridization studies (e.g., fluorescent in situ hybridization [FISH]) in chickpea were also initiated after 2000 and on the basis of information generated (repeat motifs) simple sequence repeat (SSR) markers were reported. Compared to cereals, many research gaps are present in the basic cytogenetic studies which enable chromosome manipulations. Chromosomal manipulations involve utilization of the structural (ionizing radiation–induced mutagenesis) or numerical aberrations (colchicine-induced polyploidy) of chromosomes and development of genetic stock populations in which mutant phenotypes are attributed to the chromosomal changes. Such populations reveal location of the gene responsible for a particular phenotype in a linkage group and it serves as the primary information for the development of linked SSRs or genetic linkage maps. Structural chromosomal aberrations (deletion, duplication, inversion, and translocation) originate due to destabilization of the telomeres leading to sticky ends and formation of dicentric bridges in the anaphase. Random detachment of the bridge during late anaphase perpetuates the phenomena down the generation (breakage fusion bridge cycle, McClintock, 1938, 1941) expressing an array of phenotypes. Numerical changes in chromosomes are achieved by destabilization of the spindle structure or prevention of formation of the spindle structure during cell division. Polyploidy induced by both ionizing radiations and colchicine leads to numerical aberrations (e.g., monosomy, nullisomy, trisomy types of aneuploidy). Chromosomal banding is an important approach to track the aberrations and identify the chromosomes. In chickpea, classical cytogenetics study has revealed the chromosome number, their karyotypes, banding patterns, pachytene structure, DNA content, etc. and molecular cytogenetics studies have accomplished FISH and flow cytometry; however, published reports on chromosome manipulations (informative aneuploid genetic stocks) are limited.

5.6 Cytogenetics and distant hybridization

Ploidy and diploid chromosome number of chickpea has been confirmed in the detail account of origin and evolution reported by Ladinsky and Adler (1976a). It is uniformly diploid (all the *Cicer* species) and its chromosome (2n) number is 18. Studies conducted by Ladizinsky and Adler (1976a,b) through distant hybridization between *C. arietinum* and donors *C. echinospermum*, *C. reticulatum* revealed their cytogenetic evolutionary relation. The authors reported that a reciprocal translocation has led to the evolution of *C. echinospermum* from *C. arietinum* and found their hybrids to be sterile. Later however fertile hybrids have been developed from the wide cross *C. arietinum*/*C. echinospermum* by few research groups as already discussed in this chapter. Normal meiosis and fertile hybrids were observed between *C. arietinum* and *desi* type *C. reticulatum*. The authors observed a quadrivalent, an anaphase I dicentric bridge, and an acentric fragment at meiosis in a cross between *kabuli* type *C. arietinum* and donor *C. reticulatum*, indicating they differ by a translocation and a paracentric inversion which led to hybrid with low seed fertility. The authors also reported about chromosome repatterning within *C. arietinum* and suggested cytogenetic differences between the cultivated groups. Not only inversion, *C. arietinum* and *C. reticulatum* differ in karyotypes and genome size (Ohri and Pal, 1991; Galasso et al., 1996). Wide hybridizations revealed genotypic variation for cross-compatibility and cytogenetic abnormalities in F_1s leading to pollen sterility. *C. echinospermum* accession PI 527930 was utilized by Collard et al. (2003) as donor to develop an interspecific mapping population and map QTLs for seedling resistance for ascochyta blight. This accession was used in wide hybridization in ICRISAT which set mature hybrid seeds but pollen sterility detected in F_1 generation due to formation of polyads and F_2 seeds could not be derived. When the F_1 was backcrossed with the recipient parent KAK 2 (*C. arietinum* cultivar, *kabuli* type), some mature seeds were derived (Mallikarjuna unpublished data). In the wide hybridization program of Indian Institute of Pulses Research, Kanpur, backcrossing of the F_1 of wide hybridizations has been attempted; in many cases where segregating generations have suffered from poor seed set and germination, the backcross (with recipient female parent) derived generations have thrived well. Maintaining the desirable seed characteristic as per consumer preference with the target trait is a challenge in wide hybridization, often lines with wild-type seed characteristics get selection advantage, and finally their utilization in varietal developmental program becomes limited.

5.6.1 Chromosome morphology and DNA content

Ocampo et al. (1992) used the decimal numbering (1–8) designation for the chickpea chromosomes on the basis of their length (longest metacentric chromosome numbered as 1). Alphabetical designation (A-H) has also been used for the chickpea chromosomes based on length of chromosomes in descending order (1 = A). Most recent chickpea chromosome designation on the basis of pseudomolecule has been provided by Ruperao et al. (2014) in *desi* (ICC 4958) and *kabuli* (Frontier) types. The authors have used alphabetical designation (A-H) of chromosomes and assigned the linkage groups on the basis of relative chromosome length (%) and molecular chromosome size (megabase pairs). Linkage group pseudomolecule 5 has been assigned as chromosome A by Ruperao et al. (2014), similarly chromosome B (pseudomolecule 3), chromosome C (pseudomolecule 6), chromosome D (pseudomolecule 7), chromosome E (pseudomolecule 4), chromosome F (pseudomolecule 2), chromosome G (pseudomolecule 1), and chromosome H (pseudomolecule 8).

Somatic chromosome number ($2n = 16$) of chickpea is stable across majority of the species of *Cicer* (annual and perennial), but considerable karyological variation is observed within those species. Ahmad (2000) reported small chromosomes of chickpea whose mean length during mitotic metaphase is 2.2 μm. Karyotype differences have been vividly documented by Ohri and Pal (1991) and Ahmad (2000) in the *Cicer* species and between the *desi* and *kabuli* types. Differences in chromosomal morphology with respect to the length, arm ratio, and position of constrictions (primary and secondary) have been reported between the species (Ohri and Pal, 1991; Ocampo et al., 1992; Tayyar et al., 1994; Ahmad, 2000; Ahmad and Hymowitz, 1993; Kordi et al., 2006). However, information related to karyological differences is yet to be utilized in wide crossing and development of cytogenetic stocks.

One pair of chromosome with satellite region (SAT chromosomes) or nucleolar organizer region (NOR) is present in *C. arietinum* and other species. Only in accessions of *C. reticulatum* two pairs of SAT chromosomes have been found. Ahmad (2000) contradicted this finding and suggested possibility of two cytotypes in *C. reticulatum* with one and two SAT chromosomes, respectively. Karafiátová et al. (2017) opined that during the evolution of *C. echinospermum* and *C. arietinum* from the progenitor species *C. reticulatum*, one of the two NOR was lost. Chickpea genotypes with two pairs of SAT chromosomes have been reported by Iyengar (1939) and Kutarekar and Wanjari (1983) only, such genotypes are rarely maintained under field conditions and might have been replaced with the advent of improved varieties. Tandem NORs in the same chromatid have also been reported in certain *C. arietinum* accessions; those are only visible late prophase and early metaphase (Meenakshi and Subramaniam, 1960; Ahmad, 1989, 2000; Tayyar et al., 1994; Kordi et al., 2006). Appearance of tandem NORs has been attributed to the presence of staining artifacts, movement of the NOR along the locus, and evolutionary rearrangements.

Ahmad (2000) reported the interspecific difference in location of secondary constriction site in the annual species *C. arietinum*, *C. reticulatum*, and *C. echinospermum*, where it is located in the longest chromosome, whereas in other species, secondary constriction is located in medium to small-sized chromosomes. Secondary constriction is the lone prominent cytogenetic marker on which the chromosomes can be primarily identified and classified. In mitotic metaphase, the chromosome length varies from 1.32 to 3.69 μm (Ahmad, 2000), where three chromosomes are sub-metacentric and the rest five are metacentric. Difference in chromosomal length has been reported in the *desi* and *kabuli* types, where *kabuli* types have longer length in three chromosomes and shorter length in five chromosomes with respect to their equivalent chromosomes in *desi* types. Variation in the relative chromosomal lengths ranges from 0.2% to 0.8% (Ruperao et al., 2014). Identification of the *C. arietinum* chromosomes is ambiguous and had been discussed by different authors. Kordi et al. (2006) reported distinct identification of two out of eight chromosomes of *C. arietinum*, the longest sub-metacentric chromosome (with NOR, designated as chromosome number 1 or A) and the shortest metacentric chromosome (Chromosome number 8). On the basis of karyotype symmetry (Stebbins, 1971), centromere location, and arms ratio, Ahmad (2000) reported two types of asymmetry among the annual *Cicer* species with *C. arietinum*, *C. reticulatum*, and *C. echinospermum* in one cluster which has been later validated in the molecular diversity studies as well.

Ohri and Pal (1991) from the National Botanical Research Institute, Lucknow, India, provided a detail account of karyotyping and nuclear DNA content of the *Cicer* species, where *C. judaicum* had lowest nuclear DNA content and *C. arietinum* had the highest. The nuclear DNA content was estimated to be 1.9 pg (2C value) by Bennet and Smith (1976) for *C. arietinum*. Ruperao et al. (2004) estimated nuclear DNA amounts of *kabuli* and *desi* types to be 1.80 and 1.77 pg (2C value), respectively, by the flow cytometry approach. Mean C value for *kabuli* and *desi* types were estimated to be 882 and 866 mega base pairs. In cultivars, 2C value of certain genotypes was estimated up to 3.57 pg, indicating that variation for nuclear DNA content exists within the cultivated types which may be attributed to evolution and adaptation. However, on the basis of nuclear DNA content, *Cicer* species could not be classified or clustered, rather random association or equivalence was observed in the study of Ruperao et al. (2014). Nuclear DNA content (2C) varied from 1.83 pg (*C. judaicum*) to 3.57 pg (an accession of *C. arietinum*). Even the nuclear DNA amount of perennial species (*C. songaricum*, 2C = 2.72 pg) was equivalent with certain annual species (*C. reticulatum*, 2C = 2.66 pg and *C.* echinospermum, 2C = 2.6 pg). Many of the C values mentioned above from the report of Ruperao et al. (2014) were equivalent with the reports of Galasso et al. (1996). As per reports available, the karyotype of cultivated chickpea *C. arietinum* was quite similar with its wild progenitor *C. reticulatum*, whereas the nuclear DNA content (2C value) differed significantly. Karafiátová et al. (2017) opined that this apparent anomaly between karyotype and nuclear DNA content might be due to a major change (reduction) in the nuclear DNA content in *C. arietinum* while its evolution from the progenitor *C. reticulatum* without altering the karyotype. Synonymous substitution rate in chickpea might have a role here, which was reported to be 6.05×10^{-9} per site per year (quite high) by Jain et al. (2013) in the draft genome sequence published.

5.6.2 Chromosomal manipulations, visualization, and other techniques

Researches on chromosomal manipulation in chickpea are limited to pachytene chromosome structure and colchicine-induced polyploidy. Regarding chromosome visualization, banding, fluorescent in situ hybridization, chromosome painting, and flow cytometry techniques have been utilized for chickpea characterization up to limited extent and primary results have been reported.

Ahmad and Hymowitz (1993) reported a detailed account of the chickpea pachytene chromosomes and identified them as "differentiated" type, with heterochromatin region flanking the centromere or primary constriction and distal euchromatin near the telomeres. Galasso and Pignone (1992) and Tayyar et al. (1994) developed C-banding profile of the chickpea chromosomes and characterized them. Tayyar et al. (1994) reported significant contrast between the haploid genome length and C band positive heterochromatin. Haploid genome length ranged from 20 μm (*C. judaicum*) to 28.7 μm (*C. arietinum*). The authors classified the species based on their heterochromatin content: low heterochromatin containing group (mean 41.7%) comprised *C. arietinum, C. chorassanicum, C. echinospermum, C. judaicum, C. pinnatifidum, C. reticulatum,* and *C. yamashitae,* whereas high heterochromatin (mean 59.5%) group comprised *C. bijugum* and *C. cuneatum*. Tayyar et al., (1994) concluded that there is reduction of C-heterochromatin content in the course of evolution.

Colchicine-induced polyploidy in chickpea was studied by Sohoo et al. (1970) and Phadnis and Narkhede (1972). Sterile, partially fertile, and fertile polyploids were identified. Chromosomal abnormalities due to nondisjunction (multivalents), asynaptic behavior, and univalent formation were observed in both fertile and sterile plants.

Vlacilova et al. (2002) developed the flow cytometry protocol and sorting of mitotic chromosomes in chickpea and attempted for physical mapping in the crop. *In situ* PCR with the assorted chromosome assigned certain STMS markers on chromosome 8. Zatloukalová et al. (2011) initiated work on integration of the genomic and physical maps of chickpea through flow cytometry. Linkage group 8 was conclusively assigned to chromosome H. Assigning of other linkage groups (LG) was considerably achieved by the authors. Largest chromosome A was related to LG5, whereas medium-sized chromosome E assigned to LG4 and second largest chromosome B to LG3. Chromosomes C and D were jointly assigned to LG 6 and 7, whereas LG 1 and 2 were assigned to chromosomes F and G. The authors used FISH for anchorage of probes and assignment was done on the basis of hybridization data.

5.6.2.1 Path ahead

Considerable research gaps exist in embryo rescue and chromosome manipulations in chickpea which are apparent from the above discussions. In embryo rescue, standardization of the growth regulator combinations and adjuvants, working out the parental combinations with high pod setting percentage, and standardization of growth regulator mix formulation to be applied to stalk of pollinated female flowers to avoid abscission are important areas where future research can be carried out. Modeling studies for prediction of growth stages of hybrid embryo and identification of embryogenesis triggering molecules (2,4,5-T; IAA; zeatin; BPA, etc.) have immense potential to utilize the wild alleles. Standardization for conversion of chlorotic mutant seedlings to normal green seedlings altering the illumination regimes is also an important research area. Development of genetic stocks (especially deletion stocks) and knock out studies is important for functional assignment of the genetic loci on the linkage groups. Lack of a crossing system imparting stable structural chromosomal changes in chickpea is a major bottleneck for development of genetic stock populations (which facilitate stock development in wheat, barley, and maize). Lack of natural polyploidy in progenitor and wild species of chickpea is also a limitation for cytogenetic studies. Embryo rescue and chromosome manipulation are the fields which need to be rejuvenated and addressed with novel approaches based on the technical information generated by previous researchers to harness the potential of wild alleles and broaden the genetic base of chickpea.

References

Abbo, S., Mesghenne, Y.T., Oss, H.V., 2011. Interspecific hybridization in wild *Cicer* species. Plant Breed. 130, 150–155.

Ahmad, F., Slinkard, A.E., 2004. The extent of embryo and endosperm growth following interspecific hybridization between *Cicer arietinum* L. and related annual wild species. Genet. Resour. Crop Evol. 51, 765–772.

Ahmad, F., 1989. The chromosomal architecture of *Cicer arietinum* Alef. A wild relative of chickpea. Cytologia 54, 753–757.

Ahmad, F., 2000. A comparative study of chromosome morphology among the nine annual species of *Cicer* L. Cytobious 101 (396), 37–53.

Ahmad, F., Hymowitz, T., 1993. The fine structure of chickpea (*Cicer arietinum* L.) chromosomes as revealed by pachytene analysis. Theor. Appl. Genet. 86, 637–641.

Badami, P.S., Mallikarjuna, N., Moss, J.P., 1997. Interspecific hybridization between *Cicer arietinum* and *C. pinnatifidum*. Plant Breed. 116, 393–395.

Bennett, M.D., Smith, J.B., 1976. Nuclear DNA amounts in angiosperms. Philos. Trans. R. Soc. Lond. B Biol. Sci. 274, 227–274.

Berger, F., Brownlee, C., 1995. Establishment of the apical-basal axis in multicellular plant embryos. Biol. Cell 84, 7–11.

Chaturvedi, S.K., Mondal, B., Srivastava, A.K., Tewari, Y., 2018. Chickpea breeding lines with stable green grains developed. Pulses Newsl. 29 (1), 2.

Chauhan, R., Tiwari, A., Singh, N.P., 2003. Differential requirement of mature and immature embryo of Chickpea *(Cicer arietinum L.)* for *in vitro* regeneration. Indian J. Plant Physiol. 8 (1), 28–33.

Clarke, H.J., Wilson, J.G., Kuo, I., Lulsdorf, M.M., Mallikarjuna, N., Siddique, K.H.M., 2006. Embryo rescue and plant regeneration in vitro of selfed chickpea (*Cicer arietinum* L.) and its wild annual relatives. Plant Cell Tiss. Org. Cult. 85, 197–204. https://doi.org/10.1007/s11240-005-9071-1.

Clarke, H.J., Kumari, M., Khan, T.N., Siddique, K.H.M., 2011. Poorly formed chloroplasts are barriers to successful interspecific hybridization in chickpea following in vitro embryo rescue. Plant Cell Tiss. Org. Cult. 106, 465–473.

Collard, B.C.Y., Pang, E.C.K., Ades, P.K., Taylor, P.W.J., 2003. Preliminary investigation of QTLs associated with seedling resistance to ascochyta blight from *Cicer echinospermum*, a wild relative of chickpea. Theor. Appl. Genet. 107 (4), 719–729.

Davies, A.M.R., Maxted, N., van der Maesen, L.J.G., 2007. A natural infrageneric classification for *Cicer* (Leguminosae, Cicerae). Blumea 52, 379–400.

Dixit, G.P., Srivastava, A.K., Singh, N.P., 2019. Marching towards self sufficiency in chickpea. Curr. Sci. 116 (2), 239–242.

Galasso, I., Pignone, D., Frediani, M., Maggiani, M., Cremonin, R., 1996. Chromatin characterization by banding techniques, in situ hybridization, and nuclear DNA content in *Cicer* L. Genome 39 (2), 258–265.

Galasso, I., Pignone, D., 1992. Characterization of chickpea chromosomes by banding techniques. Genet. Resour. Crop Evol. 39 (2), 115–119.

Gamborg, O.L., Miller, R.A., Ojima, K., 1968. Nutrient requirement of suspension cultures of soybean root cells. Exp. Cell Res. 50, 151–158.

Geerts, P., Toussaint, A., Mergeai, G., Baudoin, J.P., 2002. Study of early abortion in reciprocal crosses between *Phaseolus vulgaris* L. and *Phaseolus polyanthus* Greenm. Biotechnol. Agron. Soc. Environ. 6, 109–119.

Harlan, J.R., de Wet, J.M.J., 1971. Towards a rational classification of cultivated plants. Taxon 20, 509–517.

Iyengar, N.K., 1939. Cytological investigation on the genus *Cicer*. Ann. Bot. 3, 271–305.

Jain, M., Misra, G., Patel, R.K., Priya, P., Jhanwar, S., Khan, A.W., et al., 2013. A draft genome sequence of the pulse crop chickpea (*Cicer arietinum* L.). Plant J. 74 (5), 715–729. https://doi.org/10.1111/tpj.12173.

Karafiátová, M., Hřibová, E., Doležel, J., 2017. Cytogenetics of *cicer*. In: Varshney, R.K., et al. (Eds.), The Chickpea Genome, Compendium of Plant Genomes. Springer International Publishing, pp. 25–41. https://doi.org/10.1007/978-3-319-66117-9_4.

Kaur, L., Sirari, A., Kumar, D., Sandhu, J.S., Singh, S., Kapoor, K., et al., 2013. Combining Ascochyta blight and Botrytis grey mould resistance in chickpea through interspecific hybridization. Phytopathol. Mediterr. 52 (1), 157–165.

Kordi, M., Majd, A., Valizadeh, M., Sheidai, M., Sabaghpour, H., 2006. A comparative study of chromosome morphology among some genotypes of *Cicer arietinum* L. Pak. J. Biol. Sci. 9, 1225–1230.

Kumar, A.S., Gamborg, O.L., Nabors, M.W., 1988. Plant regeneration from cell suspension cultures of *Vigna aconitifolia*. Plant Cell Rep. 7 (2), 138–141. https://doi.org/10.1007/BF00270124.

Kumari, M., Clarke, H.J., des Francs-Small, C.C., Small, I., Khan, T.N., Siddique, K.H.M., 2011. Albinism does not correlate with biparental inheritance of plastid DNA in interspecific hybrids in Cicer species. Plant Sci. 180, 628–633.

Kupicha, F.K., 1981. Vicieae. In: Polhill, R.M., Raven, P.H. (Eds.), Advances in Legume Systematics. Royal Botanic Gardens, Kew, pp. 377–381.

Kutarekar, D.R., Wanjari, K.B., 1983. Karyomorphological studies in some of varieties of Bengal gram (*Cicer aretinum* Linn.). Cytologia 48, 699–705.

Ladinsky, G., Adler, A., 1976a. The origin of chickpea *Cicer arietinum* L. Euphytica 25, 211–217.

Ladinsky, G., Adler, A., 1976b. Genetic relationships among annual species of *Cicer* L. Theor. Appl. Genet. 48, 197–203.
Lersten, N.R., 1983. Suspensors in *leguminosae*. Bot. Rev. 49 (3), 233–257.
Liu, C.M., Xu, Z.H., Chua, N.H., 1993. Auxin polar transport is essential for the establishment of bilateral symmetry during early plant embryogenesis. Plant Cell 5, 621–630.
Maheshwari, P., 1937. A critical review of the types of Embryo sacs in angiosperms. New Phytol. 36 (5), 359–417.
Maheshwari, P., 1950. An Introduction to the Embryology of Angiosperms. McGraw-Hill Book Company, INC, pp. 268–312.
Mallikarjuna, N., 1999. Ovule and embryo culture to obtain hybrids from interspecific incompatible pollinations in chickpea. Euphytica 110, 1–6.
Mallikarjuna, N., Coyne, C., Cho, S., Rynearson, S., Rajesh, P.N., 2011. Cicer. In: Kole, C. (Ed.), Wild Crop Relatives: Genomic and Breeding Resources, Legume Crops and Forages. Springer, Heidelberg, pp. 63–80.
Mallikarjuna, N., Jadhav, D., Nagamani, V., Amudhavalli, C., Hoisington, D.A., 2007. Progress in interspecific hybridization between *Cicer arietinum* and wild species *C. bijugum*. SAT eJournal 5 (1). http://ejournal.icrisat.org.
McClintock, B., 1938. The production of homozygous deficient tissues with mutant characteristics by means of the aberrant mitotic behavior of ring-shaped chromosomes. For. Genet. 23 (4), 315–376. PMC 1209016. PMID 17246891.
McClintock, B., 1941. The stability of broken ends of chromosomes in *Zea mays*. For. Genet. 26 (2), 234–282. PMC 1209127. PMID 17247004.
Meenakshi, G., Subramaniam, M.K., 1960. Tandem satellites in *Cicer arietinum* linn. Curr. Sci. 29, 438–439.
Mercy, S.T., Kakar, S.N., Varghese, T.M., 1974. Embryology of *Cicer arietinum* and *C. soongaricum*. Bull. Torrey Bot. Club 101 (1), 26–30.
Mercy, S.T., Kakkar, S.K., 1975. Barriers to interspecific crossing in *Cicer*. Proc. Indian Natl. Sci. Acad. 41, 78–82.
Millan, T., Winter, P., Jungling, R., Gil, J., Rubio, J., Cho, S., et al., 2010. A consensus genetic map of chickpea (*Cicer arietinum* L.) based on 10 mapping populations. Euphytica 175, 175–189. https://doi.org/10.1007/s10681-010-0157-4.
MoAF and W, 2018. Pulses Revolution from Food to Nutritional Security. Crops Division, Department of Agriculture, Cooperation & Farmers Welfare, Ministry of Agriculture & Farmers Welfare (MoAF&W), Government of India, Krishi Bhavan, New Delhi, 110 001, 1.
Mosjidis, J.A., Wainesf, J.G., Walko, R., 1986. Prediction of the developmental stage of the embryo in chickpea (*Cicer arietinum* L.). Ann. Bot. 57, 869–887.
Murashige, T., Skoog, F.G., 1962. A revised medium for rapid growth and bioassays with tobacco tissue cultures. Physiol. Plant. 15, 473–497.
Ocampo, B., Venora, G., Errico, A., et al., 1992. Karyotype analysis in the genus *Cicer*. J. Genet. Breed. 46, 229–240.
Ohri, D., Pal, M., 1991. The origin of chickpea (*Cicer arietinum* L.): karyotype and nuclear DNA amount. Heredity 66, 367–372.
Phadnis, B.A., Narkhede, M.N., 1972. Chromosome behaviour in colchicine induced autotetraploids of *Cicer arietinum* L. Cytologia 37, 415–421.
Pigeaire, A., Abernethy, D., Smith, P.M., Simpson, K., Fletcher, N., Lu, C.Y., et al., 1997. Transformation of a grain legume (*Lupinus angustifolius* L.) via *Agrobacterium tumefaciens* mediated transfer to shoot apices. Mol. Breed. 3, 341–349.

Popov, M.G., 1929. The genus *Cicer* and its species, 21. Trudy Prikl Bot, pp. 1–254.

Pundir, R.P.S., Mengesha, M.H., 1995. Cross compatibility between chickpea and its wild relative, *Cicer echinospermum* Davis. Euphytica 83, 241–245.

Raghavan, V., 1986. Embryogenesis in Angiosperms: A Developmental and Experimental Study. Cambridge University Press, Cambridge.

Redden, R.J., Berger, J.D., 2007. History and taxonomy of chickpea. In: Yadav, S.S., Redden, R.J., Chen, W., Sharma, B. (Eds.), Chickpea Breeding and Management. CABI, Wallingford, pp. 1–13.

Reed, S.M., 2005. Embryo rescue. In: Plant Development Biotechnology, pp. 235–239.

Ruperao, P., Chan, C.K.K., Azam, S., et al., 2014. A chromosomal genomics approach to assess and validate the desi and kabuli draft chickpea genome assemblies. Plant Biotechnol. J. 6, 778–786.

Sagare, A.P., Suhasini, K., Krishnamurthy, K.V., 1995. Histology of somatic embryo initiation and development in chickpea (*Cicer arietinum* L.). Plant Sci. 109, 87–93.

Singh, A., Singh, N.P., 2012. Development of interspecific hybrid between *Cicer arietinum* and *C. judaicum* and characterization of interspecific derivatives for economic traits. Indian J. Genet. Plant Breed. 72, 174–182.

Singh, K.B., Ocampo, B., 1997. Exploitation of wild *Cicer* species for yield improvement in chickpea. Theor. Appl. Genet. 95, 418–423.

Singh, M., Bhardwaj, C., Singh, S., Panatu, S., Chaturvedi, S.K., Rana, J.C., et al., 2016. Chickpea genetic resources and its utilization in India: current status and future prospects. Indian J. Genet. Plant Breed. 76 (4), 515–529.

Singh, M., Kumar, K., Bisht, I.S., Dutta, M., Rana, M.K., Rana, J.C., et al., 2015. Exploitation of wild annual *Cicer* species for widening the gene pool of chickpea cultivars. Plant Breed. 134, 186–192.

Singh, N.P., Singh, A., Asthana, A.N., 1999. Studies on interspecific crossability barriers in chickpea. Indian J. Pulses Res. 12, 13–19.

Singh, R., Sharma, P., Varshney, R.K., Sharma, S.K., Singh, N.K., 2008. Chickpea improvement: role of wild species and genetic markers. Biotechnol. Genet. Eng. Rev. 25 (1), 267–314. https://doi.org/10.5661/bger-25-267.

Singh, R.P., Singh, B.D., 1989. Recovery of rare interspecific hybrids of gram *Cicer arietinum x Cicer cuneatum* L. through tissue culture. Curr. Sci. 58 (15), 874–875.

Smartt, J., 1984. Gene pools in grain legumes. Econ. Bot. 38 (1), 24–35.

Sohoo, M.S., Athwal, D.S., Chandra, S., 1970. Colchicine induced polyploidy in chickpeas (*Cicer arietinum* L.). Theor. Appl. Genet. 40, 163–168.

Stamigna, C., Crino, P., Saccardo, F., 2000. Wild relatives of chickpea: multiple disease resistance and problems to introgression in the cultigens. J. Genet. Breed. 54, 213–219.

Stebbins, G.L., 1971. Chromosomal Evolution in Higher Plants. Edward Arnold (Publ). Ltd, p. 216.

Suhasini, K., Sagare, A.P., Krishnamurthy, K.V., 1996. Study of aberrant morphologies and lack of conversion of somatic embryos of chickpea (*Cicer arietinum* L.) *in vitro* cellular and developmental biology. Plant 32 (1), 6–10.

Suhasini, K., Sagare, A.P., Krishnamurthy, K.V., 1994. Direct somatic embryogenesis from mature embryo axes in chickpea (*Cicer arietinum* L.). Plant Sci. 102, 189–194.

Suhasini, K., Sagare, A.P., Sainkar, S.R., Krishnamurthy, K.V., 1997. Comparative study of the development of zygotic and somatic embryos of chickpea (*Cicer arietinum* L.). Plant Sci. 128, 207–216.

Tayyar, R.I., Lukaszewski, A.J., Waines, J.G., 1994. Chromosome banding pattern in the annual species of *Cicer*. Genome 37 (4), 656–663.

USDA, 2019. Chickpeas (Garbanzo Beans, Bengal Gram), Mature Seeds, Cooked, Boiled, Without Salt. Full Report (All Nutrients) 16057. National Nutrient Database for Standard Reference. United States Department of Agriculture. https://ndb.nal.usda.gov/ndb/foods. (Accessed March 2019).

Van der Maesen, L.J.G., 1972. *Cicer* L. A Monograph of the Genus, with Special Reference to the Chickpea (*Cicer Arietinum* L.). Veenman & Sons, Wageningen, pp. 9–133.

Van der Maesen, L.J.G., 1987. Origin, history and taxonomy of chickpea. In: Saxena, M.C., Singh, K.B. (Eds.), The Chickpea. CABI, Wallingford, pp. 139–156.

Van der Maesen, L.J.G., Maxted, N., Javadi, F., Coles, S., Davies, A.M.R., 2007. Taxonomy of the genus *Cicer* revisited. In: Yadav, S.S., Redden, R.J., Chen, W., Sharma, B. (Eds.), Chickpea Breeding and Management. CAB International, pp. 14–47. ISBN-13: 9781 84593 2145.

Vlacilova, K., Ohri, D., Vrana, J., Cihalikova, J., Kubalakova, M., Kahl, G., Dolezel, J., 2002. Development of flow cytogenetics and physical genome mapping in chickpea (*Cicer arietinum* L.). Chromosome Res. 10, 695–706.

Warren Wilson, J., Warren Wilson, P.M., 1993. Mechanisms of auxin regulation of structural and physiological polarity in plants, tissues, cells and embryos. Aust. J. Plant Physiol. 20, 555–571.

Winter, P., Benko-Iseppon, A.M., Hüttel, B., Ratnaparkhe, M., Tullu, A., Sonnante, G., et al., 2000. A linkage map of the chickpea (*Cicer arietinum* L.) genome based on recombinant inbred lines from a *C. arietinum* × *C. reticulatum* cross: localization of resistance genes for *fusarium* wilt races 4 and 5. Theor. Appl. Genet. 101, 1155–1163.

Yeung, E.C., 1980. Embryogeny of Phaseolus: the role of the suspensor. Z. Pflanzenphysiol. 96, 17–28.

Yeung, E.C., Meinke, D.W., 1993. Embryogenesis in angiosperms: development of the suspensor. Plant Cell 5, 137–1381.

Zatloukalová, P., Hřibová, E., Kubaláková, M., Suchánková, P., Šimková, H., Adoración, C., et al., 2011. Integration of genetic and physical maps of the chickpea (*Cicer arietinum* L.) genome using flow-sorted chromosomes. Chromosome Res. 19, 729–739.

CHAPTER

6 Gene pyramiding and multiple character breeding

Ashutosh Kushwah[1], Soma Gupta[2], Shayla Bindra[1], Norah Johal[1], Inderjit Singh[1], C. Bharadwaj[3], G.P. Dixit[4], P.M. Gaur[5], Harsh Nayyar[6], Sarvjeet Singh[1]

[1]*Punjab Agricultural University, Ludhiana, Punjab, India;* [2]*ICAR-Indian Institute of Seed Science, Mau, Uttar Pradesh, India;* [3]*ICAR-Indian Agricultural Research Institute, New Delhi, Delhi, India;* [4]*ICAR-Indian Institute of Pulses Research, Kanpur, Uttar Pradesh, India;* [5]*International Crops Research Institute for the Semi-Arid Tropics, Patancheru, Telangana, India;* [6]*Panjab University, Chandigarh, Punjab, India*

6.1 Introduction

Chickpea (*Cicer arietinum* L.) or garbanzo bean is a temperate autogamous legume, originated from southeastern Turkey (Ladizinsky, 1975). Chickpea is a diploid ($2n = 2x = 16$) annual species having haploid genome size of 738 Mb (Varshney et al., 2013). Globally chickpea cultivation stretches over 14.56 million ha of area yielding 14.77 million tonnes (FAOSTAT, 2017). Major chickpea producing countries are India, Pakistan, Turkey, Myanmar, Australia, Iran, Ethiopia, Mexico, Canada, and the United States. India leads in terms of global production and productivity contributing about 70% of the world's total production. In India 11.16 million tonnes of chickpea has been harvested from 10.76 million ha during 2017–18 (Dixit, 2018). *C. arietinum* L. is the only cultivated species of genus *Cicer*, and *C. reticulatum* L. being interfertile with chickpea is considered to be the wild progenitor of *C. arietinum*. Chickpea is classified into two distinct types based on seed morphology—the small-seeded microsperma "*desi*" seeds having brown coat and the large-seeded macrosperma "*kabuli*" seeds enveloped under cream or beige seed coat. The *desi* type has been mostly cultivated in India, Pakistan, Bangladesh, Australia, and Myanmar, while the *kabuli* type is produced predominantly in the Mediterranean region and Central Asia.

In Indian subcontinent, chickpea is consumed as *dal* (split cotyledons) and snacks prepared from *besan* (chickpea flour), while in African regions, soups, stews, and salads are common chickpea preparations. Chickpea is also consumed in salted, roasted, boiled, raw vegetable, and fermented forms. Chickpea offers ample amount of nutrients being rich in protein (20%–30%), carbohydrates (~40%), vitamins, and several minerals (calcium, phosphorus, magnesium, potassium, manganese, iron, and zinc). Chickpea grains also contain significant proportion of essential amino acids viz. leucine, lysine, isoleucine, phenylalanine, and valine. Chickpea

Chickpea: Crop Wild Relatives for Enhancing Genetic Gains. https://doi.org/10.1016/B978-0-12-818299-4.00006-3
Copyright © 2020 Elsevier Inc. All rights reserved.

consumption helps in managing diabetes because of its lower glycemic index. Chickpea seed oil contains unsaturated fatty acids like linoleic and oleic acid and various phytosterols such as β-sitosterol, tocopherols, sterols, and tocotrienols. These phytosterols exhibit antitumoric, antibacterial, antifungal, and anti-inflammatory properties. Chickpea contains several bioactive compounds like phenolic compounds, isoflavones, and phytates which impart potential health benefits and are associated with prevention of cardiovascular diseases, cancer, blood pressure, and obesity, rendering chickpea as a functional food. Besides providing nutritive and health components to human diet, chickpea efficiently fixes atmospheric nitrogen, thereby reducing dependency on nitrogenous fertilizers. The crop fits well in crop rotation for replacing cereal-cereal cropping system and helps in maintaining soil fertility and improving soil health.

Molecular markers can effectively aid for selection of traits having low heritability, simultaneous transfer of multiple genes, cultivar/germplasm description, gene introgression from exotic/unadapted germplasm with minimum linkage drag, estimation of genetic relatedness among accessions, and identification of quantitative trait loci (QTLs) governing economically important traits. Thus, molecular markers are important tools in speeding up the process of crop improvement and in amelioration of our understanding in relation to genetic basis of traits of agronomic importance (Varshney et al., 2007). Molecular tools efficiently supplement conventional breeding schemes offering rapid yet precise alternative for improving cultivars for quantitative traits like yield, adaptability, resistance to biotic and abiotic stresses, etc.

Few decades back, chickpea used to be categorized under an orphan legume, and classical plant breeding approaches were majorly followed for its improvement. These approaches are time-consuming and less precise for improvement of complex quantitative traits. Nevertheless chickpea has now evolved as a genomic resources–rich crop. Post genome sequencing and large-scale genomic resources availability make this crop amenable for marker- and genomic-assisted breeding. Advent and subsequent advancement in molecular marker technologies have made it feasible to locate genomic regions of interest and evaluate their phenotypic effects toward various quantitative traits. This further prompted molecular breeding approaches namely, marker-assisted recurrent selection (MARS), marker-assisted backcrossing (MABC), advanced backcross quantitative trait loci (AB-QTL) analysis, and genomic-assisted breeding (GAB) activities including trait mapping as routine work in chickpea breeding programs. Next-generation sequencing (NGS) technologies led to rapid development of molecular markers in chickpea on a large scale. These advanced resources and technologies have been utilized for construction of dense linkage maps and identification of several molecular markers associated with agronomically important traits. This chapter describes about the genetic and genomic resources and their deployment for multiple trait breeding and genomics-assisted chickpea breeding.

6.2 Genetic resources in chickpea

Chickpea genetic resource comprises 99,877 accessions at global level. These accessions are safeguarded and maintained among 120 national and international gene banks located across 64 countries, out of which, 1476 are wild *Cicer* accessions (Upadhyaya et al., 2018). List of 14 major chickpea gene banks accommodating more than 1000 accessions each has been presented in Table 6.1. These 14 gene banks contain a total of 87,341 chickpea accessions representing 98.3% cultivated and 1.7% the wild *Cicer* species (Upadhyaya et al., 2018). The National Bureau of Plant Genetic Resources, India, holds 14,704 chickpea accessions including cultivated and wild species. The International Crops Research Institute for the Semi-Arid Tropics (ICRISAT) holds largest chickpea germplasm collection of 20,764

Table 6.1 List of global gene banks accommodating chickpea accessions.

Institute	Wild	Cultivated	Total
International Crops Research Institute for the Semi-Arid Tropics (ICRISAT), India	308	20,456	20,764
International Centre for Agricultural Research in the Dry Areas (ICARDA), Beirut, Lebanon	540	15,194	15,734
National Bureau of Plant Genetic Resources (NBPGR), New Delhi, India	69	14,635	14,704
Australian Temperate Field Crops Collection (ATFCC), Australia	246	8409	8655
Western Regional Plant Introduction Station, USDA-ARS, USA	194	7844	8038
National Plant Gene Bank of Iran, Seed and Plant Improvement Institute (NPGBI-SPII), Iran	–	5700	5700
N.I. Vavilov All-Russian Scientific Research Institute of Plant Industry (VIR), Russian Federation	–	2767	2767
Plant Genetic Resources Program (PGRP), Pakistan	89	2057	2146
Plant Genetic Resources Department, Aegean Agricultural Research Institute (AARI), Turkey	21	2054	2075
Institute of Plant Production nd. a. V. Ya. Yuryev of NAAS, Ukraine	–	1760	1760
Estación de Iguala, Instituto Nacional de Investigaciones Agrícolas, Iguala, Mexico	–	1600	1600
Institute of Biodiversity Conservation (IBC), Ethiopia	–	1173	1173
Institute for Agrobotany (RCA), Hungary	9	1161	1170
Uzbek Research Institute of Plant Industry (UzRIPI), Uzbekistan	–	1055	1055
Total	1476	85,865	87,341

Source: Upadhyaya, H.D., Dwivedi, S.L., Sharma, S., 2018. Managing and Discovering Agronomically Beneficial Traits in Chickpea Germplasm Collections. The Chickpea Genome, vols. 43–52. Springer International Publishing, Cham, Switzerland, pp. 43–52.

accessions (20,456 cultivated and 308 wild) representing 59 countries of origin including the United States, Syria, Nepal, Azerbaijan, and Bulgaria.

Despite the availability of enormous global wild and cultivated chickpea accessions, utilization of these accessions in chickpea breeding programs is meager. Scanty information on agronomically important traits, cost, and time required for screening huge sets of germplasm lines prevent their utilization in breeding programs. As many as 16,990 chickpea accessions were evaluated at ICRISAT for 13 traits to form a core collection comprising 1956 accessions so as to promote significance of global chickpea genetic resources in genomics and breeding (Upadhyaya et al., 2001). Furthermore, a mini-core collection of 211 chickpea accessions has also been developed (Upadhyaya and Ortiz, 2001). ICRISAT and ICARDA with their joint efforts have developed a reference set of 300 lines under the Generation Challenge Program (GCP) of the Consortium of International Agricultural Research Centres depicting genetic variability available in the germplasm maintained at aforementioned institutions (Upadhyaya et al., 2008). These manageable numbers of accessions representing mini-core, core, and reference sets of germplasm contribute ideal resource for association genetics, gene mapping and cloning, allele mining, and applied breeding for the development of elite cultivars.

6.3 Genomic resources

Conventional approaches being practiced in chickpea inflated crop yield, but narrow genetic base hampered the desired improvement in chickpea productivity. Genomic studies aim toward the identification of genes and metabolic pathways affecting crop production, thus accelerating the realization of enhanced genetic gain. This triggered international platforms to initiate developing and further deploying chickpea genomic resources in breeding.

Isozymes are biochemical markers catalyzing the same chemical reaction but differentiated on the basis of electrophoretic mobility. In chickpea, development and segregation of isozyme markers was reported in the F_2 hybrids obtained from interspecific crosses of *C. arietinum* L. with *C. echinospermum* and *C. reticulatum* (Gaur and Slinkard, 1990a,b). Kazan and Muehlbauer (1991) categorized chickpea species into four categories based on isozyme profile analysis of annual (nine) and perennial (one) chickpea accessions, which was further supported by several studies (Ahmad and Slinkard, 1992; Tayyar and Waines, 1996; Labdi et al., 1996).

RFLP (restriction fragment length polymorphism) markers have also been extensively used in *desi* and *kabuli* type chickpea for assessing polymorphism (Udupa et al., 1993), diversity analysis, identifying center of genetic diversity (Serret et al., 1997), and linkage map construction (Simon and Muehlbauer, 1997). The RAPD (random amplified polymorphic DNA) markers have been used to assess the polymorphism (Banerjee et al., 1999), trait mapping (Tullu et al., 1998), genetic diversity assessment, and phylogenetic analysis (Sant et al., 1999; Iruela et al., 2002;

Singh et al., 2003). The AFLP (amplified fragment length polymorphism) markers have been utilized in genetic diversity assessment, phylogenetic classification of germplasm lines (Nguyen et al., 2004; Shan et al., 2005; Talebi et al., 2008), and construction of genetic maps (Winter et al., 2000) in chickpea. Microsatellite markers in chickpea have been mainly developed following: using oligonucleotide repeats to probe genomic libraries (Winter et al., 1999; Hüttel et al., 1999), sequencing of microsatellite-enriched libraries, and sequencing of bacterial artificial chromosome (BAC) clones (Nayak et al., 2010; Thudi et al., 2011). Later BAC and BIBAC (binary BAC) have been instrumental in development of SSR markers (Lichtenzveig et al., 2005). DArT (diversity arrays technology) markers are profoundly used for constructing diversity analysis and linkage maps. DArT arrays with 15,360 clones have been developed by ICRISAT in a collaborative effort with DArT Pty Ltd. (Thudi et al., 2011). DArT arrays followed similar trend of narrow genetic diversity present in gene pool of cultivated genotypes in comparison to wild *Cicer* species (Roorkiwal et al., 2014b).

Transcriptome profiles in various stress challenged plant tissues, and stress-responsive expressed sequence tags (ESTs) have also been instrumental in identification of gene of interest and thus, development of functional markers further being utilized into breeding program. Buhariwalla et al. (2005) initiated efforts toward developing functional markers from ESTs, while Varshney et al. (2009) used drought- and salinity-responsive ESTs for developing 177 new EST-SSRs. Several studies were conducted on transcriptome profiling of different stress-responsive tissues (Garg et al., 2011a,b; Hiremath et al., 2011; Singh et al., 2013; Kudapa et al., 2014; Afonso-Grunz et al., 2014) that led to EST libraries sequencing, resulting in a plethora of EST sequences in public domains.

SNP (single nucleotide polymorphism) genotyping platforms have their applications for genetic diversity analysis, association mapping, fine mapping of genes, evolutionary studies, and genomic selection. In due course of time SNP markers became marker of choice owing to their cost-effectiveness, genome-wide abundance, and high-throughput genotyping. The identification of 184 putative SNPs in 19 contigs constructed with 1499 ESTs from various *Cicer* species has been accessible in public domain (Varshney et al., 2007). Advancement in NGS technologies generates sample quantity of sequencing data. More than 20,000 ESTs have been generated from salinity and drought stress-challenged tissues in chickpea using Sanger sequencing technology (Varshney et al., 2009). Additional sequencing data were generated on more than 20 tissues differing and representing varietal developmental stages (Hiremath et al., 2011). They analyzed the pooled data through NGS transcripts and Sanger ESTs to generate first transcript assembly with 103,215 tentative unique sequences (TUSs) and further identification of few thousand SNPs. Several thousand SNPs were also identified through Illumina sequencing (Varshney et al., 2013), allele-specific sequencing (Gujaria et al., 2011; Roorkiwal et al., 2014a), and 454 transcriptome sequencing (Deokar et al., 2014). A total of 6698 SNPs were mapped on eight linkage groups extending 1083.93 cM in an interspecific RIL population utilizing genomic and transcriptomic SNPs (Gaur et al.,

2015) and 70,463 nonredundant SNPs were selected from a pool of 4.9 million SNPs and further on the basis of p-convert score, 50,590 SNPs were tiled out of 61,174 selected SNPs on Affymetrix Axiom array (Roorkiwal et al., 2017).

NGS technologies are time- and cost-effective that enable sequence-based trait mapping. Genotyping by sequencing (GBS), skim sequencing, and whole-genome sequencing approaches furnish genome-wide large-scale marker data enabling high-resolution trait mapping (Pandey et al., 2016). GBS has been employed for "*QTL-hotspot*" refinement which was identified through an intraspecific cross between genotypes ICC 4958 × ICC 1882 (Jaganathan et al., 2015), whereas skim sequencing identified 84,963 SNPs utilizing same parental cross, of which 76.01% were distributed over eight pseudomolecules (Kale et al., 2015). More than 40,000 and 80,000 genome-wide SNPs have been identified using integrated reference genome–based and *de novo*-based genotyping by sequencing approach from 93 wild and cultivated chickpea accessions by Kujur et al. (2015) and Bajaj et al. (2015), respectively. Availability of large-scale SNP information imposed challenges to make use of these markers routinely in breeding programs. Different SNP genotyping platforms such as Golden Gate and Vera Code (Roorkiwal et al., 2013) and KASP markers (Hiremath et al., 2012) were developed to meet automation, and time- and cost-effective for data turnaround so as to deploy this large-scale SNP marker information in breeding programs.

6.3.1 Transcriptomics resources

Identification and isolation of stress-responsive genes can be carried out with the help of transcriptomic approaches such as gene expression profiling based on RNA sequencing. Further identified stress-responsive genes would contribute toward unveiling the underlying molecular mechanisms of stress tolerance and help in development of stress-tolerant chickpea cultivars. Several genes or ESTs involved in various stress responses have been reported on the basis of several transcriptomic and proteomic studies (Mantri et al., 2007; Pandey et al., 2008; Varshney et al., 2009a; Molina et al., 2011). But very limited studies have been carried out in chickpea, and cloning of few candidate genes has been validated functionally (Shukla et al., 2009; Peng et al., 2010; Kudapa et al., 2013).

Advancement in NGS technologies has paved the way to genome sequencing and transcriptome sequencing. Large-scale transcriptome sequencing and NGS approaches have led to requisite amount of information regarding the chickpea gene content. A large amount of ESTs have been generated from stage-specific plant tissues, including various developmental phases as well as stress-challenged tissues in chickpea. For example, 20,162 ESTs have been identified from plants exposed under salt and drought stress (Varshney et al., 2009), and 103,215 tentative unique sequences (TUSs) based on 435,018 FLX/454 reads and 21,491 Sanger ESTs have been developed by Hiremath et al. (2011) constructing transcriptome assembly. An array of 53,409 contigs representing ~28 Mb of unique transcriptome sequence using Illumina sequencing were assembled (Garg et al., 2011a). Illumina sequencing and FLX/454 defined another set of 34,760 contigs which represents ~4.8% (35.5 Mb) of the total chickpea genome (Garg et al., 2011b). A hybrid assembly of 46,369 contigs from RNA has been developed showcasing range of different developmental stages and eight tissues exposed to various stresses (Kudapa et al., 2014). Jain and Chattopadhyay (2010) investigated the differences in transcriptome profiling of chickpea varieties during drought stress and concluded that highly expressed ESTs involved in drought tolerance encoded proteins responsible for signal transduction, cellular organization, and protein metabolism. Deokar et al. (2011) identified 3062 unigenes by studying the upregulated and downregulated genes in drought-tolerant and drought-susceptible genotypes, out of which about 51% were novel and 2185 genes were found similar with nonredundant database in the NCBI (National Center for Biotechnology Information).

In silico differential expression studies were conducted from Sanger ESTs that were further generated from drought-stressed tissues of drought-tolerant (ICC 4958) and drought-sensitive (ICC 1882) genotypes (Varshney et al., 2009). Hiremath et al. (2011) aligned 37 million Illumina short sequence tags characterized from root tissues of drought-stressed plants and identified 2974 TUSs with significant expression changes. Differential range in the levels of expression has been observed by mapping all reads onto a nonredundant set of chickpea transcripts, and the number of reads corresponding to each transcript ranged from 14 to 270,894 with an average of 1617 (Garget al., 2011a). Another resilient approach i.e., gene cloning relating to isolation of candidate genes functionally related to the trait of interest is also coming into play. A one-step forward genetics approach for gene identification related to a trait is positional cloning. Discovery of trait-associated genes by any of the aforementioned approaches renders to the next step, that is, their functional validation. Certain approaches such as overexpression, RNAi, virus-induced gene silencing, and TILLING have been utilized for this task. Optimized protocols with maximal output are readily available in chickpea (Acharjee et al., 2010). Genetic transformation and further validation through RNAi or virus-induced gene silencing possessing inefficient transformation systems is a time-consuming process. This situation has been combated by implementation of TILLING to study gene function.

Various approaches of functional genomics like suppression subtractive hybridization (SSH), microarray and EST sequencing, serial analysis of gene expression (SAGE), and their modifications such as super serial analysis of gene expression (Super SAGE) and deep Super SAGE are used in identification of transcripts of various abiotic stress-responsive genes in chickpea to facilitate quantification of global gene expression (Matsumura et al., 2005; Buhariwalla et al., 2005; Molina et al., 2008). For example, Kahl et al. (2007) investigated salt- and drought-stress-responsive transcripts in chickpea and analyzed these transcripts which represented 40,000 unique mRNAs and led to identification of 3000 associated transcripts using Super SAGE. Similarly deep Super SAGE led to identification of 80,238 tags representing 17,493 unique transcripts from drought-stressed and nonstressed control roots in chickpea (Molina et al., 2008, 2011).

6.3.2 Genetic and physical maps

In the process of marker-assisted selection and marker-assisted backcross program, identification the gene/QTL-specific markers related with particular trait of interest merges as the initial step. Two approaches, viz., biparental mapping and genome-wide association mapping (GWAS) can be used for identification of associated markers. Genetic or linkage maps are constructed before the discovery of markers related to trait of interest. These linkage maps assist the identification of molecular markers closely linked with major gene/QTL and allowed the gene tagging in molecular breeding. Linkage maps also assist in the characterization of genomic regions controlling the traits governed by several genes, decoding the gene action including epistatic effects for determining the specific phenotype (Tanksley, 1993). After the major gene/QTL mapping, closely linked markers should be further validated for use in marker-assisted selection. These closely linked markers can be serving as starting point for positional cloning of the underlying functional gene/QTL. Linkage map in chickpea dramatically progressed from morphological markers ranging to NGS techniques like SNPs, InDels, and DArT (Table 6.2). Recent advances in genotyping using NGS techniques have enabled large-scale simultaneous genome-wide SNP discovery leading to the development of high-resolution linkage maps in chickpea (Deokar et al., 2014; Kujur et al., 2015; Jaganathan et al., 2015). Now, these highly saturated genetic linkage maps and large-scale whole-genome sequences generated using NGS technologies–based SNP markers facilitate in mapping of QTLs as well as positional cloning of QTLs to reveal the underlying candidate genes.

6.3.3 Functional or transcript maps

Transcript map is of immense use for targeted QTL mapping, QTL cloning, and identification of candidate genes for important agronomic traits in chickpea. Procurement of large-scale mRNA sequences as EST-SSR and EST-SNP markers has been practiced to develop transcript maps in chickpea. First large-scale transcript map of chickpea employing SSRs, SNPs, and intron spanning region markers in an interspecific mapping population was developed by Gujaria et al. (2011) spanning about 767 cM of the genome size having intermarker distance of 2.5 cM. The same mapping population was used by Choudhary et al. (2012) with a new set of EST-derived genic molecular markers (GMMs) to generate a large-scale transcript map of 1498 cM with intermarker distance of 3.7 cM. Further a high-resolution transcript map was generated using tentative orthologous genes (TOGs)-SNPs spanning about 788.6 cM distance (Hiremath et al., 2012). This comprehensive and highly saturated interspecific transcript map was applied to anchor the first draft version of chickpea genome of CDC Frontier (Varshney et al., 2013). SSR and SNP markers derived from the transcription factors of candidate genes led to development of highly saturated interspecific genetic linkage map spanning map length of 949 cM (Saxena et al., 2014). Now, these transcription factors of candidate genes–derived SSRs and SNPs markers associated with QTL can act as efficient markers in chickpea for marker-assisted genetic improvement.

Table 6.2 List of various genetic linkage maps constructed in chickpea employing phenotypic and DNA-based molecular markers.

Type of cross	Mapping population	Markers mapped	Number of linkage groups	Distance (cM)	Average marker density (cM)	References
Interspecific	F_2	16 isozyme markers	7	200	–	Gaur and Slinkard (1990a,b)
Interspecific	F_2	28 morphological and isozyme markers	8	257	–	Kazan et al. (1993)
Interspecific	F_2, F_3	9 morphological, 27 isozyme, 10 RFLP, and 45 RAPD markers	10	550	–	Simon and Muehlbauer (1997)
Interspecific	RIL	120 STMS	11	613	–	Winter et al. (1999)
Interspecific	RIL	89 RAPD, 17 ISSR, 9 isozyme, and 1 morphological markers	9	981.6	8.4	Santra et al. (2000)
Interspecific	RIL	118 STMS, 96 DAF, 70 AFLP, 37 ISSR, 17 RAPD, 8 isozyme, 3 cDNA, 2 SCAR, and 3 morphological markers	16	2077.9	6.8	Winter et al. (2000)
Intraspecific (ICCV-2 × JG-62)	RIL	55 STMS, 20 RAPD, 3 ISSR, and 2 phenotypic markers	14	297.5	3.7	Cho et al. (2002)
Interspecific	RIL	44 STMS, 89 RAPD, 17 ISSR, 9 isozyme, and 2 morphological markers	9	1174.5	7.0	Tekeoglu et al. (2002)
Interspecific	RIL	47 gene-specific markers along with previously reported STMS, DAF, RAPD, and AFLP markers	12	2483.3	–	Pfaff and Kahl (2003)

Continued

Table 6.2 List of various genetic linkage maps constructed in chickpea employing phenotypic and DNA-based molecular markers.—*cont'd*

Type of cross	Mapping population	Markers mapped	Number of linkage groups	Distance (cM)	Average marker density (cM)	References
Intraspecific (JG62 × Vijay) and (Vijay × ICC 4958)	RIL	44 RAPD, 16 ISSR, 2 RGA, 165 SSR, and 1 ASAP markers	8	739.6	3.2	Radhika et al. (2007)
Intraspecific (ICC 4991 × ICCV 04516)	F_2	82 SSR and 2 EST markers	10	724.4	8.6	Kottapalli et al. (2009)
Interspecific	Segregating population	135 STMS and 33 cross-genome markers along with other markers	8	652.67	–	Millan et al. (2010)
Intraspecific	Segregating population	99 STMS, 3 SCAR, 1 ASAP, fusarium resistance gene, 5 morphological traits, RAPD, and ISSR	8	426.96	–	
Interspecific	RIL	47 SSR (ICCM), 46 SSR (H-series), 71 gene-based SNP along with 357 legacy markers	8	2602.1	4.99	Nayak et al. (2010)
Intraspecific (ICCV 2 × JG 62)	RIL	144 markers including SSRs, RAPD, DAF, MP-PCR, RMMFP, AFLP, ISSR, and morphological markers	11	442.8	3.3	Anuradha et al. (2011)
Intraspecific (ICCV 2 × JG 62)	RIL	138 STMS	8	630.9	4.57	Gaur et al. (2011)
Intraspecific (BGD 112 × FLIP 90–166)	F_2	33 STMS	8	471.1	14.2	Bharadwaj et al. (2011)

Interspecific	RIL	157 SSR, 621 DArT loci, 145 GMM, and 368 legacy markers	8	845.56	0.65	Thudi et al. (2011)
Interspecific	RIL	47 EST-SSR, 49 ITP, 22 ESTP, 57 SNP, 2 MtEST, 126 gSSR, and 103 STMS	8	1497.7	3.68	Choudhary et al. (2012)
Interspecific	RIL	625 CKAM, 314 TOG-SNP, 389 GMM, and legacy markers	8	788.6	0.59	Hiremath et al. (2012)
Intraspecific (ILC 3279 × ICCV 2)	$F_{2:3}$	27 STMS, 17 RAPD, and 13 ISSR markers	8	379.47	–	Jamalabadi et al. (2013)
Intraspecific	ICCRIL03 (ICC 4958 × ICC, 1882)	214 SSRs, 6 GMMs, and 21 DArT	8	621.51	2.71	Varshney et al. (2014b)
	ICCRIL04 (ICC 283 × ICC 8261)	151 SSRs, 10 GMMs, and 7 DArT	8	533.06	3.27	
Intraspecific (ICCV 96029 × CDC Frontier)	RIL	1336 SNPs	8	653	0.5	Deokar et al. (2014)
Intraspecific (K850 × WR315)	RIL	23 SSR	4	690	30	Jingade and Ravikumar (2015)
Interspecific	RIL	6698 SNPs	8	1083.93	0.16	Gaur et al. (2015)
Intraspecific (ICC 4958 × ICC 1882)	RIL	743 SNP, 232 SSR, 21 DArT, 7 EST-SSR, and 4 GMM	8	727.29	1.30	Jaganathan et al. (2015)
Intraspecific (ILC 3279 × ILC 588)	$F_{2:3}$ $F_{3:4}$	10 RAPD, 8 URP, 15 STMS 3 morphological markers	7	285.3	8.6	Karami et al. (2015)
Interspecific	RIL	1063 Genic EST-SSR and genomic SSR, SNP, ITP, and 634 transcript-derived and genomic SSR	8	1061.16	0.625	Khajuria et al. (2015)

Continued

Table 6.2 List of various genetic linkage maps constructed in chickpea employing phenotypic and DNA-based molecular markers.—*cont'd*

Type of cross	Mapping population	Markers mapped	Number of linkage groups	Distance (cM)	Average marker density (cM)	References
Intraspecific (SBD377 × BGD112)	RIL	3363 SNP	8	1006.98	0.33	Verma et al. (2015)
Interspecific	F5 (Pusa 1103 × ILWC 46)	1059 InDel	8	978.21	0.92	Srivastava et al. (2016)
	F5 (Pusa 256 × ILWC 46)	594 InDel	8	603.26	1.01	
	Consensus map	1479 InDel	8	978.61	0.66	
Intraspecific (ICC 4567 × ICC 15614)	RIL	271 GBS-based SNPs	8	529.11	1.95	Paul et al. (2018b)

Source: Roorkiwal, M., Jain, A., Thudi, M., Varshney, R.K., 2018. Advances in chickpea genomic resources for accelerating the crop improvement. In: The Chickpea Genome, pp 53-67, Springer International Publishing, Cham, Switzerland.

6.3.4 Genome sequencing and resequencing

Draft genome sequence will help to understand the molecular mechanism of functional scenario of genes as well as response mechanism of plants against various biotic and abiotic stresses. NGS technology is highly useful in identification and development of remarkably stable molecular markers which allowed the researchers in genomic-assisted breeding. Considering the utility of genome sequence, ICRISAT-led International Chickpea Genome Sequencing Consortium (ICGSC) decoded the chickpea genome sequence. A *kabuli* chickpea variety (CDC Frontier) was sequenced through Illumina sequencing and assembled 544.73 Mb of genome which represented 74% of the chickpea genome (Varshney et al., 2013). Improving the assemblies is vital, for which chromosomes from chickpea (*desi* and *kabuli*) varieties have been sequenced by NGS-based sequencing technology to further validate the assemblies. This will help to detect the misassembled regions. A small region in *kabuli* and a large region in *desi* were found to be misassembled (Ruperao et al., 2014). Parween et al. (2015) generated additional sequence data to advance the assembly of *desi* variety and reported an improved assembly.

Accessions constitute tremendous genetic diversity and are the reservoir of novel alleles. To exploit these genetic variations toward the identification of novel alleles, large-scale sequencing and resequencing efforts were initiated. Illumina HiSeq 2500 has been instrumental in resequencing of 100 chickpea genotypes across 14 countries. Resequencing data on these 100 elite varieties along with 29 prior resequenced lines were useful in gaining the importance relating to effect on temporal trends in chickpea diversity via breeding on genetic diversity (Thudi et al., 2016). First-generation Hap Map of chickpea was developed utilizing resequencing data on 100 elite chickpea varieties. In addition, resequencing of chickpea reference set (300 lines) using whole-genome resequencing approach generated 1.8 Tb raw sequence data which helped to identify 4.9 million SNPs. These resequencing data obtained from reference set along with multilocation and multiseason phenotyping data were exploited for GWAS analysis for the identification of trait-related markers of interest. Post draft genome sequencing, ICGSC also resequenced chickpea accessions of both cultivated and wild chickpea accessions (90 in total) utilizing NGS-based whole-genome resequencing and restriction site–associated DNA (RAD) technology (Varshney et al., 2013). Further genome of ICC 4958, a *desi* chickpea genotype, was sequenced for assembling the draft chickpea genome assembly. Chickpea genome ~520 Mb was assembled using NGS technology and BAC end sequencing (Jain et al., 2013). A draft assembly of a wild progenitor of chickpea was developed by Gupta et al. (2016), and 78% of this assembly has been anchored to eight linkage groups. "The 3000 Chickpea Genome Sequencing Initiative" has been launched by ICRISAT to resequence the chickpea germplasm conserved in gene banks for unveiling novel alleles (Varshney, 2016).

6.3.5 Trait mapping

The wide access of DNA-based genetic markers related to trait of interest helps to infer genotypic constitution of plants and to overcome the problems of stage dependency, confounding effects of genotype × environment interactions, and operational difficulties. Chickpea molecular breeding has gone miles away by mapping several desirable traits for various biotic and abiotic stress tolerance and yield-related traits. The molecular markers closely linked to these complex traits have been applied in several marker-assisted breeding programs. Recently, several markers have been identified using SUPER genome-wide association study (GWAS) significantly related to yield and yield-related traits against drought-stressed conditions.

A "*QTL-hotspot*" of 12 major QTLs imparting drought stress tolerance has been identified on LG 4 and the estimated size of this "*QTL-hotspot*" was 29 cM on the linkage map and 7.74 Mb on the physical map of chickpea genome (Varshney et al., 2014b). This "*QTL-hotspot*" was further refined to 14 cM on linkage map from 29 cM corresponding to ~4 Mb of chickpea genome on the physical map by GBS approach (Jaganathan et al., 2015). Now this "*QTL-hotspot*" region was further refined by Kale et al. (2015) using a combination of GWAS based gene enrichment analysis of skim sequenced data approach and sliding window–based bin mapping approach and spilt into two subregions, i.e., "*QTL-hotspot a*" of size of 139.22 Kb and "*QTL-hotspot b*" of size of 153.36 Kb.

In another study, 312 markers related to drought and heat stress tolerance–related traits using whole-genome sequencing and candidate gene–based association mapping approach in chickpea have been identified (Thudi et al., 2014a,b). Likewise, two genomic regions harboring four QTLs and a total of 25 putative candidate genes responsible for heat tolerance–related traits were identified on LG5 and LG6 in chickpea (Paul et al., 2018a). Several QTLs have been also identified for salinity tolerance in chickpea. Markers closely associated with salt stress resistance–associated traits have been identified on LG1, LG2, LG3, and LG7 using RIL mapping population developed from the cross between ICC6263 × ICC1431 (Saminemi, 2010). In another study, QTLs for yield and yield component traits associated with salt stress tolerance have been described on LG3 and LG6 using RIL mapping population derived from a cross between ICCV2 (salinity sensitive) × JG62 (salinity tolerant) (Vadez et al., 2012). Further, Pushpavalli et al. (2015) reported 27 QTLs governing yield and yield-related traits for salinity stress tolerance harbored on LG5, LG7, and LG8 using RIL mapping population developed from the cross between ICCV 2 (salinity sensitive) x JG 11 (salinity tolerant).

QTLs for ascochyta blight resistance have been mapped on LG2, LG3, LG4, and LG8 by Millán et al. (2013) on the chickpea linkage map and different genetic backgrounds of chickpea have been validated by employing the different mapping populations. In another study, QTLs for ascochyta blight resistance have been mapped by Anbessa et al. (2009) on LG6 using the CDC Frontier as resistant source of blight. Recently, Deokar et al. (2019) mapped ascochyta blight–resistant QTLs on chromosomes 1, 2, 4, 6, and 7 using two RIL mapping populations and identified a total of 11 and 66 QTLs from both of the populations, respectively, through NGS-based bulked segregant analysis approach.

The first fusarium wilt resistance gene H1 (*foc*1) providing resistance to race 1 has been mapped by Mayer et al. (1997). RAPD markers (UBC-170550, CS-27700) were found to be closely linked with fusarium wilt resistance gene to race 4 (Tullu et al., 1999). In another study, ISSR markers UBC-855500 and CS-27700 were tagged with fusarium wilt resistance gene to race 4 by Ratnaparkhe et al. (1998). Several SSR markers like TR59 and $OPJ20_{600}$ that were linked to the resistance gene *foc*0 (Cobos et al., 2005), TA110 and H3A12 linked to resistance gene *foc*1, H3A12, and TA96 linked to resistance gene *foc*2 (Gowda et al., 2009), TA96 and TA194 linked to resistance gene *foc*3 (Sharma et al., 2004; Gowda et al., 2009), TA96 and CS27 linked to resistance gene *foc*4 (Winter et al., 2000; Sharma et al., 2004), and TA59 and TA96 linked to resistance gene *foc*5 (Sharma et al., 2005; Cobos et al., 2009) have been mapped which are responsible for providing resistance against *Fusarium* race 0, 1, 2, 3, 4, and 5, respectively. Molecular markers closely linked to various biotic and abiotic stresses in chickpea are presented in Table 6.3.

6.3.6 Genomic-assisted breeding

GAB in breeding practices is integration of genomic tools for the development of tolerant lines with escalated biotic or abiotic stress tolerance resulting in the margin of existing yield. The intent of GAB is establishment and exploitation of interaction between genotype and related phenotype resulting in crop improvement. It includes different strategies such as genomics, transcriptomics, and proteomics for identification of molecular markers related with agronomically important characteristics that help in prediction of phenotype from the genotype to assist breeding. Advent of NGS technologies and high-throughput genotyping technologies has made possible to use genome-wide marker data to predict phenotype of hybrids for selection in another breeding cycle. MABC approach is found to be effective for breeding traits governed by major QTL/genes. MABC is helpful in holding requisite gene pyramiding of QTLs together in a given genetic background. However, majority of agronomically desirable traits are governed by several genes/QTLs, thus rendering MABC approach futile. Under such conditions MARS has been considered a better option. Genome-wide marker data generated by the technique of high-throughput genotyping or NGS approaches have led emergence of genome-wide selection or genomic selection (GS) as a powerful approach for identifying desirable progenies from favorable crosses (Jannink et al., 2010). Wild species are the storehouse of several desirable/superior alleles responsible for resistance to several biotic and abiotic stresses. AB-QTL approach (Tanksley and Nelson, 1996) is used for simultaneous identification and further transfer of desirable alleles from wild accessions into elite ones for development of improved lines/cultivars. This approach has been used for introgressing productivity-enhancing traits and resistance to diseases from *C. reticulatum* (Singh et al., 2005).

Table 6.3 List of molecular markers closely associated with various desirable chickpea breeding traits for utilization in marker-assisted breeding.

Trait mapped	Source	QTL/ gene	Associated marker	Genetic effects	References
Abiotic stresses					
Drought stress	ICCRIL 03	LGV1 (Cluster 1)	NCPGR184, ICCM0009b	3.60 −10.67	Varshney et al. (2013)
		LGV1 (Cluster 2)	cpPb 679915, CaM 0393	14.36	
		LG 8	NCPGR 164, CaM 1918, CaM 2187	9.11 −26.87	
	ICCRIL 04	LGV1 (Cluster 3)	TA103II, TA122, TA113	6.08 −15.57	
		LG 5	CaM 029, TA 11	14.04 −16.79	
		LG 6	CaM 1257, ICCeM 15, NCPGR 200, CaM 0317	2.95 −12.06	
		LG8	CaM 0772, TA118, TS 45, NCPGR 138, GA 6	8.77 −31.32	
	ILC 588 × ILC 3279	LG3	TA 125, TA 6, NCPGR 12, NCPGR 10	9–33	Rehman et al. (2011)
		LG 6	GA 21, CaSTMS 2	13	
		LG 7	TA 28, CaSTMS 25	5–6	
	ILC 588 × ILC 3279	LG2	TA 96	8.5	Hamweih et al. (2013)
	ICCRIL 03	LG4	NCPGR 127, NCPGR 21, TAA170, ICCM 0249, TR 11, GA 24, STMS 11, ICCM 0065	10.90 −58.20	Varshney et al. (2013); Jaganathan et al. (2015); Kale et al. (2015)
Salinity tolerance	ICCV 2 × JG 11	LG 1	CaM 1301, CKAM 1971	66.75	Pushpavalli et al. (2015)
		LG 4	CKAM 0003, CKAM 1003	8.5	

Table 6.3 List of molecular markers closely associated with various desirable chickpea breeding traits for utilization in marker-assisted breeding.—*cont'd*

Trait mapped	Source	QTL/ gene	Associated marker	Genetic effects	References
		LG 5	CaM 0463, ICCM 272	24.98 −40.69	
		LG 8	CKAM 1903, CKAM 0343	37.75 −56.87	
	ICCV 2 × JG 62	LG 3	TA 106-Podnode	19.2	Vadez et al. (2012)
		LG 6	TR 20s, TA 46, TA 186, GA137, GA 25	18.2 −43.2	
		LG 7	TA 11, TA 42	27.6	
	ICCV 6263 × ICC 1431	LG 2	TR 37, TR 14	7.5	Saminemi (2010)
Biotic stresses					
Ascochyta blight	ILC3279, ILC72	LG2	GA16, TA194, TR19, Ein3	44.3	Udupa and Baum (2003); Cobos et al. (2006); Iruela et al. (2007); Anbessa et al. (2009); Madrid et al. (2014)
	ICC1, CDC Frontier, Amit	LG3	STMS28, TS12, TA64, TR58	18	Flandez-Galvez et al. (2003); Tar'an et al. (2007); Anbessa et al. (2009); Kottapalli et al. (2009)
	FLIP84-92C, Hadas, ILC3279	LG4	GAA47, GA24, H3C041, CaETR	16.9	Santra et al. (2000); Tekeoglu et al. (2002); Lichtenzveig et al. (2006); Madrid et al. (2013)
	FLIP84-92C, ILC3279, Hadas,	LG4	SCAR733b, TA2, TA146, TA72, TA130, TA132, TS54,	29.4	Santra et al. (2000); Udupa and Baum (2003); Cho

Continued

Table 6.3 List of molecular markers closely associated with various desirable chickpea breeding traits for utilization in marker-assisted breeding.—*cont'd*

Trait mapped	Source	QTL/ gene	Associated marker	Genetic effects	References
	CDC Frontier		H1G20, SCY17		et al. (2004); Iruela et al. (2007); Lichtenzveig et al. (2006); Tar'an et al. (2007); Anbessa et al. (2009); Kottapalli et al. (2009)
	CDC Frontier	LG6	TA176, TA80, TA22, TA40	11.6	Tar'an et al. (2007); Anbessa et al. (2009)
	CDC Corinne	LG8	TA3, H3C11a, TS46, TS45	16	Flandez-Galvez et al. (2003); Lichtenzveig et al. (2006); Anbessa et al. (2009)
	ICCV 05530	LG4	CKAM0847, CKAM0964	2.88	Garg et al. (2018)
Fusarium wilt	JG62	LG5	TR59, TS35	37.8	Cobos et al. (2005)
	ICCL81001	LG2	TA59	46.5	Cobos et al. (2009)
	WR315	LG6	CaM1402, CaM1101	16.4	Sabbavarapu et al. (2013)
	ICCV 05530	LG2 LG4	TR19, H2B061 TA132, TA46	24.56	Garg et al. (2018)
Botrytis gray mold	ICCV2	LG6	SA14, TS71	12.8	Anuradha et al. (2011)
		LG8	TA25, TA144 TA159, TA118	9.5–48	
Rust	Cr5-10	LG7	TA18, TA180	73.7	Madrid et al. (2008)

6.3.7 Marker-assisted backcrossing

Molecular breeding utilizing MABC approach is appropriate for introgression of few gene combinations for which elite varieties are deficient. MABC involves introgression of one or two major genes from a donor line into the genetic background of a recipient (usually elite variety) with the assistance of molecular markers (Hospital, 2005). Availability of a number of markers, high-throughput genotyping, and fine mapping has made possible for any major QTL to be introgressed into a variety with least linkage drag and without changing the desirable agronomic characteristics. Thus, the outcome of MABC is a new line similar to recurrent parent in terms of whole-genome constitution introgressed with few major genes/QTLs from the donor parent. Since this approach transfers restricted number of loci from one genetic background to other, the same can also be instrumental in generating near-isogenic lines (NILs) or chromosome segment substitution lines (CSSLs). NILs and CSSLs are the populations which are usually used for genetic analysis of genes/QTLs as well as alien gene introgressions. Gene pyramiding is a vital application of MABC curtailing to the same trait or for different traits controlled by few genes that are mosaiced in one genetic background with the assistance of molecular markers. However, clubbing large number of genes via backcross breeding or introgressing all the genes by forward breeding with the help of cost-effective genotyping system happens to be a daunting task.

Resistance to ascochyta blight (AB) has recessive phenotype showing complex inheritance pattern. MABC has been exploited to introgress QTLs resistant to AB and QTLs responsible for double podding simultaneously in elite chickpea cultivars through backcrossing of moderately resistant donors and adapted cultivars (Tar'an et al., 2013). For this purpose, two backcrosses following the selection for two QTLs for AB resistance and a locus associated with double podding were carried out which led to 92% recovery of recipient parent genome by BC_2F_1 except for the segments of LG4, LG6, and LG8 possessing targeted QTLs.

As AB and FW are major constraints to chickpea production, Varshney et al. (2014b) opted stepwise MABC approach for the development of FW- and AB-resistant line. An elite chickpea cultivar, C 214 was used as recipient parent in which two QTLs for AB and *foc*1 locus for FW were introgressed. Eight markers were linked to QTLs for AB resistance and six markers were linked to *foc*1 locus for foreground selection, while 40 evenly distributed SSR markers for background selection were used to obtain lines having high recurrent parent genome with targeted QTLs. Twenty-two MABC lines for FW and 14 lines for AB in BC_3F_4 generation were generated following three backcrosses and three rounds of selfing (Varshney et al., 2014a). Phenotypic screening led to the identification of three resistant lines for FW and seven resistant lines for AB. MABC approach has been utilized for introgressing resistance to two races (*foc*2 and *foc*4) individually and pyramiding of resistance to two races (*foc*1 and *foc*3) for FW and two QTLs conferring resistance to AB in chickpea (Varshney et al., 2014b). More recently, five germplasm lines showing resistance against *foc2* race have been introgressed into the background of Pusa

256 using foreground selection having SSR markers TA 37 and TA110 (Pratap et al., 2017). Similar efforts to introgress resistance to AB and FW have also been initiated by Punjab Agricultural University, Ludhiana, ICAR-Indian Agricultural Research Institute, New Delhi, and ICAR-Indian Institute of Pulse Research, Kanpur.

Apart from resistance to diseases, introgression of QTL(s)/genomic region(s) for yield under rainfed condition has also been performed. Introgression of genomic region LG04 labeled as "*QTL-hotspot*" (responsible up to 58% phenotypic variation for root traits) that maintained yield under rainfed condition was carried out into elite chickpea genotype JG 11 using MABC approach (Varshney et al., 2014b). Introgression lines thus generated after three backcrosses and two rounds of selfing (BC_3F_3) marked enhanced upheaval of 12% (under rainfed) to 24% (under irrigated) in yields. "*QTL-hotspot*" harboring several drought tolerance–related traits has been transferred in chickpea varieties as well using MABC approach (Thudi et al., 2014a,b).

6.3.8 Marker-assisted recurrent selection

The recurrent-selection method utilizing molecular markers is termed "marker-assisted recurrent selection." Marker technology identifies individual plants possessing favorable alleles within large populations due to the higher number of loci of interest and genotypes having a combination of favorable alleles present in the population. Utility of MARS in breeding programs is dependent on the totality of genetic gain achieved by selection accuracy, selection efficiency, marker-trait associations, and distribution of favorable alleles across the parents selected for crossing. MARS involves phenotyping of $F_{2:4}$ or $F_{2:5}$ progenies along with estimation of marker effects from genotyping of F_2 and F_3 population. This is followed by two or three recombination cycles on the basis of presence of marker alleles for small-effect QTLs (Eathington et al., 2007). MARS is useful for capturing large number of major as well as minor QTLs; hence the genetic gain achieved by MARS is higher than that by MABC. In MARS, the identification of QTLs is de novo in the breeding population developed from good × good crosses. Subsequently, selected hybrids having desirable alleles for maximum QTLs are crossed so that these alleles get pyramided in single genetic background. The recombinants are phenotyped for identifying promising lines for multilocation evaluation to be released as varieties.

Multinational companies have been using MARS mainly in maize and soybean, whereas only a few public sectors have been practicing MARS in crops like rice (Grenier et al., 2012), wheat (Charmet et al., 2001), and sorghum (Abdallah et al., 2009). Recently in chickpea, MARS has been exploited for pyramiding favorable set of alleles against drought stress using ICCV 05107 × ICCV 94954 and ICCV 04112 × ICCV 93954 crosses (Thudi et al., 2014a,b). The crosses JG 11 × ICCV 04112 and JG 130 × ICCV 05107 were carried out in chickpea to assemble favorable alleles for QTLs governing yield (Thudi et al., 2014a,b). About 200 F_3 plants each were genotyped in two crosses using SSR markers, and $F_{3:5}$ progenies were investigated at multilocations. QTL analysis revealed few major and several minor QTLs relating to yield and component traits. Based on the QTL(s) information on several yield-related parameters in different F_5 progenies, four lines from the cross JG 11 × ICCV 04112 and three lines from the cross JG 130 × ICCV 05107 were selected with various combinations of favorable alleles for recombination cycle. Shortlisted lines were subjected further to two recombination cycles. F_1 plants with favorable alleles of yield-related traits were identified from both the crosses, and those having favorable alleles in homozygous condition were grown. Shortlisted F_1 plants were advanced to F_4 generation for evaluation. Thus numerous cycles of MARS escalated the frequency of favorable marker alleles related with agronomically important traits.

6.3.9 Genomic-assisted breeding through MAGIC lines

Insufficient genetic diversity in chickpea is one of the constraints, hurdling the attempts for higher yield gains and resistance to various biotic and abiotic stresses. Survey regarding chickpea germplasm from "Fertile Crescent" revealed by GIS technology has initiated a wide-scale introgression breeding program to achieve wide genetic-based populations that have potential traits which will result in chickpea improvement (von Wettberg et al., 2018). Inclusion of multiple parents would serve better to predict allelic effects in diverse genetic backgrounds. Thus, development of MAGIC (multiparent advanced generation Intercrossing) population for highlighting the point regarding tapering of genetic diversity is an excellent approach. MAGIC lines detect QTLs with high accuracy and precision. Involvement of multiple parents ensures the segregation of the population for multiple QTLs associated with multiple traits. MAGIC lines can be well employed for gene discovery, gene characterization, and deployment of genes for understanding complex traits. MAGIC population can be utilized in maize to comprehensively understand the genetic architecture of complex traits (Buckler et al., 2009; Poland et al., 2011).

Development of MAGIC lines in chickpea from eight different genotypes from South Asia and Sub-Saharan Africa provides sufficient recombination and higher genetic diversity. The lines developed can directly be implemented further in breeding programs for developing cultivars suited to diverse agro-ecological zones. Some research groups have evaluated the MAGIC lines developed using multiparents at field level exposed to rainfed and irrigated treatments. As a result, lines having significantly higher yield were identified in comparison to the best parents. Parents used for these MAGIC lines were genotyped using SSRs and SNPs. Parental lines were resequenced and SNPs and InDels were identified among the parental lines by aligning the same to CDC Frontier used as a reference genome. MAGIC lines would offer an opportunity for establishment of marker-trait association using GWAS for unveiling several complex chickpea-related traits.

6.3.10 Advanced backcross QTL analysis—based breeding

Aforesaid molecular breeding approaches (MABC, MARS, and GS) play a major role toward chickpea breeding provided the superior alleles for desired trait are accessible in the breeding germplasm collection. However, the absence of genetic variability relative to a particular trait in the primary gene pool (e.g., resistance to pod borer in chickpea) may direct breeders to resort to wild relatives (secondary or higher gene pools) of the crop. Gene transfer from wild into the cultivars imposes difficulties in terms of efficient tracking for desirable and nondesirable alleles in breeding lines. AB-QTL-based breeding approach proposed by Tanksley and Nelson (1996) is convenient for introducing novel alleles from unadapted germplasm into the cultivars. In AB-QTL-based breeding approach, selected set of wild species is backcrossed to elite lines and selection is practiced in segregating BC_2F_2 or BC_2F_3 population to identify and regain individuals with resistant traits in the population. Segregating (BC_2F_2 or BC_2F_3) populations serves for generating both phenotypic and genotyping data which are further subjected to QTL analysis. AB-QTL analysis will be effective in identifying additive, dominant, partially dominant, or overdominant QTLs and QTL-associated markers and also in finding out involvement of QTL in trait improvement in the preserved progenies. Thus, AB-QTL strategy necessitates discovery and transfer of required QTLs from unadapted germplasm into shortlisted breeding lines (Tanksley and Nelson, 1996). In addition, AB-QTL strategy deferred the QTL mapping up to BC_2 or BC_3 generations to avoid any complication regarding pollen fertility and incompatibility in the prior backcross populations and to ensure utmost genome recovery from the recurrent parent.

The accuracy of QTL identification enhances with a backcross population like BC_2 or BC_3 and it acts as a suitable statistical power to ensure sufficient similarity with the recurrent parent. Identification of QTLs using near isogenic lines (NILs) population is performed in a short span of time by AB-QTL-based breeding. These NILs can serve either as parents to be used in future hybridization programs or as improved varieties and for studies relating to heterosis as well as the QTL effects can be established. AB-QTL-based breeding has been initially exercised in tomato where crosses between wild species with the elite lines were made for identification and introgression of QTLs governing fruit characteristics (Fulton et al., 2000). The resultant progenies obtained from the crosses using cultivated and wild chickpea (*C. arietinum* × *C. reticulatum*) paved the way to increased seed yield (Jaiswal et al., 1986; Singh et al., 2005). Singh and Ocampo (1997) identified transgressive segregation for agronomic traits in an F_2 population of a cross between *C. arietinum* and *C. reticulatum*. The results offered options in exploiting genetic variation available in wild species of chickpea, and hence, fresh initiatives were initiated at ICRISAT to introgress stress tolerance through AB-QTL-based breeding approach. Low genetic diversity in chickpea renders AB-QTL-based breeding of utmost importance for diversification in the primary gene pool with favorable alleles to alleviate the chances of crop improvement. Accessibility of densely saturated linkage maps ensures limiting linkage drag through stringent background selection and following the presence of nondesirable genomic region from the wild relatives of chickpea.

6.3.11 Genomic selection

Genome-wide selection (GWS) or genomic selection (GS) targets for identification of superior lines having higher breeding value on the basis of genome-wide marker profile data. Breeding values in GS approach are based on genome-wide marker data and hence referred as genomic-estimated breeding values (GEBVs). GS approach exploits two populations, i.e., training population and candidate population. Training population usually comprises breeding lines for which phenotyping data for overall performance in terms of yield and yield-related traits are available across different environments. Candidate population comprises lines that are in current use by the breeder or derived from the parental lines present in the training population. All the individuals are genotyped in training population with considerable number of markers considering linkage disequilibrium (LD) in the germplasm collection. Historical phenotyping and genotyping data are considered to develop statistical models for estimation of GEBVs of the lines.

GEBVs are calculated on the progenies of the candidate population based on marker genotyping data employing different models. Superior lines from candidate population are selected as per their GEBVs for making the next cycle of crosses. Genotyping data on candidate population train the model developed on the basis of training populations. Progenies having higher GEBVs from the screened lines of the candidate population are further genotyped with the assistance of same set of markers. GEBVs calculated on the basis of marker data using the model provide another list of suitable progenies. These progenies can be utilized for next cycle of crossing, can be selfed for field evaluation in particular set of environments, and can be advanced for multilocation field trials.

Genomic selection thus minimizes time-span and expenditure by lowering the frequency of extensive phenotyping, compresses selection cycle length of a breeding program, and bypasses the need for QTL mapping. However, implementation of appropriate statistical model is very essential for estimating the GEBVs with higher accuracy. GEBVs predicted using either best linear unbiased prediction (BLUP) or Bayesian methods are more effective statistical models according to simulation studies (Bernardo and Yu, 2007). Additionally Bayes B and wBSR (weighed Bayesian Shrinkage Regression) are also considered effective methods for genomic selection (Takeshi and Hiroyoshi, 2010).

Historical data on phenotyping of several chickpea breeding lines, existence of larger linkage disequilibrium (LD) blocks in breeding populations, and availability of genome-wide marker genotyping system like DArT and SNP markers make GS a deployable approach for chickpea in immediate future. Looking into the perspective, ICRISAT has commenced efforts to use GS in chickpea breeding program using a set of 320 elite chickpea lines genotyped by DArT markers. Phenotyping has been carried out at Patancheru and New Delhi for yield and yield-related traits. Six different statistical GS models utilizing phenotyping and genotyping data provided promising results with higher prediction accuracies (up to 0.91) for yield-related traits (Roorkiwal et al., 2016). GS considering multiple variables simultaneously assisted to obtain higher prediction accuracies through inclusion of G × E effects in chickpea breeding programs (Roorkiwal et al., 2018). Incorporation of information regarding the significant markers from GWAS results to different GS models might increase prediction accuracies (Li et al., 2018). Crossa et al. (2017) suggested that prebreeding programs will be highly profited from GS models that will help in screening the accessions for subsequent introgression.

A tentative outline of sequence-based breeding using genomic selection has been proposed by Varshney et al. (2018). According to the proposed outline firstly all possible parental lines of a particular breeding program have to be sequenced at higher depth. Founder genotypes can be sequenced to develop GWAS or available Hap Map that can further be used to select suitable parental combinations having higher frequency of desirable alleles. Large number of crosses using higher number of lines has to be made followed by early generation screening with existing 10 SNP panels for a given crop. Genomic analysis approach can be performed on selected lines from such crosses derived from the germplasm set for developing the training and candidate populations. Fixed SNP array has been considered the best genotyping platform for GS, though this may not be feasible to large-scale breeding programs. Alternatively segregating progenies at F_6/F_7 generations can be sequenced at lower coverage using skim sequencing or 384-plex-based genotyping. Sequencing data of parental lines and of other available germplasm lines can be used for developing practical haplotype graph (PHG) which can help in the selection of SNP loci. Consequently, using sequence-based approaches, these SNPs can be evaluated using rhAmp SNP genotyping technology (Integrated DNA Technology) or DArT-seq SNP genotyping technology. GS-based breeding approach can be deployed on these segregating populations and elite lines with higher GEBVs can be selected.

6.4 Speed breeding

Global food security for ever-growing human population necessitates accelerated breeding and research programs so as to meet future agro-based demands. Pace between time required and development of improved varieties has to be optimum to meet breeding challenges. Longer generation time required by crops slows down the progression toward fast-track research and development of improved varieties. Watson et al. (2018) present a method called "Speed breeding" which significantly shortens the generation time of crops. Speed breeding is a promising breeding technique to achieve up to 6 generations of chickpea per year compared to two to three generations under normal glasshouse conditions. Speed breeding utilizes fully enclosed, controlled-environment growth chambers for hastened plant development. Supplemented lighting through LED in glasshouse environments promotes rapid generation advancement with enhanced photoperiod. A photoperiod of 22 h with controlled temperature regime has been demonstrated to produce 4.5 generations per year in chickpea (Watson et al., 2018) though the results are cultivar dependent.

Prolonged light exposure leads to photosynthesis overdrive and prompts plant growth. Quality of light, photoperiod duration, and temperature are other parameters influencing generation time.

Speed breeding being an indoor plant cultivation technique can be well adapted for large-scale crop improvement programs. Speed breeding can be easily integrated with other modern chickpea breeding technologies, such as genome editing, genomic selection, mutant analysis, transformation, trait phenotyping, and high-throughput genotyping. Plant density can also be manipulated to obtain homozygous plant populations for single seed descent programs in shorter span. Genomic-assisted breeding amplifies genetic gains by favorable gene action, enhancing diversity, heritability, and intensity of selection (Moose and Mumm, 2008). Genetic gains can further be improved through speed breeding by narrowing of the breeding cycles and quick generation advancement by means of controlled photoperiod, temperature, germination of immature seeds, humidity, and harvesting (Watson et al., 2018; Chiurugwi et al., 2018). Phenotyping protocols adapted to hasten breeding strategies have been developed to enable characterization and selection of agronomically important traits. Phenotyping and selection can be well integrated for single as well as multiple traits in the same plant generation grown under speed breeding conditions. In durum wheat, phenotyping method for characterization of fixed lines out of season and providing selection pressure during early generations has been designed and implemented for novel multitraits (Alahmad et al., 2018). Large F_2 population was selected for multiple traits to evaluate the shift in phenotypic response and to calculate the chances to hasten pyramiding of multiple target traits. Being time and resource saving, speed breeding will accelerate agricultural research, improve stability, and increase chickpea production to meet the demands of rising population.

6.5 Conclusion

In chickpea, breeding for multiple traits including yield and its component traits, quality, and resistance/tolerance to various abiotic and biotic stresses is of paramount importance in order to develop cultivars that meet consumer's demands and thrive well under variable climatic and farming conditions. Integrating multiple traits into single genetic background through conventional breeding approaches is feasible but challenging. Chickpea breeders must prioritize target traits for optimized use of resources, cost, and time to breed for multiple traits. Molecular breeding technologies including marker-assisted selection, MABC, AB-QTL strategy, gene stacking, gene pyramiding, and genomic selection can greatly supplement conventional breeding to gain momentum in multiple trait breeding and gene integration. Recent advancement and forthcoming molecular breeding approaches present immense opportunities in chickpea. Genomic selection, a future molecular breeding tool, has great potential in chickpea breeding to increase the efficiency of selection and improve multiple traits simultaneously. Chickpea, a previously considered orphan crop, has metamorphosed into a genomic resource–rich crop. In summary, chickpea improvement is heading toward integration of modern genomics approach with existing breeding programs, leading to resultant enhanced yield.

References

Abdallah, A.A., Ali, A.M., Geiger, H.H., Parzies, H.K., 2009. Marker-assisted recurrent selection for increased out crossing in Caudatum–race Sorghum. In: Proceedings of the International Conference on Applied Biotechnology, 28–30 September 2009, Khartoum, Sudan.

Acharjee, S., Sarmah, B.K., Kumar, P.A., Olsen, K., Mahon, R., Moar, W.J., et al., 2010. Transgenic chickpeas (*Cicer arietinum* L.) expressing a sequence-modified *cry2Aa* gene. Plant Sci. 178, 333–339.

Afonso-Grunz, F., Molina, C., Hoffmeier, K., Rycak, L., Kudapa, H., et al., 2014. Genome-based analysis of the transcriptome from mature chickpea root nodules. Front. Plant Sci. 5, 325.

Ahmad, F., Slinkard, A.E., 1992. Genetic relationship in the genus *Cicer* L. as revealed by polyacrylamide gel electrophoresis of seed storage protein. Theor. Appl. Genet. 84, 688–692.

Alahmad, S., Dinglasan, E., Leung, K.M., Riaz, A., Derbal, N., Voss-Fels, K.P., et al., 2018. Speed breeding for multiple quantitative traits in durum wheat. Plant Methods 14, 36.

Anbessa, Y., Taran, B., Warkentin, T.D., Tullu, A., Vandenberg, A., 2009. Genetic analyses and conservation of QTL for ascochyta blight resistance in chickpea (*Cicer arietinum* L.). Theor. Appl. Genet. 119, 757–765.

Anuradha, C., Gaur, P.M., Pande, S., Gali, K.K., Ganesh, M., Kumar, J., Varshney, R.K., 2011. Mapping QTL for resistance to botrytis grey mould in chickpea. Euphytica 182, 1–9.

Bajaj, D., Das, S., Badoni, S., et al., 2015. Genome-wide high-throughput SNP discovery and genotyping for understanding natural (functional) allelic diversity and domestication patterns in wild chickpea. Sci. Rep. 5, 12468.

Banerjee, H., Pai, R.A., Sharma, R.P., 1999. Restriction fragment length polymorphism and random amplified polymorphic DNA analysis of chickpea accessions. Biol. Plant. 42, 197–208.

Bernardo, R., Yu, J., 2007. Prospects for genomewide selection for quantitative traits in maize. Crop Sci. 47 (3), 1082–1090.

Bharadwaj, C., Srivastava, R., Chauhan, S.K., Satyavathi, C.T., Kumar, J., Faruqui, A., Yadav, S., Rizvi, A.H., Kumar, T., 2011. Molecular diversity and phylogeny in geographical collection of chickpea (*Cicer* sp.) accessions. J. Genet. 90, 94–100.

Buckler, E.S., Holland, J.B., Bradbury, P.J., Acharya, C.B., Brown, P.J., et al., 2009. The genetic architecture of maize flowering time. Science 325, 714–718.

Buhariwalla, H.K., Jayashree, B., Eshwar, K., et al., 2005. Development of ESTs from chickpea roots and their use in diversity analysis of the *Cicer* genus. BMC Plant Biol. 5, 16.

Charmet, G., Robert, N., Perretant, M.R., Gay, G., Sourdille, P., Groos, C., Bernard, S., Bernard, M., 2001. Marker assisted recurrent selection for cumulating QTLs for bread-making related traits. Euphytica 1 (19), 89–93.

Chiurugwi, T., Kemp, S., Powell, W., Hickey, L.T., 2018. Speed breeding orphan crops. Theor. Appl. Genet. 132: 607-616. https://doi.org/10.1007/s00122-018-3202-7.

Cho, S., Chen, W., Muehlbauer, F.J., 2004. Pathotype-specific genetic factors in chickpea (*Cicer arietinum* L.) for quantitative resistance to ascochyta blight. Theoretical and Applied Genetics 109, 733–739.

Cho, S., Kumar, J., Shultz, J.L., Anupama, K., Tefera, F., Muehlbauer, F.J., 2002. Mapping genes for double podding and other morphological traits in chickpea. Euphytica 128 (2), 285–292.

Choudhary, S., Gaur, R., Gupta, S., Bhatia, S., 2012. EST-derived genic molecular markers: development and utilization for generating an advanced transcript map of chickpea. Theor. Appl. Genet. 124, 1449–1462.

Cobos, M.J., Fernández, M.J., Rubio, J., Kharrat, M., Moreno, M.T., et al., 2005. A linkage map in chickpea (*Cicer arietinum* L.) in two populations from Kabuli x Desi crosses: location of a resistance gene for fusarium wilt race 0. Theor. Appl. Genet. 110, 1347–1353.

Cobos, M.J., Rubio, J., Strange, R.N., Moreno, M.T., Gil, J., Millan, T., 2006. A new QTL for ascochyta blight resistance in an RIL population derived from an interspecific cross in chickpea. Euphytica 149 (1–2), 105–111.

Cobos, M.J., Winter, P., Kharrat, M., Cubero, J.I., Gil, J., Millan, T., Rubio, J., 2009. Genetic analysis of agronomic traits in a wide cross of chickpea. Field Crop Res. 111 (1–2), 130–136.

Crossa, J., Pérez-Rodríguez, P., Cuevas, J., Montesinos-López, O., Jarquín, D., et al., 2017. Genomic selection in plant breeding: methods, models, and perspectives. Trends Plant Sci. 22, 961–975.

Deokar, A., Sagi, M., Daba, K., Tar'an, B., 2019. QTL sequencing strategy to map genomic regions associated with resistance to ascochyta blight in chickpea. Plant Biotechnol. J. 17 (1), 275–288.

Deokar, A.A., Kondawar, V., Jain, P.K., Karuppayil, M., Raju, N.L., Vadez, V., et al., 2011. Comparative analysis of expressed sequence tags (ESTs) between drought-tolerant and -susceptible genotypes of chickpea under terminal drought stress. BMC Plant Biol. 11, 70.

Deokar, A.A., Ramsay, L., Sharpe, A.G., Diapari, M., Sindhu, A., Bett, K., et al., 2014. Genome wide SNP identification in chickpea for use in development of a high density genetic map and improvement of chickpea reference genome assembly. BMC Genomics 15, 708.

Dixit, G.P., 2018. Project Coordinator's Report of All India Coordinated Research Project on Chickpea. Indian Institute of Pulses Research, Kanpur.

Eathington, S.R., Crosbie, T.M., Edwards, M.D., Reiter, R.D., Bull, J.K., 2007. Molecular markers in a commercial breeding program. Crop Sci. 47, 154–163.

FAOSTAT, 2017. http://www.fao.org/faostat/.

Flandez-Galvez, H., Ades, P.K., Ford, R., Pang, E.C.K., Taylor, P.W.J., 2003. QTL analysis for ascochyta blight resistance in an intraspecific population of chickpea (*Cicer arietinum* L.). Theor. Appl. Genet. 107, 1257–1265.

Fulton, T.M., Grandillo, S., Beck-Bunn, T., Fridman, E., Frampton, A., Lopez, J., et al., 2000. Advanced backcross QTL analysis of a *Lycopersicon esculentum* × *Lycopersicon parviflorum* cross. Theor. Appl. Genet. 100, 1025–1042.

Garg, R., Patel, R.K., Jhanwar, S., Priya, P., Bhattacharjee, A., et al., 2011b. Gene discovery and tissue-specific transcriptome analysis in chickpea with massively parallel pyrosequencing and web resource development. Plant Physiol. 156, 1661–1678.

Garg, R., Patel, R.K., Tyagi, A.K., Jain, M., 2011a. De novo assembly of chickpea transcriptome using short reads for gene discovery and marker identification. DNA Res. 18, 53–63.

Garg, T., Mallikarjuna, B.P., Thudi, M., Samineni, S., Singh, S., Sandhu, J.S., et al., 2018. Identification of QTLs for resistance to Fusarium wilt and Ascochyta blight in a recombinant inbred population of chickpea (*Cicer arietinum* L.). Euphytica 214 (3). https://doi.org/10.1007/s10681-018-2125-3.

Gaur, P.M., Slinkard, A.E., 1990a. Genetic control and linkage relations of additional isozyme markers in chickpea. Theor. Appl. Genet. 80, 648–656.

Gaur, P.M., Stinkard, A.E., 1990b. Inheritance and linkage of isozyme coding genes in chickpea. J. Hered. 81, 455–461.

Gaur, R., Jeena, G., Shah, N., Gupta, S., Pradhan, S., et al., 2015. High density linkage mapping of genomic and transcriptomic SNPs for synteny analysis and anchoring the genome sequence of chickpea. Sci. Rep. 5, 13387.

Gaur, R., Sethy, N.K., Choudhary, S., Shokeen, B., Gupta, V., Bhatia, S., 2011. Advancing the STMS genomic resources for defining new locations on the intra-specific genetic linkage map of chickpea (*Cicer arietinum* L.). BMC Genomics 12, 117.

Gowda, S.J.M., Radhika, P., Kadoo, N.Y., Mhase, L.B., Gupta, V.S., 2009. Molecular mapping of wilt resistance genes in chickpea. Mol. Breed. 24, 177–183.

Grenier, C., Chatel, M.H., Ospina, Y., Cao, T., Guimaraes, E.P., Martinez, C.P., et al., 2012. Population improvement through recurrent selection in rice. Prospects for maker assisted recurrent selection and genome-wide selection. In: 'Plant and Animal Genome XX Conference', 14–18 January 2014, San Diego.

Gujaria, N., Kumar, A., Dauthal, P., et al., 2011. Development and use of genic molecular markers (GMMs) for construction of a transcript map of chickpea (*Cicer arietinum* L.). Theor. Appl. Genet. 122, 1577–1589.

Gupta, S., Nawaz, K., Parween, S., Roy, R., Sahu, K., et al., 2016. Draft genome sequence of *Cicer reticulatum* L., the wild progenitor of chickpea provides a resource for agronomic trait improvement. DNA Res. 24 (1), 1–10. https://doi.org/10.1093/dnares/dsw042.

Hamwieh, A., Imtiaz, M., Malhotra, R.S., 2013. Multi-environment QTL analyses for drought-related traits in a recombinant inbred population of chickpea (*Cicer arientinum* L.). Theor. Appl. Genet. 126, 1025–1038.

Hiremath, P.J., Farmer, A., Cannon, S.B., Woodward, J., Kudapa, H., Tuteja, R., et al., 2011. Large-scale transcriptome analysis in chickpea (*Cicer arietinum* L.), an orphan legume crop of the semi-arid tropics of Asia and Africa. Plant Biotechnol. J. 9 (8), 922–931.

Hiremath, P.J., Kumar, A., Penmetsa, R.V., et al., 2012. Large-scale development of cost-effective SNP marker assays for diversity assessment and genetic mapping in chickpea and comparative mapping in legumes. Plant Biotechnol. J. 10, 716–732.

Hospital, F., 2005. Selection in backcross programmes. Philos. Trans. R. Soc. Lond. B Biol. Sci. 360 (1459), 1503–1511.

Hüttel, B., Winter, P., Weising, K., Choumane, W., Weigand, F., Kahl, G., 1999. Sequence-tagged microsatellite site markers for chickpea (*Cicer arietinum* L.). Genome 42, 210–217.

Iruela, M., Castro, P., Rubio, J., Cubero, J.I., Jacinto, C., Millán, T., Gil, J., 2007. Validation of a QTL for resistance to *ascochyta* blight linked to resistance to *fusarium* wilt race 5 in chickpea (*Cicer arietinum* L.). Eur. J. Plant Pathol. 119, 20–37.

Iruela, M., Rubio, J., Cubero, J.I., Gil, J., et al., 2002. Phylogenetic analysis in the genus Cicer and cultivated chickpea using RAPD and ISSR markers. Theor. Appl. Genet. 104, 643–651.

Jaganathan, D., Thudi, M., Kale, S., Azam, S., Roorkiwal, M., Gaur, P.M., Kishor, P.B., 2015. Genotyping-by-sequencing based intra-specific genetic map refines a "*QTL-hotspot*" region for drought tolerance in chickpea. Mol. Genet. Genomics 290, 559–571.

Jaiswal, H.K., Singh, B.D., Singh, A.K., Singh, R.M., 1986. Introgression of genes for yield and yield traits from C. reticulatum x C. arietinum. Int. Chickpea Newsl. 14, 5–8.

Jain, D., Chattopadhyay, D., 2010. Analysis of gene expression in response to water deficit of chickpea (*Cicer arietinum* L.) varieties differing in drought tolerance. BMC Plant Biol. 10, 24.

Jain, M., Misra, G., Patel, R.K., et al., 2013. A draft genome sequence of the pulse crop chickpea (*Cicer arietinum* L.). Plant J. 74 (5), 715–729.

Jamalabadi, J.G., Saidi, A., Karami, E., Kharkesh, M., Talebi, R., 2013. Molecular mapping and characterization of genes governing time to flowering, seed weight, and plant height in an intraspecific genetic linkage map of chickpea (*Cicer arietinum* L.). Biochem. Genet. 51 (5–6), 387–397.

Jannink, J.L., Lorenz, A.J., Iwata, H., 2010. Genomic selection in plant breeding: from theory to practice. Briefings Funct. Genomics 9, 166–177.

Jingade, P., Ravikumar, R.L., 2015. Development of molecular map and identification of QTLs linked to Fusarium wilt resistance in chickpea. J. Genet. 94 (4), 723–729.

Kahl, G., Molina, C., Udupa, S.M., 2007. Super SAGE: exploring the stress transcriptome in chickpea. In: 'Plant and Animal Genome XV Conference,' 13–17 January 2007, San Diego.

Kale, S.M., Jaganathan, D., Ruperao, P., Chen, C., Punna, R., Kudapa, H., et al., 2015. Prioritization of candidate genes in '*QTL-hotspot*' region for drought tolerance in chickpea (*Cicer arietinum* L.). Sci. Rep. 5, 15296.

Karami, E., Talebi, R., Kharkesh, M., Saidi, A., 2015. A linkage map of chickpea (*Cicer arietinum* L.) based on population from ILC3279 X ILC588 crosses: location of genes for time to flowering, seed size and plant height. Genetika 47, 253–263.

Kazan, K., Muehlbauer, F.J., 1991. Allozyme variation and phylogeny in annual species of *Cicer* (Leguminosae). Plant Syst. Evol. 175, 11–21.

Kazan, K.M.F.J., Muehlbauer, F.J., Weeden, N.E., Ladizinsky, G., 1993. Inheritance and linkage relationships of morphological and isozyme loci in chickpea (*Cicer arietinum* L.). Theor. Appl. Genet. 86 (4), 417–426.

Khajuria, Y.P., Saxena, M.S., Gaur, R., Chattopadhyay, D., Jain, M., Parida, S.K., Bhatia, S., 2015. Development and integration of genome-wide polymorphic microsatellite markers onto a reference linkage map for constructing a high-density genetic map of chickpea. PLoS One 10 (5), e0125583.

Kottapalli, P., Gaur, P.M., Katiyar, S.K., Crouch, J.H., Buhariwalla, H.K., et al., 2009. Mapping and validation of QTLs for resistance to an Indian isolate of Ascochyta blight pathogen in chickpea. Euphytica 165, 79–88.

Kudapa, H., Ramalingam, A., Nayakoti, S., Chen, X., Zhuang, W.J., Liang, X., Kahl, G., Edwards, D., Varshney, R.K., 2013. Functional genomics to study stress responses in crop legumes: progress and prospects. Funct. Plant Biol. 40, 1221–1233.

Kudapa, K., Azam, S., Sharpe, A.G., Taran, B., Li, R., et al., 2014. Comprehensive transcriptome assembly of chickpea (*Cicer arietinum* L.) using Sanger and next generation sequencing platforms: development and applications. PLoS One 9, e86039.

Kujur, A., Bajaj, D., Upadhyaya, H.D., et al., 2015. Employing genome-wide SNP discovery and genotyping strategy to extrapolate the natural allelic diversity and domestication patterns in chickpea. Front. Plant Sci. 6, 162.

Labdi, M., Robertson, L.D., Singh, K.B., Charrier, A., 1996. Genetic diversity and phylogenetic relationships among the annual Cicer species as revealed by isozyme polymorphism. Euphytica 88, 181–188.

Ladizinsky, G., 1975. A new *Cicer* from Turkey. Notes R. Bot. Gard. Edinb. 34, 201–202.
Li, Y., Ruperao, P., Batley, J., Edwards, D., Khan, T., Colmer, T.D., et al., 2018. Investigating drought tolerance in chickpea using genome-wide association mapping and genomic selection based on whole-genome resequencing data. Front. Plant Sci. 9, 190.
Lichtenzveig, J., Bonfil, D.J., Zhang, H.B., Shtienberg, D., Abbo, S., 2006. Mapping quantitative trait loci in chickpea associated with time to flowering and resistance to *Didymella rabiei* the causal agent of ascochyta blight. Theor. Appl. Genet. 113 (7), 1357–1369.
Lichtenzveig, J., Scheuring, C., Dodge, J., Abbo, S., Zhang, H.B., 2005. Construction of BAC and BIBAC libraries and their applications for generation of SSR markers for genome analysis of chickpea, *Cicer arietinum* L. Theor. Appl. Genet. 110, 492–510.
Madrid, E., Chen, W., Rajesh, P.N., Castro, P., Millán, T., Gil, J., 2013. Allele-specific amplification for the detection of ascochyta blight resistance in chickpea. Euphytica 189, 183–190.
Madrid, E., Rubiales, D., Moral, A., Moreno, M.T., Millán, T., Gil, J., Rubio, J., 2008. Mechanism and molecular markers associated with rust resistance in a chickpea interspecific cross (*Cicerarietinum* × *Cicer reticulatum*). Eur. J. Plant Pathol. 121, 43–53.
Madrid, E., Seoane, P., Claros, M.G., Barro, F., Rubio, J., Gil, J., Millan, T., 2014. Genetic and physical mapping of the QTLAR3 controlling blight resistance in chickpea (*Cicer arietinum* L). Euphytica 198, 69–78.
Mantri, N.L., Ford, R., Coram, T.E., Pang, E.C., 2007. Transcriptional profiling of chickpea genes differentially regulated in response to high-salinity, cold and drought. BMC Genomics 8, 303.
Matsumura, H., Ito, A., Saitoh, H., Winter, P., Kahl, G., Reuter, M., Kruger, D.H., Terauchi, R., 2005. Super SAGE. Cell Microbiol. 7, 11–18.
Mayer, M.S., Tullu, A., Simon, C.J., Kumar, J., Kaiser, W.J., Kraft, J.M., Muehlbauer, F.J., 1997. Development of a DNA marker for fusarium wilt resistance in chickpea. Crop Sci. 37, 1625–1629.
Millán, T., Madrid, E., Imtiaz, M., Kharrat, M., Chen, W., 2013. Disease Resistance in Chickpea. Translational Genomics for Crop Breeding. Wiley, Hoboken.
Millàn, T., Winter, P., Jungling, R., Gil, J., Rubio, J., Cho, S., et al., 2010. A consensus genetic map of chickpea (*Cicer arietinum* L.) based on 10 mapping populations. Euphytica 175, 175–189.
Molina, C., Rotter, B., Horres, R., Udupa, S.M., Besser, B., Bellarmino, L., Baum, M., Matsumura, H., Terauchi, R., Kahl, G., Winter, P., 2008. Super SAGE: the drought stress-responsive transcriptome of chickpea roots. BMC Genomics 9, 553.
Molina, C., Zaman-Allah, M., Khan, F., Fatnassi, N., Horres, R., Rotter, B., et al., 2011. The salt-responsive transcriptome of chickpea roots and nodules via deepSuperSAGE. BMC Plant Biol. 11, 31.
Moose, S.P., Mumm, R.H., 2008. Molecular plant breeding as the foundation for 21st century crop improvement. Plant Physiol. 147, 969–977.
Nayak, S.N., Zhu, H., Varghese, N., et al., 2010. Integration of novel SSR and gene-based SNP marker loci in the chickpea genetic map and establishment of new anchor points with *Medicago truncatula* genome. Theor. Appl. Genet. 120, 1415–1441.
Nguyen, T.T., Taylor, P.W.J., Redden, R.J., Ford, R., 2004. Genetic diversity estimates in *Cicer* using AFLP analysis. Plant Breed. 123, 173–179.
Pandey, A., Chakraborty, S., Datta, A., Chakraborty, N., 2008. Proteomics approach to identify dehydration responsive nuclear proteins from chickpea (*Cicer arietinum* L.). Mol. Cell. Proteom. 7, 88–107.

Pandey, M.K., Roorkiwal, M., Singh, V.K., Ramalingam, A., Kudapa, H., et al., 2016. Emerging genomic tools for legume breeding: current status and future prospects. Front. Plant Sci. 7, 455.

Parween, S., Nawaz, K., Roy, R., Pole, A.K., Venkata, Suresh, B., Misra, G., et al., 2015. An advanced draft genome assembly of a desi type chickpea (*Cicer arietinum* L.). Sci. Rep. 5, 12806.

Paul, P.J., Samineni, S., Sajja, S.B., Rathore, A., Das, R.R., Chaturvedi, S.K., et al., 2018a. Capturing genetic variability and selection of traits for heat tolerance in a chickpea recombinant inbred line (RIL) population under field conditions. Euphytica 214, 27.

Paul, P.J., Samineni, S., Thudi, M., Sajja, S.B., Rathore, A., Das, R.R., et al., 2018b. Molecular mapping of QTLs for heat tolerance in chickpea. Int. J. Mol. Sci. 19 (8), E2166.

Peng, H., Yu, X., Cheng, H., Shi, Q., Zhang, H., Li, J., Ma, H., 2010. Cloning and characterization of a novel NAC family gene *CarNAC1* from chickpea (*Cicer arietinum* L.). Mol. Biotechnol. 44, 30–40.

Pfaff, T., Kahl, G., 2003. Mapping of gene-specific markers on the genetic map of chickpea (*Cicer arietinum* L.). Mol. Genet. Genomics 269, 243–251.

Poland, J.A., Bradbury, P.J., Buckler, E.S., Nelson, R.J., 2011. Genome-wide nested association mapping of quantitative resistance to northern leaf blight in maize. Proc. Natl. Acad. Sci. U. S. A. 108, 6893–6898.

Pratap, A., Chaturvedi, S.K., Tomar, R., Rajan, N., Malviya, N., Thudi, M., Saabale, P.R., Prajapati, U.,, Varshney, R.K., Singh, N.P., 2017. Marker assisted introgression of resistance to fusarium wilt race 2 in Pusa 256, an elite cultivar of *desi* chickpea. Molecular Genetics and Genomics 292, 1237–1245.

Pushpavalli, R., Krishnamurthy, L., Thudi, M., Gaur, P.M., Rao, M.V., et al., 2015. Two key genomic regions harbour QTLs for salinity tolerance in ICCV 2 × JG 11 derived chickpea (*Cicer arietinum* L.) recombinant inbred lines. BMC Plant Biol. 15, 124.

Radhika, P., Gowda, S.J.M., Kadoo, N.Y., Mhase, L.B., Jamadagni, B.M., Sainani, M.N., Chandra, S., Gupta, V.S., 2007. Development of an integrated intraspecific map of chickpea (*Cicer arietinum* L.) using two recombinant inbred line populations. Theor. Appl. Genet. 115, 209–216.

Ratnaparkhe, M., Santra, D.K., Tullu, A., Muehlbauer, F.J., 1998. Inheritance of inter simple sequence repeat polymorphisms and linkage with a fusarium wilt resistance gene in chickpea. Theor. Appl. Genet. 96, 348–353.

Rehman, A.U., Malhotra, R.S., Bett, K., Tar'an, B., Bueckert, R., Warkentin, T.D., 2011. Mapping QTL associated with traits affecting grain yield in Chickpea (*Cicer arietinum* L.) under terminal drought stress. Crop Sci. 51 (2), 450–463.

Roorkiwal, M., Jain, A., Kale, S.M., Doddamani, D., Chitikineni, A., et al., 2017. Development and evaluation of high density SNP array (Axiom®CicerSNP Array) for high resolution genetic mapping and breeding applications in chickpea. Plant Biotechnol. J. 16 (4), 890–901. https://doi.org/10.1111/pbi.12836.

Roorkiwal, M., Jain, A., Thudi, M., Varshney, R.K., 2018. Advances in chickpea genomic resources for accelerating the crop improvement. In: The Chickpea Genome. Springer International Publishing, Cham, Switzerland, pp. 53–67.

Roorkiwal, M., Nayak, S.N., Thudi, M., Upadhyaya, H.D., Brunel, D., et al., 2014a. Allele diversity for abiotic stress responsive candidate genes in chickpea reference set using gene based SNP markers. Front. Plant Sci. 5, 248.

Roorkiwal, M., Rathore, A., Das, R.R., Singh, M.K., Jain, A., Srinivasan, S., et al., 2016. Genome-enabled prediction models for yield related traits in chickpea. Front. Plant Sci. 7, 1666.

Roorkiwal, M., Sawargaonkar, S.L., Chitikineni, A., Thudi, M., Saxena, R.K., et al., 2013. Single nucleotide polymorphism genotyping for breeding and genetics applications in chickpea and pigeonpea using the BeadXpress platform. Plant Genome 6, 2.

Roorkiwal, M., Von Wettberg, E.J., Upadhyaya, H.D., Warschefsky, E., Rathore, A., et al., 2014b. Exploring germplasm diversity to understand the domestication process in *Cicer* spp. using SNP and DArT markers. PLoS One 9, e102016.

Ruperao, P., Chan, C.K.K., Azam, S., Karafiatova, M., Hayashi, S., Cizkova, J., et al., 2014. A chromosomal genomics approach to assess and validate the desi and kabuli draft chickpea genome assemblies. Plant Biotechnol. J. 12, 778–786.

Sabbavarapu, M.M., Sharma, M., Chamarthi, S.K., Swapna, N., Rathore, A., et al., 2013. Molecular mapping of QTLs for resistance to Fusarium wilt (*race 1*) and Ascochyta blight in chickpea (*Cicer arietinum* L.). Euphytica 193, 121–133.

Saminemi, S., 2010. Physiology, Genetics and Molecular Mapping of Salt Tolerance in Chickpea (*Cicer arietinum* L) (Ph D thesis). The University of Western Australia.

Sant, V.J., Patankar, A.G., Sarode, N.D., Mhase, L.B., Sainani, M.N., et al., 1999. Potential of DNA markers in detecting divergence and in analysing heterosis in Indian elite chickpea cultivars. Theor. Appl. Genet. 98, 1217–1225.

Santra, D.K., Tekeoglu, M., Ratnaparkhe, M., Kaiser, W.J., Muehlbauer, F.J., 2000. Identification and mapping of QTLs conferring resistance to ascochyta blight in chickpea. Crop Sci. 40, 1606–1612.

Saxena, M.S., Bajaj, D., Das, S., Kujur, A., Kumar, V., Singh, M., et al., 2014. An integrated genomic approach for rapid delineation of candidate genes regulating agro-morphological traits in chickpea. DNA Res. 21, 695–710.

Serret, M.D., Udupa, S.M., Weigand, F., 1997. Assessment of genetic diversity of cultivated chickpea using microsatellite-derived RFLP markers: implications for origin. Plant Breed. 116, 573–578.

Shan, F., Clarke, H.C., Plummer, J.A., Yan, G., Siddique, K.H.M., 2005. Geographical patterns of genetic variation in the world collections of wild annual *Cicer* characterized by amplified fragment length polymorphisms. Theor. Appl. Genet. 110, 381–391.

Sharma, K.D., Chen, W., Muehlbauer, F.J., 2005. Genetics of chickpea resistance to five races of Fusarium wilt and a concise set of race differentials for *Fusarium oxysporum* f. sp. *Ciceris*. Plant Dis. 89, 385–390.

Sharma, K.D., Winter, P., Kahl, G., Muehlbauer, F., 2004. Molecular mapping of *Fusarium oxysporum* f. sp. *cicerisrace 3* resistance gene in chickpea. Theor. Appl. Genet. 108, 1243–1248.

Shukla, R.K., Tripathi, V., Jain, D., Yadav, R.K., Chattopadhyay, D., 2009. CAP2 enhances germination of transgenic tobacco seeds at high temperature and promotes heat stress tolerance in yeast. FEBS J. 276, 5252–5262.

Simon, C.J., Muehlbauer, F.J., 1997. Construction of a chickpea linkage map and its comparison with maps of pea and lentil. J. Hered. 88, 115–119.

Singh, B.K., Ocampo, B., 1997. Exploitation of wild *Cicer* species for yield improvement in chickpea. Theor. Appl. Genet. 95, 418–423.

Singh, R., Prasad, C.D., Singhal, V., Randhawa, G.J., 2003. Assessment of genetic diversity in chickpea cultivars using RAPD, AFLP and STMS markers. J. Genet. Breed. 57, 165–174.

Singh, S., Gumber, R.K., Joshi, N., Singh, K., 2005. Introgression from wild *Cicer reticulatum* to cultivated chickpea for productivity and disease resistance. Plant Breed. 124, 477–480.

Singh, V.K., Garg, R., Jain, M., 2013. A global view of transcriptome dynamics during flower development in chickpea by deep sequencing. Plant Biotechnol. 11, 691–701.

Srivastava, R., Singh, M., Bajaj, D., Parida, S.K., 2016. A high-resolution InDel (Insertion–Deletion) markers-anchored consensus genetic map identifies major QTLs governing pod number and seed yield in chickpea. Front. Plant Sci. 7, 1362.

Takeshi, H., Hiroyoshi, I., 2010. EM algorithm for Bayesian estimation of genomic breeding values. BMC Genet. 11, 3.

Talebi, R., Naji, A.M., Fayaz, F., 2008. Geographical patterns of genetic diversity in cultivated chickpea (*Cicer arietinum* L.) characterized by amplified fragment length polymorphism. Plant Soil Environ. 54, 447–452.

Tanksley, S.D., 1993. Mapping polygenes. Annu. Rev. Genet. 27, 205–233.

Tanksley, S.D., Nelson, J.C., 1996. Advanced backcross QTL analysis: a method for simultaneous discovery and transfer of valuable QTL from unadapted germplasm into elite breeding lines. Theor. Appl. Genet. 92, 191–203.

Tar'an, B., Warkentin, T.D., Tullu, A., Vandenberg, A., 2007. Genetic mapping of ascochyta blight resistance in chickpea (*Cicer arietinum* L.) using a simple sequence repeat linkage map. Genome 50, 26–34.

Tar'an, B., Warkentin, T.D., Vandenberg, A., 2013. Fast track genetic improvement of ascochyta blight resistance and double podding in chickpea by marker-assisted backcrossing. Theor. Appl. Genet. 126, 1639–1647.

Tayyar, R.I., Waines, J.G., 1996. Genetic relationships among annual species of *Cicer* (Fabaceae) using isozyme variation. Theor. Appl. Genet. 92, 245–254.

Tekeoglu, M., Rajesh, P.N., Muehlbauer, F.J., 2002. Integration of sequence tagged microsatellite sites to the chickpea genetic map. Theor. Appl. Genet. 105 (6–7), 847–854.

Thudi, M., Bohra, A., Nayak, S.N., Varghese, N., Shah, T.M., et al., 2011. Novel SSR markers from BAC-end sequences, DArT arrays and a comprehensive genetic map with 1,291 marker loci for chickpea (*Cicer arietinum* L.). PLoS One 6, e27275.

Thudi, M., Gaur, P.M., Krishnamurthy, L., Mir, R.R., Kudapa, H., Fikre, A., et al., 2014a. Genomic assisted breeding for drought tolerance in chickpea. Funct. Plant Biol. 41, 1178–1190.

Thudi, M., Khan, A.W., Kumar, V., Gaur, P.M., Katta, A.V.S.K., Garg, V., et al., 2016. Whole genome re-sequencing reveals genome wide variations among parental lines of mapping populations in chickpea (*Cicer arietinum* L.). BMC Pant Biol. 16 (1), 10.

Thudi, M., Upadhyaya, H.D., Rathore, A., Gaur, P.M., Krishnamurthy, L., et al., 2014b. Genetic dissection of drought and heat tolerance in chickpea through genome-wide and candidate gene-based association mapping approaches. PLoS One 9, e96758.

Tullu, A., Kaiser, W.J., Kraft, J.M., Muehlbauer, F.J., 1999. A second gene for resistance to *race 4* of Fusarium wilt in chickpea and linkage with a RAPD marker. Euphytica 109, 43–50.

Tullu, A., Muehlbauer, F.J., Simon, C.J., Mayer, M.S., Kumar, J., et al., 1998. Inheritance and linkage of a gene for resistance to *race 4* of fusarium wilt and RAPD markers in chickpea. Euphytica 102, 227–232.

Udupa, S.M., Baum, M., 2003. Genetic dissection of pathotype-specific resistance to ascochyta blight disease in chickpea (*Cicer arietinum* L.) using microsatellite markers. Theor. Appl. Genet. 106, 1196–1202.

Udupa, S.M., Sharma, A., Sharma, R.P., Pai, R.A., 1993. Narrow genetic variability in *Cicer arietinum* L. as revealed by RFLP analysis. J. Plant Biochem. Biotechnol. 2, 83–86.
Upadhyaya, H.D., Bramel, P.J., Singh, S., 2001. Development of a chickpea core subset using geographic distribution and quantitative traits. Crop Sci. 41, 206–210.
Upadhyaya, H.D., Dwivedi, S.L., Baum, M., Varshney, R.K., Udupa, S.M., Gowda, C.L.L., et al., 2008. Genetic structure, diversity, and allelic richness in composite collection and reference set in chickpea (*Cicer arietinum* L.). BMC Plant Biol. 8, 106.
Upadhyaya, H.D., Dwivedi, S.L., Sharma, S., 2018. Managing and discovering agronomically beneficial traits in chickpea germplasm collections. In: The Chickpea Genome, vols. 43–52. Springer International Publishing, Cham, Switzerland.
Upadhyaya, H.D., Ortiz, R., 2001. A mini core subset for capturing diversity and promoting utilization of chickpea genetic resources in crop improvement. Theor. Appl. Genet. 102, 1292–1298.
Vadez, V., Krishnamurthy, L., Thudi, M., Anuradha, C., Colmer, T., Turner, N.C., et al., 2012. Assessment of ICCV 2 × JG 62 chickpea progenies shows sensitivity of reproduction to salt stress and reveals QTL for seed yield and yield components. Mol. Breed. 30 (1), 9–21.
Varshney, R.K., 2016. Exciting journey of 10 years from genomes to fields and markets: some success stories of genomic-assisted breeding in chickpea, pigeon pea and groundnut. Plant Sci. 242, 98–107.
Varshney, R.K., Chabane, K., Hendre, P.S., Aggarwal, R.K., Graner, A., 2007. Comparative assessment of EST-SSR, EST-SNP and AFLP markers for evaluation of genetic diversity and conservation of genetic resources using wild, cultivated and elite barleys. Plant Sci. 173, 638–649.
Varshney, R.K., Hiremath, P.J., Lekha, P., Kashiwagi, J., Balaji, J., et al., 2009. A comprehensive resource of drought- and salinity responsive ESTs for gene discovery and marker development in chickpea (*Cicer arietinum* L.). BMC Genomics 10, 523.
Varshney, R.K., Mohan, S.M., Gaur, P.M., Chamarthi, S.K., Singh, V.K., et al., 2014a. Marker-assisted backcrossing to introgress resistance to *Fusarium* wilt (FW) *race 1* and ascochyta blight in C 214, an elite cultivar of chickpea. Plant Genome 7 (1), 1–11.
Varshney, R.K., Song, C., Saxena, R.K., Azam, S., Yu, S., et al., 2013. Draft genome sequence of chickpea (*Cicer arietinum*) provides a resource for trait improvement. Nat. Biotechnol. 31, 240–246.
Varshney, R.K., Thudi, M., Nayak, S.N., Gaur, P.M., Kashiwagi, J., et al., 2014b. Genetic dissection of drought tolerance in chickpea (*Cicer arietinum* L.). Theor. Appl. Genet. 127, 445–462.
Varshney, R.K., Thudi, M., Pandey, M.K., Tardieu, F., Ojiewo, C., Vadez, V., et al., 2018. Accelerating genetic gains in legumes for the development of prosperous smallholder agriculture: integrating genomics, phenotyping, systems modelling and agronomy. J. Exp. Bot. 69, 3293–3312.
Verma, S., Gupta, S., Bandhiwal, N., Kumar, T., Bharadwaj, C., Bhatia, S., 2015. High-density linkage map construction and mapping of seed trait QTLs in chickpea (*Cicer arietinum* L.) using Genotyping-by-Sequencing (GBS). Sci. Rep. 5, 17512.
Von Wettberg, E.J.B., Chang, P.L., Başdemir, F., Carrasquila-Garcia, N., Korbu, L.B., et al., 2018. Ecology and genomics of an important crop wild relative as a prelude to agricultural innovation. Nat. Commun. 9, 649.
Winter, P., Benko-Iseppon, A.M., Hüttel, B., Ratnaparkhe, M., Tullu, A., Sonnante, G., et al., 2000. A linkage map of the chickpea (*Cicer arietinum* L.) genome based on recombinant

inbred lines from a *C. arietinum* × *C. reticulatum* cross; localization of resistance genes for fusarium wilt races 4 and 5. Theor. Appl. Genet. 101 (7), 1155–1163.

Winter, P., Pfaff, T., Udupa, S.M., Hüttel, B., Sharma, P.C., Sahi, S., Arrequin- Espinoza, R., et al., 1999. Characterization and mapping of sequence-tagged microsatellite sites in the chickpea (*C. arietinum* L.). Mol. Gen. Genet. 262, 90–101.

Watson, A., Ghosh, S., Williams, M., Cuddy, W.S., Simmonds, J., et al., 2018. Speed breeding: a powerful tool to accelerate crop research and breeding. Nat. Plants 4, 10.

CHAPTER

Molecular markers and marker trait associations

7

Kamal Dev Sharma, Surinder Singh Chandel, Rajeev Rathour

Department of Agricultural Biotechnology, CSK Himachal Pradesh Agricultural University, Palampur, Himachal Pradesh, India

7.1 Introduction

Chickpea (*Cicer arietinum* L.) is one of the most important pulse crops in the world and is grown and consumed primarily in the developing and underdeveloped countries where it is the primary source of protein to human beings. World population, especially in developing and underdeveloped countries, where chickpea is largely grown and consumed, is expanding rapidly and demand for chickpea is expected to rise considerably in near future. To meet this increased demand, there is a need to enhance productivity of this crop. Failure to increase productivity will reduce per capita availability of chickpea, resulting in protein deficiency in the people if alternative protein sources are not available, thereby forfeiting the objective of nutritional sufficiency in near future. The crop fits well into the crop rotations where it enriches the soils by fixing the atmospheric nitrogen and reducing the amount of nitrogen fertilization to chickpea as well as next crop in crop rotation. Major factors responsible for low yield of chickpea are low yield potential of present varieties, biotic stresses such as wilt caused by *Fusarium oxysporum* f. sp. *ciceris* and blight caused by *Ascochyta rabiei*, and abiotic stresses such as drought, cold, and heat. While sources of complete resistance to wilt are available in chickpea and are being exploited to develop wilt-resistant cultivars of chickpea, the resistance to ascochyta blight is incomplete and governed by several genes, and quantitative trait loci (QTLs) for blight resistance are present on different chromosomes. Transfer of blight resistance is, thus, tedious and may require transfer of all the QTLs into the recurrent parent. Tolerance to abiotic stresses, i.e., drought, heat, and cold, is also available in chickpea, but like ascochyta blight resistance (ABR), tolerance to abiotic stresses is incomplete and is governed by complex, partially known mechanisms involving several genes. DNA-based markers can be used to tag genes governing different traits, map QTLs, introgress genes into recurrent parents, and estimate background of recurrent parent genome in advanced progeny lines.

DNA marker technologies starting from restriction fragment length polymorphism (RFLP; Botstein et al., 1980) are making inroads into various fields of human, animal, and plant sciences. RFLPs, though highly robust, reveal limited

Chickpea: Crop Wild Relatives for Enhancing Genetic Gains. https://doi.org/10.1016/B978-0-12-818299-4.00007-5
Copyright © 2020 Elsevier Inc. All rights reserved.

polymorphism and the procedure is tedious and time-consuming. Marker technologies frog leaped after the invention of random amplified polymorphic DNA (RAPD; Williams et al., 1990) that was based on polymerase chain reaction (PCR) developed in 1983 (Mullis et al. 1986). RAPD does not require prior knowledge of DNA sequences, it can reveal adequate polymorphism, and the procedure is simple. Consequently, RAPD remained the basic tool to study diversity and tag genes for a long time until more robust markers were developed. Due to use of 10-mer primers, RAPD lacked robustness and reproducibility and is no more used for precise scientific studies. RAPD was followed by the development of inter–simple sequence repeat (ISSR) markers that were based on simple sequence repeats (SSRs) and amplify the region between two closely spaced microsatellites. Like RAPD, ISSRs do not require prior sequence information but display limited polymorphism and are rarely used. Major breakthrough in marker technologies was development of SSR markers where primers designed from regions flanking the microsatellites were used to amplify the microsatellites and variation in microsatellite repeat units depicts polymorphism. The SSR markers are versatile because regions flanking the microsatellites, which are used for primer design, are conserved within the species as well as genus and to some extent among genera, and hence, primers once designed can be used within and across species. Because of robustness, high polymorphism, and co-dominant nature, SSRs have become a tool of choice in all biological systems including plants. Ease of sequencing introduced us to another important source of detection of polymorphism i.e., single nucleotide polymorphisms (SNPs).

Markers are important genomic resources for a variety of studies such as paternity issues (hybridity), forensic studies, diversity/relatedness, mapping the genes, and developing linkage maps. In plant systems, markers can also be used to introgress genes, especially for pyramiding of genes. Construction of well-saturated maps, i.e., maps with markers at short intervals on all the chromosomes, is desirable to meet our objectives of genetic relationships among individuals, gene/QTL mapping, and evolution. In chickpea, development of saturated linkage maps remained a challenge owing to the narrow genetic base of the crop that allowed limited polymorphism. Less dense maps have limited markers and pose problems in narrowing down the boundaries of the QTLs or develop markers tightly linked to monogenic traits in chickpea. SNPs, due to their abundance in chickpea, can be used to increase marker density on chickpea linkage maps. These markers have been useful in narrowing down the boundaries of QTL regions. In addition to hybridization-based mapping populations, genome-wide association studies (GWAS) using a panel of genotypes to map genes are also gaining popularity. Scientists aim to develop designer crops wherein a genotype will be modified to generate crops with phenotype best suited to our requirements. With developments in marker and sequencing technologies, the possibility of prediction of agronomic traits of a plant based on its genotype is within the realm of the scientists.

7.2 DNA-based markers for chickpea

7.2.1 Simple sequence repeat markers

At first, RAPD and ISSR markers were used in chickpea but with little success as these markers could not reveal adequate polymorphism (Ahmad and Slinkard, 1992) and were not sufficiently useful for linkage map development, gene mapping, and elucidation of genetic diversity. The reason for low polymorphism by these markers was the narrow genetic base of chickpea (Ahmad and Slinkard, 1992). Initial studies demonstrated that there was abundance of repetitive (di-, tri-, and tetra-) motifs in chickpea genome, and probes developed for these motifs revealed variability in chickpea genotypes (Weising et al., 1992; Sharma et al., 1995a). Sharma et al. (1995a) studied 38 different microsatellite motifs in four accessions of chickpea and detected 35 hybridization signals. The study also demonstrated the possibility of use of SSRs in elucidating polymorphism in chickpea, especially the interspecific polymorphism. Soon after this, a PCR-based method called "microsatellite primed polymerase chain reaction" (MP-PCR), wherein microsatellite oligonucleotides complementary to SSRs were used as single PCR primers to detect polymorphism in chickpea and two wild annual *Cicer* species, was developed (Sharma et al., 1995b). The touchdown MP-PCR generated highly reproducible banding pattern indicating that PCR-based microsatellite analysis can be employed to identify interspecific polymorphism. Subsequently, Huttel et al. (1999) generated two small-insert genomic libraries in chickpea followed by identification of microsatellites and development of primers from regions flanking the 28 microsatellites. Twenty-two primer pairs produced amplicons of expected size, sixteen of which were polymorphic (Huttel et al., 1999). The markers were named "*Cicer arietinum* sequence-tagged microsatellite site (CaSTMS) markers." In further elaboration of this work and using the similar methodology, Winter et al. (1999) developed 174 primer pairs flanking chickpea microsatellites and used those for their ability to reveal polymorphism in chickpea. One hundred twenty of these markers were subsequently used to develop linkage map of chickpea (Winter et al., 1999). The STMS developed by Winter et al. (1999) remained the primary markers for mapping disease resistance genes (Sharma et al., 2005; Sharma and Muehlbauer, 2007a,b) and identifying QTLs linked to ABR (Anbessa et al., 2009) or agronomic traits such as double podding (Cho et al., 2002). SSR development was boosted by the development of bacterial artificial chromosome (BAC) library in chickpea (Rajesh et al., 2004). In the beginning, SSRs were developed from BAC ends because sequencing of whole BAC clones was tedious. When sequencing came into the ambit of some laboratories, BAC clones were sequenced to fish out the SSRs.

At present, SSRs are one of the most commonly used marker systems in chickpea primarily due to their distribution throughout the genome (Weising et al., 1997), hypervariability, co-dominant nature, and ease of PCR-based assay. Chickpea breeders, agronomists, and biotechnologists prefer SSR markers as these could reveal considerably high polymorphism in chickpea as compared to other markers.

The paucity of SSR markers was the major bottleneck in saturating chickpea linkage maps, and development of additional SSR markers remained a favorite area in chickpea breeding. In contrast to their use, the SSR markers (synonym: STMS markers) development is tedious, is costly, and involves genomic library preparation and enrichment of genomic library for SSRs (first two steps usually summed in one as construction of SSR-enriched libraries) followed by microsatellite identification and primer development. Finally, the primers are tested against a set of chickpea cultivars/lines for evaluation of their polymorphic potential. During the initial years of development of SSR markers, DNA sequencing was a costly affair and limited to labs with adequate resources. Consequently, the rate of development of SSR markers was slow and limited markers were being added per year. The sequencing of human genome using Sanger sequencing technology (International Human Genome Sequencing Consortium, 2001; Bentley et al., 2008) followed by development of high-throughput cost-effective technologies for whole-genome sequencing (WGS; Reuter et al., 2015) and parallel development of bioinformatics tools to analyze the whole-genome sequence data propelled the development of SSR markers in crops including chickpea (Sethy et al., 2006a; Gaur et al., 2011). Not only SSR development but these technologies also led to decoding of chickpea genome where draft genome sequences were released into public domain by separate groups from India (Varshney et al., 2013, Jain et al., 2013, and Parween et al., 2015). With enhancement in chickpea DNA sequence data, these sequences were also exploited to fish out SSR markers in chickpea (Qadir et al., 2007). Primers were designed from sequences flanking 64 SSRs fished from DNA databases, of which 45 (70.3%) were polymorphic between two chickpea cultivars (Qadir et al., 2007). India is the largest producer of chickpea in the world; however, the country was a late entrant to development of STMS markers. In one of the preliminary studies in India, Sethy et al. (2006a) identified 92 new microsatellites and developed 74 functional STMS primer pairs. STMS marker development in chickpea was further enriched by Gaur et al. (2011) who identified 181 new chickpea STMS markers. Gaur et al. (2011) constructed a microsatellite-enriched library of chickpea (enriched for (GT/CA)n and (GA/CT)n repeats) and identified 387 putative microsatellite-containing clones. A total of 254 STMS primers were designed from these clones, out of which 48 (18.8%) primer pairs yielded no amplicons, 25 (9.8%) yielded inconsistent amplicons, and remaining 181 primer pairs amplified fragments of anticipated sizes. One of the major boosts to STMS (SSR) marker development in chickpea was by Thudi et al. (2011) who developed 1344 polymorphic SSR markers and 5397 polymorphic DArT markers. Sequences in databases were also used to develop SSR primers, e.g., development of a nonredundant set of 202 *in silico* SSR markers (Agarwal et al., 2015) and 1494 SSR markers (1016 genomic, 478 transcript-derived) (Khajuria et al., 2015). MIcroSAtellite (MISA) search module (Thiel et al., 2003; http://pgrc.ipk-gatersleben.de/misa/) was used to search SSRs (6845) in the BAC end sequences. Primers from these SSRs were designed using standalone Primer3 (http://frodo.wi.mit.edu/) program using MISA-generated Primer3 input file (Thiel et al., 2003).

Efficacy of SSRs for revealing polymorphism was also studied. At interspecific (*C. arietinum* ICC4958 and *C. reticulatum* PI489777) level, the polymorphism was 59%, whereas at intraspecific level it was 48% (Khajuria et al., 2015). A set of markers was used to analyze multiparent advanced generation intercross (MAGIC) as well as near isogenic lines, and the markers indicated their usefulness in diversity analysis in chickpea (Agarwal et al., 2015). Identity of hybrids in chickpea was traditionally established using morphological markers. Of late, STMS markers have been shown to be valuable tools to confirm hybridity in chickpea (Caballo et al., 2018).

Not only chickpea, but the wild progenitor of chickpea i.e., *Cicer reticulatum* was also exploited to develop STMS markers for use in chickpea; however, the number of markers developed is only 11 (Sethy et al., 2006b). The wild relatives of chickpea are sources of some useful genes such as those possessing resistance to abiotic and biotic stresses. These traits can be transferred to chickpea to develop tolerant plants. Draft genome of *C. reticulatum* has also been released (Gupta et al., 2016), and *in silico* analysis of sequences of the genome is expected to identify complete array of SSRs in this wild relative and, hence, more STMS markers specific to *C. reticulatum*.

7.2.2 Expressed sequence tag—simple sequence repeat markers

Sequencing of expressed sequence tags (ESTs) and further analysis of those sequences revealed that SSRs may be present in the transcribed regions of the genome. This knowledge provided a new source of SSR markers in chickpea. Prior to development and extensive utilization of modern DNA sequencing technologies, subtractive suppressive hybridization (SSH) and differential display RT-PCR were the major tools to detect ESTs from the given tissue. SSH was used to develop a database of about 2800 chickpea ESTs from root tissues of two chickpea genotypes (ICC16349 and Annigeri) (Jayashree et al., 2005), whereas cDNA-AFLP was used to generate 4800 transcript-derived fragments from chickpea leaves (Dinari et al., 2013) and differential display RT-PCR (DDRT-PCR) to generate 9205 EST bands from chickpea anthers (Sharma and Nayyar, 2014). Fifty-four ESTs from leaf and 102 ESTs from anthers were sequenced and added to public domains (Dinari et al., 2013; Sharma and Nayyar, 2014). EST-SSR markers (106) for chickpea were also developed from 477 sequences of which 44 were polymorphic across the *Cicer* species (Buhariwalla et al., 2005). These techniques, however, could detect only a fraction of ESTs expressed in the cells. Second-generation sequencing technologies enabled sequencing of whole transcriptomes in the given tissues and consequently the information generated was many fold than that obtained from SSH, cDNA-AFLP, or DDRT-PCR. The automated DNA sequencing was used to generate 822 chickpea ESTs from immature seeds (Choudhary et al., 2009). These ESTs and previously described 1309 chickpea ESTs contained 246 SSRs, from which 183 primer pairs were designed and 60 were validated as functional markers (Choudhary et al., 2009). EST database from chickpea seed was further enriched by identification of 2496 (Choudhary et al., 2012) and 2029

(Varshney et al., 2009) ESTs. These ESTs were also used to develop 125 (Choudhary et al., 2012) and 177 EST-SSR functional markers (Varshney et al., 2009). Subsequently, 187 EST-SSR markers were developed by Gupta et al. (2015) and 143 by Stephens et al. (2014). Considerably high number of chickpea EST resources has been added/are being added that can be exploited to identify SSRs in ESTs and develop EST-SSR markers.

In addition to *C. arietinum*, the transcriptome of wild chickpea, *C. reticulatum* (PI489777) was sequenced by GS-FLX 454 technology (Jhanwar et al., 2012). A total of 37,265 transcripts with an average length of 946 bp were generated. A total of 4072 SSRs could be identified in these transcript sequences, of which at least 561 EST-SSRs were polymorphic between *C. arietinum* and *C. reticulatum*.

Above descriptions adequately demonstrate that the number of SSRs is increasing rapidly in case of chickpea, and for effective utilization of these SSRs as markers by breeders and geneticists, a storage and retrieval database that may also act as repository of new SSRs is need to be developed. To fulfill this gap, ICRISAT, Hyderabad, developed an open-access database named "Chickpea Microsatellite Database (CicArMiSatDB http://cicarmisatdb.icrisat.org)" that provided detailed information on SSRs along with their features in the genome (Doddamani et al., 2014). The database has web interface to mine and filter the SSR markers. Usually it is believed that polymorphism detection by SSRs is due to variation in repeat units. In *Cicer* species, size variation in the EST-SSR marker bands, however, cannot be attributed solely to variation in number of repeat units. In addition to variation in microsatellite repeat units, insertions, deletions, and point mutations in SRRs as well as flanking sequences contributed toward size variation in SSR bands (Choudhary et al., 2009).

7.2.3 Transcription factor gene–derived microsatellite markers

Like other genes, the transcription factor (TF) genes also contain SSRs. These SSRs can be used for development of genic markers called "transcription factor gene–derived microsatellite (TFGMS)" markers. Functional domains of TF genes have been used to develop genic microsatellite markers in tomato and pepper (Yu et al., 2010). The transcriptomics studies in chickpea led to identification of genes encoding TFs (Garg et al., 2011; Kujur et al., 2013). In the first study, 1773 TF-encoding genes from transcript sequences of chickpea desi genotype ICC 4958 (Chickpea Transcriptome Database, CTDB release 1.0, http://www.nipgr.res.in/ctdb.html) were identified (Garg et al., 2011), whereas in the second study 707 TFs from the same genotypes were identified (Kujur et al., 2013). From these 707 TF-genes, 1108 TFGMS markers were identified and 161 transcription factor functional domain–associated microsatellite (TFFDMS) markers were developed (Kujur et al., 2013). These markers had 96.5% amplification efficiency and high intraspecific polymorphic potential (34%). Thus, these markers are additional genomic resources for genotyping, developing linkage maps, and diversity analysis in chickpea.

7.3 DNA-based markers other than SSRs

7.3.1 Genic markers

7.3.1.1 Intron targeted primers

Intron targeted primers (ITPs) are designed from the exon boundaries of the introns and hence, amplify the introns. The polymorphism displayed by ITPs is based on variations in introns and is called as intron polymorphisms (IPs). There are only two reports on development of ITP markers for chickpea (Choudhary et al., 2012; Gupta et al., 2015). In the first study, 151 ITPs were developed from chickpea seed ESTs (Choudhary et al., 2012) whereas in the second, 130 potential intron polymorphisms (PIPs) were identified from chickpea embryos (Gupta et al., 2015).

7.3.1.2 Expressed sequence tag polymorphisms

ESTs may also have certain deletions and additions. For example, a gene encoding the enzyme thymidylate synthase has 6-bp variation (deletion) at bp 1494 in the 3′-untranslated region (Ulrich et al., 2000). This variation provides an additional mean to study polymorphism (called expressed sequence tag polymorphisms, ESTPs) and elucidation of variability. To study ESTPs, usually, ESTs are amplified and run on denaturing gradient gel electrophoresis (DGGE), and variation in distance covered by the bands denotes polymorphism (Temesgen et al., 2001). In chickpea, 159 ESTPs have been identified, 109 from seed and 50 from embryo (Choudhary et al., 2012; Gupta et al., 2015).

7.3.2 Nongenic markers

7.3.2.1 Diversity arrays technology

Diversity arrays technology (DArT) searches for SNPs and InDels across the genome. The DArT technology works effectively against the polyploids; however, it can also be used in diploids such as chickpea. DArT technology does not require prior sequence information and can be accomplished with much less resources compared to SNP assay. In chickpea, 5397 polymorphic DArT markers have been developed (data based on 94 genotypes tested) (Thudi et al., 2011).

7.3.2.2 Insertion-Deletion markers

Insertion-deletion (InDel) markers detect polymorphism based on insertions and deletions in genomic region of plants and are reported to be "An extended marker resource for molecular breeding in chickpea" (Jain et al., 2019). In chickpea, the length of insertions or deletions varies from 1 to 54 bp as InDel markers displayed 1−18 bp (Srivastava et al., 2016) or 2−54 bp polymorphism (Das et al., 2015a). Though studies on development of InDel markers are limited in chickpea, considerably high number of InDel markers has been developed e.g., 21,499 genome-wide InDel markers with 2 to 54 bp fragment length polymorphism (Das et al., 2015a), 8628 InDel markers with 1−18 bp polymorphism (Srivastava et al., 2016), 7523 InDel markers with >20 bp polymorphism (Jain et al., 2019). Srivastava et al.

(2016) also mapped the InDel markers onto eight chromosomes of chickpea. Further evaluation of polymorphic potential of the InDel markers established that these markers reveal high polymorphism wherein 3414 markers generated 6849 alleles in a panel of 24 chickpea and wild chickpea accessions (Srivastava et al., 2016). The InDels were also detected in genic and regulatory regions. Of the 21,499 InDel markers, 4566 were from genic and regulatory regions (Das et al., 2015a), indicating the importance of these markers in revealing trait associations. Detection of polymorphism using agarose gels is relatively easier and faster compared to polyacrylamide gels and does not require sophisticated instruments or chemicals and can be accomplished in ordinary laboratories. To facilitate detection of InDel polymorphism in agarose gels, Jain et al. (2019) suggested the use of InDel markers with >20 bp polymorphism. From the 231,658 InDels, they selected 8307 InDels that had more than 20 bp size difference, out of these 8307, 7523 were used for InDel marker development (Jain et al., 2019). The rate of amplification by these primers was 80% and polymorphism varied from 46.06% to 58.01%. Since the InDels are detected from sequencing data, the identification of InDels require computer-based software, and one of the commonly used software for detection of InDels is Dindel software (Albers et al., 2011).

7.3.2.3 Single nucleotide polymorphisms

SNPs are the most common genetic variation in all the living organisms and depict variation in single nucleotides. SNPs are mostly biallelic (Krawczak, 1999), co-dominantly inherited, and sequence tagged and occur at high density within genomes (Xing et al., 2005). Next-generation sequencing technologies can sequence genomes in a short span of time, and subsequent reduction in cost of sequencing has brought SNPs into the mainstay of genotyping of plant systems. There are several studies on identification of SNPs in chickpea; e.g., Hiremath et al. (2012) detected 2486 SNPs from 70 different chickpea genotypes, Stephens et al. (2014) searched 1430 SNP markers, Kujur et al. (2015) identified 44,844 genome-wide SNPs using 92 accessions, Bajaj et al. (2015a) identified 82,489 genome-wide SNPs from 93 wild and cultivated *Cicer* accessions, Basu et al. (2018) discovered 2150 desi and 2199 kabuli accessions–derived SNPs from 92 chickpea accessions, and Deokar et al. (2019a) identified 15,581 SNPs from RILs (135) and parents (ICCV 96029 and Amit). In addition to sequencing, EcoTILLING was also used to detect SNPs in chickpea (Bajaj et al., 2016). Illumina GoldenGate SNP array of 1536 genic SNPs in chickpea was also developed (Deokar et al., 2014). Comparison of genome sequences of *C. arietinum* with *C. reticulatum* led to identification of new SNPs (842,104 genomic SNPs with average SNP density of 1.13 SNPs per kb) (Gaur et al., 2015). The sequenced reads of *C. reticulatum* were compared with draft genome sequence of *C. arietinum* (desi chickpea) to identify new genic SNPs (36,446; Gaur et al., 2015). Using EST libraries from roots of chickpea, 21,405 high-confidence SNPs in 742 contigs were also identified (Varshney et al., 2009). In addition to this, recognition sites for restriction enzymes were identified for 7884 SNPs in 240 contigs (Varshney et al., 2009).

A combination of four approaches, i.e., Solexa/Illumina sequencing, amplicon sequencing of tentative orthologous genes, mining of ESTs, and sequencing of candidate genes, was used to mine SNPs (Hiremath et al., 2012). For easy, flexible, and cheap analysis by future researchers, the assay for 2005 SNPs was converted into allele-specific PCR (KASPar) assays named "Chickpea KASPar Assay Markers" (CKAMs) (Hiremath et al., 2012). 1341 CKAMs showed polymorphism in 70 genotypes of chickpea and clustered these accessions into groups based on geographical origin. The tools for development of allele-specific PCR assay such as "KASP-Hedgehog" that can be used to design primers from multidisciplinary data types are available (Alsamman et al., 2019).

The SNPs developed through high-throughput genotyping platforms such as Infinium or iScan may be costly and out of reach of common laboratories. Roorkiwal et al. (2013) developed cost-effective SNP platform (Illumina BeadXpress) for chickpea and pigeonpea. Using this platform, 96 chickpea SNPs were developed and tested by genotyping 288 genotypes of chickpea (Roorkiwal et al., 2013). Of the 96 SNPs used, 94 (97.9%) displayed polymorphism on 288 genotypes. While studying efficacy of these markers vis-à-vis SSR markers for phylogenetic analysis, these markers generated four clusters in comparison to two clusters by SSR markers. However, grouping was not based on geographic or biological origin or seed type (Roorkiwal et al., 2013).

Detection of SNPs in wild progenitor of chickpea, i.e., *C. reticulatum*, was also accomplished, though the work is limited to a few studies. Rajesh and Muehlbauer (2008) studied SNPs in coding and genomic regions between FLIP 84-92C (*C. arietinum*) and PI599072 (*C. reticulatum*) and found SNP frequency of 1 in 66 bp in coding regions and 1 in 71 bp in genomic regions. They observed 110 single base changes or substitutions in their study. Subsequently, 1022 SNPs were identified from sequences of *C. arietinum* (genotype: ICC4958) and *C. reticulatum* (accession: PI489777), of which 697 were useful for genotyping the mapping population derived from cross of ICC4958 and PI489777 (Gaur et al., 2012). The transcriptome of *C. reticulatum* (PI489777) was sequenced by GS-FLX 454 technology, and 37,265 transcripts with an average length of 946 bp were generated followed by detection of 36,446 SNPs (Jhanwar et al., 2012). The estimation of synonymous substitution rates of orthologous transcript pairs suggested that the speciation event for divergence of *C. arietinum* and *C. reticulatum* may have happened approximately 0.53 million years ago (Jhanwar et al., 2012).

7.4 Genotyping-by-sequencing

Genotyping-by-sequencing (GBS) is a technique where genomes of the parents and SNPs are sequenced followed by detection of the SNPs. The SNPs are subsequently used for genotyping the mapping populations, developing linkage maps, or identifying SNPs tightly linked to the phenotype. Such large-scale SNP detection by sequencing the genomes is possible only through the use of next-generation sequencing technologies. In chickpea, GBS is limited. However, GBS approach was applied for SNP calling between the parents and recombinant inbred lines derived from cross of ICC 4958 and ICC 1882 (Jaganathan et al., 2015). Polymorphic loci were detected based on consensus sequences from all samples and used for subsequent analysis. Finally, the data were used to develop linkage map and QTL analysis (Jaganathan et al., 2015).

7.4.1 Transferability of chickpea SSRs to other taxa and vice versa

The DNA sequences flanking the microsatellites are conserved within the same species. Several studies revealed that such conservation of microsatellite flanking regions extends beyond the species (cross-species amplification), but such conservation reduces as the diversity increases. Same is true for legumes as several of the chickpea SSR markers amplified microsatellites in other genera or SSR markers from other genera amplified microsatellites in chickpea. In one of the preliminary studies in legumes, primer sequences were shown to be conserved among Leguminosae tropical trees (Dayanandan et al., 1997). Later on, Peakall et al. (1998) found up to 65% cross-amplification within the genus *Glycine* and lower rates (3% −13%) outside the genus.

STMSs are conserved in the genus *Cicer* (Choumane et al., 2000) and in different taxa of family Leguminosae (Choumane et al., 2004; Datta et al., 2010). The degree of conservation of primer binding sites depends on the closeness of the species/genus to chickpea. For example, transferability was 92.2% in *C. reticulatum* (potential progenitor of chickpea) and 50% for *C. cuneatum* (Choumane et al., 2000). The transferability rate of chickpea STMS primers in lentil and *Pisum* was 54.4% (Choumane et al., 2004). Considerable variation in cross genera transferability rates was, however, reported in different studies e.g. transferability of chickpea STMS was 5% (Pandian et al., 2000) to 85.71% (Agrawal and Katiyar, 2008) in lentil and 18% in field pea (Pandian et al., 2000). Agrawal and Katiyar (2008) evaluated 42 STMS of chickpea, all of which generated amplicons in lentil and 36 of these were polymorphic among lentil genotypes. Not only lentil, chickpea STMS primers also amplified regions of *Cajanus* species with a transferability rate of 46% (Datta et al., 2010). The primers also exhibited extensive polymorphism in *Cajanus* species (4.1 alleles per primer). Cross-species transferability of 172 genic and 29 genomic SSRs of chickpea developed earlier at the National Institute of Plant Genome Research, New Delhi (Jhanwar et al., 2012; Kujur et al., 2013; Garg et al., 2011) was studied for important legumes viz., horsegram, blackgram, cowpea, pigeonpea, and soybean (Jingade et al., 2014). The transferability to these genera was very high (66.7%: blackgram, 62.2%: horsegram, 61.7%: redgram, 54.7%: cowpea, 62.7%: soybean). The study led to identification of 134, 133, 126, 124, and 110 new SSR primers for blackgram, horsegram, soybean, redgram, and cowpea, respectively.

Similar to cross-species transferability of chickpea STMSs to other genera, STMSs from other genera amplified chickpea genomic regions. *Medicago truncatula*, *Lotus japonica*, *Glycine max*, and *Pisum sativum* are four well-studied model systems in legumes, and marker resources from these crops can be exploited in other legumes due to cross-species transferability. The cross-species transferability of 209 EST-based and 33 bacterial artificial chromosome (BAC)-based microsatellites from *M. truncatula* to chickpea was demonstrated by Gutierrez et al. (2005). The transferability of *M. truncatula* microsatellites to chickpea was 36.3%. Not only chickpea, *M. truncatula* microsatellites also showed transferability to faba bean (40%) and pea (37.6%). The study further demonstrated that transferability rates were almost twofold higher for EST-SSRs (39%–43%) compared to genomic SSRs (21%–24%). The transferability was still higher in another study i.e., 62% of pea EST-SSR primer pairs (out of 58 primer pair analyzed) and 60% of Medicago EST-SSR primer pairs (out of 25 primer pair analyzed) amplified in chickpea (Mishra et al., 2012). Similarly, transferability of field pea STMS primers across chickpea/Cicer sp. accessions was also very high (62%; Pandian et al., 2000). Cereal SSR primers also demonstrated transferability to chickpea at lower rates. Of the 42 rice SSR primers, 9 (21.4%) showed transferability to chickpea (Gupta et al., 2016).

The studies clearly demonstrate high rates of cross transferability of chickpea STMS to other legumes and these STMSs can be used to reveal polymorphism for map construction and gene mapping in other crops. Almost similar rates of transferability were observed from other legumes to chickpea. The rates of transfer of SSRs from taxa that were distantly related to chickpea such as rice (a cereal) were lower. Thus, the SSRs from other taxa can be evaluated in chickpea for amplification as well as polymorphism and those can be valuable addition to already existing marker resources in chickpea. Chickpea EST-SSR cross transferability to other taxa has not been evaluated so far; the transferability of EST-SSRs were reported to be higher than SSRs from the repository of already available markers in other legumes.

7.5 Development of linkage maps

Densely saturated linkage maps are critical for identification of markers linked to new traits or to study evolutionary relationship with other closely related species as these can provide knowledge of inversions, deletions, additions, translocations, etc. or point mutations if maps are generated using SNPs. Well-saturated maps are also prerequisite for marker-assisted selection (MAS) and marker assisted breeding (MAB) using foreground and background selection. Map development in chickpea was initiated in 1990s, but development of saturated maps took more than two decades as chickpea being a species of narrow genetic base lacked adequate polymorphism for map development. Consequently, map development in the beginning involved use of interspecific mapping populations with its progenitor *C. reticulatum* as one of the parents and *C. arietinum* as the other parent. First linkage map in chickpea was developed using isozyme markers and populations derived from interspecific cross between *C. arietinum* and *C. reticulatum* (Gaur and Slinkard (1990a,b). These studies used 13 isozyme loci which were allocated to four linkage groups (LGs). Chickpea has eight chromosomes and the map has only four LGs. Isozyme markers are limited in number and not adequate to develop linkage maps in chickpea. In contrast to protein or isozyme markers, DNA-based marker systems such as RFLP (Botstein et al., 1980) and RAPD (Williams et al., 1990) are more in number, but still insufficient to reveal adequate polymorphism in chickpea (Simon and Muehlbauer, 1997). Integrating 9 morphological, 27 isozyme, 10 RFLP, and 45 RAPD markers covering 550 cM, a genetic linkage map of chickpea was developed using three interspecific population derived from *C. arietinum* and *C. reticulatum* (Simon and Muehlbauer, 1997). The map consisted of 10 linkage groups (LGs) for eight chromosomes of chickpea.

Much needed impetus for map development in chickpea was provided by the STMS markers developed for chickpea for the first time by Winter et al. (1999). They mapped 120 STMS markers into 11 linkage groups covering a distance of 613 cM. To develop the map, an interspecific mapping population consisting of 90 RILs derived from cross of *C. arietinum* and *C. reticulatum* was used. Due to lack of adequate polymorphism, development of linkage maps in the initial years was based on interspecific mapping populations (Winter et al., 1999, 2000; Tekeoglu et al., 2002; Nayak et al., 2010). Development of adequate STMS markers that could detect polymorphism even in intraspecific populations led to the development of intraspecific linkage maps in chickpea (Cho et al., 2002, 2004; Udupa and Baum, 2003; Flandez-Galvez et al., 2003; Radhika et al., 2007; Taran et al., 2007; Anbessa et al., 2009), however, intraspecific map developed in the initial years had sparse marker densities. The marker densities in intraspecific map improved considerably afterward, especially when several new SSRs were added to the existing repository of chickpea SSRs and new marker systems were developed in chickpea. The intraspecific maps have advantages over interspecific maps as these could depict more accurately the genetic distances between the markers or the markers and the genes. Hunt for new SSR markers is still continued and these markers are being added continuously to the interspecific (Aryamanesh et al., 2010; Khajuria et al., 2015; Das et al., 2015a) and the intraspecific (Gaur et al., 2011; Thudi et al., 2011; Stephens et al., 2014; Bajaj et al., 2015b; Mallikarjuna et al., 2017) linkage maps in chickpea and are considered to be the cost-effective marker system to map genes or mine QTLs in chickpea. EST-SSR, InDel, DArT arrays, and SNPs are other most common marker systems developed for chickpea. These markers are being used to develop more saturated linkage maps in chickpea and those maps that have high density of markers are usually being termed as "ultra-high-density linkage maps" (Thudi et al., 2011, Gaur et al., 2012, Hiremath et al., 2012, Stephens et al., 2014, Jaganathan et al., 2015, Bajaj et al., 2015a,b, Kujur et al., 2015, 2016, Verma et al., 2015, Gaur et al., 2015, Das et al., 2015a, Srivastava et al., 2016, Amalraj et al., 2018, Roorkiwal et al., 2018, Jain et al., 2019).

An interspecific linkage map termed as "second-generation linkage map" was developed by Hiremath et al. (2012) in which 1328 marker loci (625 CKAMs, 314 tentative orthologous genes (TOGs)-SNPs, and 389 published marker loci) were placed on eight different LGs. The LGs were named CaLG01 to CaLG08 where Ca denotes *C. arietinum*, LG the linkage group, and numeral the number of LG. LG01 to LG08 denote 8 linkage groups for eight chromosomes of chickpea. In later studies, additional markers were placed on the linkage maps of chickpea e.g., 1494 markers (1016 genomic and 478 EST SSRs) (Khajuria et al., 2015), 143 EST-SSRs and 768 SNPs (Stephens et al., 2014), and 6698 SNP markers (Gaur et al., 2015). In some of the studies, only one type of the markers were used to develop linkage maps, whereas in others different types of markers were used; e.g., Thudi et al. (2011) developed map using BES-SSR markers and DArT arrays, Stephens et al. (2014) used EST-SSRs and 768 SNPs, Bajaj et al. (2015b) used SNP and SSR, and Amalraj et al. (2018) used SNP and DArT markers. Of late, several maps were developed using SNPs detected through GBS approach (Gaur et al., 2012; Kujur et al., 2015, 2016; Verma et al., 2015; Roorkiwal et al., 2018; Deokar et al., 2019a). Among these, the prominent ones are by Bajaj et al. (2015a) (1536 SNPs), Verma et al., (2015) (3228 SNPs), Kujur et al., (2016) (3625 SNPs), Deokar et al., (2014) (51,632 SNPs), Deokar et al., (2019a) (15,581 SNPs), and Roorkiwal et al., (2018) (13 679 SNPs). The map developed by Roorkiwal et al. (2018) is presented here (see Fig. 7.1).

Similar to SNPs, several InDel markers have been mapped onto the physical maps of the chickpea reference genomes. Das et al. (2015a) physically mapped 21,499 InDel markers on eight chromosomes of desi chickpea genome. Maximum marker density was on chromosome 3 (1468 markers with map density of 15.9 kb) and minimum on chromosome 7 (488 markers with map density of 17.3 kb). In another study, Srivastava et al. (2016) added 5263 InDel markers to physical maps of chickpea chromosomes and Jain et al. (2019) added 7523 InDels. Jain et al. (2019) used whole genome resequencing of five parental lines of an interspecific and two intraspecific mapping populations to fish the InDels. Primers were developed for those InDels (7523) which were more than 20 bp in size and these InDels were assessed by gel assay followed by construction of the linkage map. The map had marker density of 1038 InDels/LG (maximum on CaLG04: 1952 InDels, minimum on CaLG08: 360 InDels) (Jain et al., 2019). The use of InDels with more than 20 bp size facilitated the evaluation of polymorphism using agarose gels, whereas InDels with smaller size did not resolve in agarose gels and hence, failed to reveal polymorphism. Due to ease of agarose gel—based assays for >20 bp InDels, use of these markers for map development or diversity studies is expected to expand to common labs across the globe. Moreover, InDels with smaller size can also be assayed by using slightly tedious polyacrylamide gels. Since gel-based analysis does not require genome sequencing, the cost of InDel-based assays will be considerably lower if the polymorphism is analyzed through gels.

GWAS where marker analysis is carried using a panel of diverse accessions/genotypes to study marker trait or marker-marker linkages has also been attempted in chickpea. In these studies, GWAS were combined with whole-genome

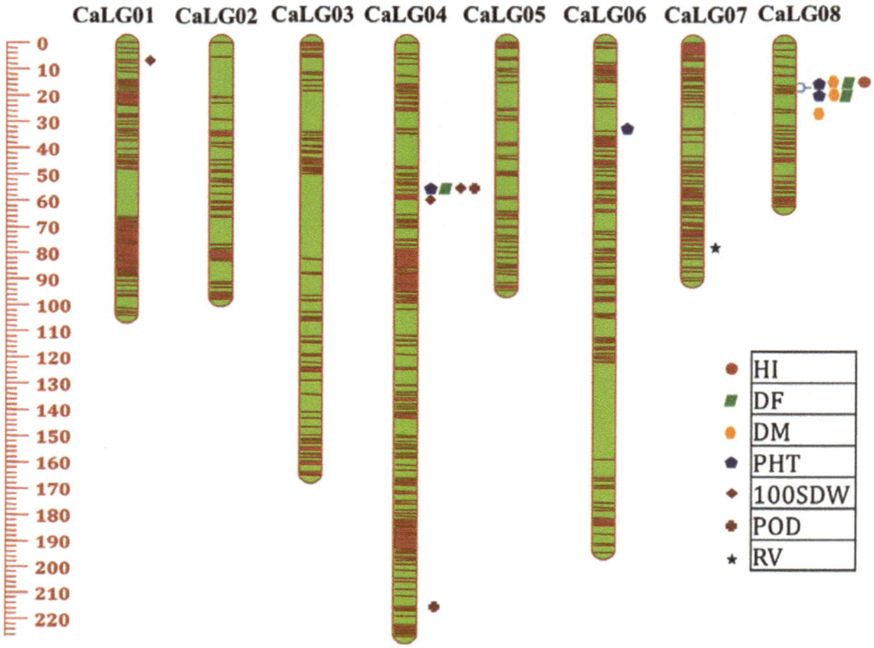

FIGURE 7.1

High-resolution linkage map of chickpea developed using Axiom CicerSNP Array (13,679 SNP markers). The intraspecific mapping population used to develop the map was derived from cross of ICC 4958 and ICC 1882. Major QTLs are shown on the right-hand side of linkage groups. The QTLs are for HI, height; DF, days to 50% flowering; DM, days to maturity; PHT, plant height; 100SDW, 100 seed weight; POD, pods per plant; and RV, root volume.

The map reproduced from Roorkiwal, M., Jain, A., Kale, S.M., Doddamani, D., Chitikineni, A., Thudi, M., Varshney, R.K., 2018. Development and evaluation of high-density Axiom® Cicer SNP array for high-resolution genetic mapping and breeding applications in chickpea. Plant Biotechnol. J. 16 (4), 890–901.

resequencing (WGRS) methods. In a first step, WGRS was carried out to identify SNPs on a panel of accessions/genotypes followed by GWAS where the SNPs were ordered and subsequently placed on physical maps of chickpea. Using 69 chickpea genotypes for WGRS, 827,411 SNPs were detected among the genotypes. The linkages among the SNPs were detected using GWAS and finally these SNPs were placed on map of chickpea. Maximum SNPs (170,747) were found on Ca4 and minimum (28,624) on Ca8 (Li et al., 2017). In an earlier study, Bajaj et al. (2015a) used similar methodology on a set of kabuli genotypes and physically mapped 11,989 SNPs on eight kabuli chromosomes with a frequency of 28.9 kb. SNP map density was maximum on chromosome 4 (20.3 kb) and minimum on 5 (39.1 kb) (Bajaj et al., 2015a).

7.6 Genetic mapping of disease resistance and agro-morphological traits in chickpea

7.6.1 Disease resistance

7.6.1.1 Genetic mapping of wilt resistance

Wilt caused by *Fusarium oxysporum* (F) sp. *ciceris* (*Foc*) is an important disease of chickpea in almost all chickpea growing areas of the world, and search for markers linked to genes governing resistance to this disease is well studied. The pathogen is known to have eight well-defined races (0, 1A, 1B, 2, 3, 4, 5, and 6), and resistance to each of these races is either monogenic or polygenic (Upadhyaya et al., 1983a,b, Sharma et al., 2004a,b, Sharma et al., 2005, Sharma and Muehlbauer, 2007b, Anupama et al., 2018). For example, resistance to race 1 is governed by three genes, all of which provide partial resistance and combination of any two of these provide complete resistance (Upadhyaya et al., 1983a, b). Markers CS-27_{700} (RAPD), UBC-170_{550}, were reported to be linked to gene *h1* (syn. *foc-1*) that govern resistance to race-1 with 7% recombination (Mayer et al., 1997) whereas ISSR marker, UBC-855_{500} situated on LG 6, at a distance of 5.2 cM was linked to race 4 resistance gene (Ratnaparkhe et al., 1998a). The UBC-855_{500} was at 0.6 cM to CS-27_{700} (Ratnaparkhe et al., 1998a) indicating that both the genes lie close to each other. This was followed by mapping of several wilt resistance genes in chickpea e.g., *foc-0$_1$* (Rubio et al., 2003; Cobos et al., 2005), *foc-1* (Sharma and Muehlbauer, 2007b), *foc-2* (Sharma and Muehlbauer, 2007b; Gowda et al., 2009), *foc-3* (Sharma et al., 2004b; Sharma and Muehlbauer, 2007b), *foc-4* (Ratnaparkhe et al., 1998a,b; Tullu et al., 1998; Tullu et al., 1999; Tekeoglu et al., 2000; Winter et al., 2000; Benko-Iseppon et al., 2003; Sharma and Muehlbauer, 2007a,b), the second resistance gene for race 4 (Tullu et al., 1999), and *foc-5* (Ratnaparkhe et al., 1998b; Tekeoglu et al., 2000; Winter et al., 2000; Benko-Iseppon et al., 2003; Sharma and Muehlbauer, 2007a,b) (see Table 7.1 for list of markers linked to wilt resistance genes).

These studies also showed that fusarium wilt resistance genes in chickpea exist on two separate LGs i.e., *Foc-0$_1$/foc-0$_1$* on LG3 [Cobos et al., (2005); syn. LG 5 (chromosome C [or D]) of Winter et al. (2000)] and other five genes (*foc-1*, *foc-2*, *foc-3*, *foc-4*, *foc-5*) on LG 2 (chromosome F [or G]) (Sharma and Muehlbauer, 2007b). Gene *foc-5* that confers resistance to race 5 of *Foc* is linked to STMS marker TA59 (Sharma and Muehlbauer, 2007b), and taking this marker as starting point, 820 kb region of the genome associated with *foc-5* resistance was identified on draft genome sequence of chickpea (Caballo et al., 2019). The NILs were used to narrow down the genomic region followed by physical map construction and saturated region contained 26 candidate genes. Additional markers were also added to this region (Caballo et al., 2019) which can be useful for MAS for *foc-5*. Similarly, chromosome five region surrounding the *Foc0* resistance gene that govern resistance to race 0 of *Foc* was also physically delineated using RIl populations and draft genome sequences of chickpea (Jendoubi et al., 2016). With this study, six STMS

Table 7.1 Markers linked to wilt resistance genes in chickpea.

Wilt resistance genes	Markers linked to the gene	Linage group (LG)
$Foc0_1/foc0_1$	H2120, CaGM20820, CaGM20889, TS43 (Jendoubi et al., 2016); $OPJ20_{600}$ (Rubio et al., 2003); TR59 (Cobos et al., 2005)	LG 5
$Foc0_2/foc0_2$	TA59, TS47 (Halila et al., 2009)	LG 2
foc-1 (syn h1)	$CS27_{700}$, UBC-170_{550} (Mayer et al., 1997); CS27A, TA59, TA96, TA27 (Sharma and Muehlbauer, 2007b); H3A12, TA110 (Gowda et al., 2009)	LG 2
foc-2	TA96, TA27, TR19, TA59, CS27A (Sharma and Muehlbauer, 2007b); H3A12 (Gowda et al., 2009)	LG 2
Foc-3/foc-3	CS27A, TA59, TA96, TA27, CS27A (Sharma et al., 2004a,b; Sharma et al., 2005; Sharma and Muehlbauer 2007b); H1B06y, TA194 (Gowda et al., 2009)	LG 2
foc-4	UBC-855_{500}, CS-27_{700} (Ratnaparkhe et al., 1998a, b, Tullu et al., 1998, 1999); CS27A (Tekeoglu et al., 2000); TR19 (Sharma and Muehlbauer, 2007b); TA96, CS 27, EAAMCTA12 (Winter et al., 2000; Sharma and Muehlbauer, 2007b); R-2609-01, OP-U17-1 (Benko-Iseppon et al., 2003)	LG 2
foc-5	CS27A (Tekeoglu et al., 2000); TA59 (Sharma and Muehlbauer, 2007a,b); TA96 (Sharma and Muehlbauer, 2007b; Cobos et al., 2009); TA27, ECAMCTA07 (Winter et al., 2000; Sharma and Muehlbauer, 2007a,b); SPA, PRP-RGA1 (Caballo et al., 2019); OP-M20-2/3 (Benko-Iseppon et al., 2003);	LG 2

and one genic markers were added to chromosome 5 and marker-gene distances were reduced. The *foc-0₁* governs resistance to yellowing syndrome caused by race 0_1. There is another unmapped gene (*foc-1B/C*) that also provides resistance to yellowing syndrome. Similarly, a wilt syndrome resistance gene *foc-6* has also not been mapped so far. The possibility of occurrence of *foc-1B/C* at LG5 or that of *foc-6* at LG2, where other genes governing resistance to yellowing syndrome and wilting syndrome are situated, exists as has been hypothesized by Sharma and Muehlbauer (2007a,b). It appears that LG2 is a hot spot for disease resistance genes as, in addition to wilt resistance genes, QTLs (ar1 and ar2a) governing resistance to *Ascochyta rabiei*, some genes with putative function in plant defense against diseases (Udupa and Baum, 2003; Pfaff and Kahl, 2003) and a few RGAs also lie on the same LG. In a recent study, five QTLs for resistance to fusarium wilt were detected (Garg et al., 2018). Race 1 resistance was governed by three QTLs (a major QTL on CaLG02 and a minor QTL each on CaLG04 and CaLG06), whereas race 3 resistance was governed by two major QTLs each on CaLG02 and CaLG04 (Garg et al., 2018). Earlier, wilt resistance to race 1 in WR315 was reported to be governed by two QTLs (Sabbavarapu et al., 2013), whereas late wilting was governed by five QTLs (Jingade and Ravikumar 2015).

7.6.1.2 Genetic mapping of QTLs governing ABR

Ascochyta blight is one of the most destructive foliar diseases of chickpea worldwide and is caused by a fungus called *Didymella rabiei* (syn.: *Ascochyta rabiei*). ABR in chickpea is partial and polygenic, and at least three recessive and complementary genes along with modifiers are involved in ABR (Tekeoglu et al., 2000; Millan et al., 2006, 2010; Tar'an et al., 2007). In a study based on 15 chickpea accessions, Labdi et al. (2013) found race 4 resistance in different accessions to be governed by single recessive gene, two recessive complementary genes, two dominant complementary genes, and two recessive genes with epistasis interaction. Identification of QTLs governing resistance to this disease was pursued since the use of DNA-based markers in chickpea. The QTLs governing resistance to AB under field as well as glasshouse conditions were identified (Santra et al., 2000, Rajesh et al., 2002; Tekeoglu et al., 2002, Collard et al., 2003). Since, *(A.) rabiei* infection is influenced by the prevailing environment, especially the humidity and temperature in addition to host and inoculum, adequate disease development needs specialized individuals for phenotyping of mapping populations as accurate phenotyping is key to identify the ABR QTLs. The ABR to pathotypes I, II, and III of *(A.) rabiei* or unspecified pathotypes was shown to be governed by 2 to 11 QTLs (Santra et al., 2000, Rajesh et al., 2002, Tekeoglu et al., 2002, Collard et al., 2003, Flandez-Galvez et al., 2003, Aryamanesh et al., 2010, Sabbavarapu et al., 2013, Stephens et al., 2014, Daba et al., 2016, Deokar et al., 2019a, b). The ABR QTLs situated on chromosomes 1, 2, 3, 4, 5, 6, and 7 were identified using inter- as well as intra-specific mapping populations (Udapa and Baum, 2003, Anbessa et al., 2009; Lichtenzveig et al., 2006; Millan et al., 2010; Radhika et al., 2007; Tar'an et al., 2007, Garg et al., 2018, Deokar et al., 2019a, b). These QTLs explained small to medium amounts of the phenotypic variation and act in additive manner (Cho et al., 2004; Anbessa et al., 2009). Madrid et al. (2013) developed a co-dominant marker, CaETR, using allelic variation in sequences of a gene called ethylene receptor-like gene that is positioned in an ABR QTL (QTL_{AR1}). A combination of this marker (linked to QTL_{AR1}) with another co-dominant SCAR marker (SCY17590; linked to another ABR QTL called QTL_{AR2}) successfully identified AB resistant and susceptible genotypes of chickpea (Madrid et al., 2013). Further confirmation of evaluation of genotypes for ABR in chickpea using these QTLs is still awaited; the study, however, points toward use of combination of major QTLs for estimation of phenotype of a plant.

Availability of reference genome sequences in chickpea allowed delineation of ABR QTL regions. Daba et al. (2016) following identification of eight QTLs delineated QTL regions on chromosome 3 (interval 8.6–23.11 cM) and on chromosome 8 (interval 53.88–62.33 cM). WGS was also used to identify ABR QTLs. WGS revealed more QTLs (11 QTLs in first population and 6 QTLs in second population)

(Deokar et al., 2019b) than those mapped previously. The QTLs were on five chromosomes (Ca1, Ca2, Ca4, Ca6, Ca7) and those on Ca1, Ca4, Ca6, and Ca7 in first population overlapped with the previously detected QTLs (Deokar et al., 2019b). Moreover, the QTL on Ca4 was present on both the populations. Using in silico analysis of this QTL region, six candidate genes were identified and association of these genes was validated for blight resistance in the second population (Deokar et al., 2019b). In an earlier study, Deokar et al. (2019a) also identified ABR QTLs on Ca3 and Ca5. The Fst genome-scan and GWAS narrowed down the ABR QTL region on chromosome 4 to a 100 kb region (AB4.1) wherein 12 predicted genes were identified (Li et al., 2017). The genes belonged to four categories i.e., NBS-LRR receptor like kinase, wall-associated kinase, zinc finger protein, and serine/threonine protein kinases (Li et al., 2017). Kumar et al. (2018) used multiple QTL-sequencing (mQTL-seq) approach and identified three ABR QTLs (two major QTLs: *qABR4.1* and *qABR4.2* and a minor QTL: *qABR4.3*). The *qABR4.1* region was further delineated to 4.568−4.618 Mb and physical map of the delineated region was constructed. *In silico* analysis revealed 4 genes in this region wherein a gene, *AT-HOOK MOTIF CONTAINING NUCLEAR LOCALIZED* (*AHL*) named *CaAHL18* for chickpea, overexpressed in AB-resistant parent after inoculation with *A. rabiei* (Kumar et al., 2018) rendering it a potential gene associated with blight resistance in chickpea. Further analysis revealed that allelic variation was associated with higher expression (Kumar et al., 2018).

In brief, ABR in chickpea is governed by several QTLs, up to 11 and these QTLs are situated on all chromosomes except chromosome 5 depicting the complexity of ABR in chickpea. Despite such complexity, focus remained on two major QTLs (*qABR4.1*, syn: QTLAR1 [Cho et al., 2004; Iruela et al., 2006; Madrid et al., 2012] and *qABR4.2*, syn: QTL ar2b [Madrid et al., 2012; Udupa and Baum, 2003]), since *A. rabiei* populations comprise several pathotypes and boundaries of QTLs appear to change with the pathotype used to identify the QTLs as has been seen in studies of Li et al. (2017) and Deokar et al. (2019b). In these studies, different pathotypes were used and QTL intervals for *qABR4.1* did not overlap (Deokar et al., 2019b). Inconsistencies in genome assemblies in different versions of the chickpea draft genomes have been reported (Deokar et al., 2014; Parween et al., 2015; Ruperao et al., 2014). Such inconsistencies may be one of the reasons of nonoverlapping of QTL regions among different studies, e.g., QTLs, CPR01-qAB6.1, and CPR01-qAB6.2 were mapped to two locations on chromosome 6 in genome assembly v2.6 whereas to one location in genome assembly version 1 (Deokar et al., 2019b). Nonetheless, multiple QTLs on different chromosomes depict the complexity of ABR in chickpea. At the same time, multiple QTLs also point toward the difficulties a breeder may face in transferring blight resistance using MAS to agronomically superior but blight susceptible cultivars. In an earlier study, a combination of major QTLs (QTLAR1, syn: *qABR4.1* and QTL ar2b, syn: *qABR4.2*) was used to predict the blight resistant phenotype (Madrid et al., 2013). The findings of Madrid et al. (2013), if confirmed, indicate that out of several QTLs, major QTLs can be used to accurately identify the resistant genotypes. Use of limited QTLs for blight resistant phenotyping may pave the way for foreground selection using markers for QTLs in chickpea.

7.6.1.3 Genetic mapping of QTLs governing resistance to other diseases

Chickpea is also infected by other pathogens such as *Phytophthora medicaginis* (disease: phytophthora root rot) and *Botrytis cinerea* Pers. ex. Fr. (disease: botrytis gray mold). The resistance to *P. medicaginis* is available in *C. arietinum* (moderate field resistance) and in *C. echinospermum* (higher level of resistance than *C. arietinum*). QTLs linked to phytophthora root rot resistance were identified in *C. arietinum* as well as *C. echinospermum* (Amalraj et al., 2019). The QTLs for phytophthora root rot resistance in *C. arietinum* were present on chromosomes 5 and 6 and those in *C. echinospermum* on chromosomes 3 and 6 (Amalraj et al., 2019). It appears that resistance sources are different on both the species of *Cicer* (Amalraj et al., 2019) and for introgression of better resistance to phytophthora root rot in chickpea, resistance from *C. arietinum* as well as *C. echinospermum* must be combined into the recurrent parent. Three QTLs which together accounted for 43.6% of the resistance to *Botrytis cinerea* were identified in a moderately resistant genotype ICCV 2 (Anuradha et al., 2011). One QTL was present on LG 6A whereas other two QTLs were present on LG 3.

7.6.1.4 Genetic mapping of agro-morphological traits

Plant growth habits determine the yield potential of crops as well as ease of field operations such as mechanical harvesting. It has been demonstrated adequately that the growth habit traits in plants such as number of pods and branches are correlated to yield and, hence, development of crops with desirable growth habits is always a priority area in breeding. With advancements in gene and marker technologies, efforts were diverted to ascertain genetic bases of growth traits, map the genes (in case of mono- or di-genic traits) and QTLs (in case of traits governed by several genes) governing different traits, and isolate the genes. Though chickpea is considered the poor man crops, information, especially on the mapping of genes/QTLs related to several growth-related traits, is available at present.

7.6.1.5 Monogenic/digenic traits

Chickpea usually bear one flower and one pod per node. but some desi lines of chickpea possess double podding character wherein two flowers and two pods develop at each node. Double podding has been reported to increase productivity in chickpea; however, some studies disagree with this claim. This character resulted from a natural mutation and is inherited by a recessive gene called "*s*" or "*sfl*." The *sfl* was mapped for the first time on LG6 at 4.84 cM apart from the STMS marker TA80 (Rajesh et al., 2002) and subsequently on the same LG by other workers (Cho et al., 2002; Radhika et al., 2007; Ali et al., 2016). The region surrounding *sfl* was enriched by addition of eight new STMS markers and the locus was delineated between markers, CAGM27819 and CAGM27777, in an interval of 2 cM. On physical map, the target region of the QTL spanned 92.6 Kb. *In silico* analysis revealed the presence of seven annotated candidate genes for double podding trait in the QTL region of chickpea (Ali et al., 2016).

Domestication process has favored the selection of upright types of chickpeas as the cultivated chickpeas are erect and semierect (bushy), whereas annual wild species of *Cicer* are prostrate in growth habit. The erect and semierect traits in chickpea are governed by two major genes called *Hg/hg* and *Hg2/hg2* (Kazan et al., 1993; Ali et al., 2015). The gene *Hg/hg* was mapped on CaLG 3 (Kazan et al., 1993; Cobos et al., 2009; Aryamanesh et al., 2010) whereas *Hg2/hg2* onto CaLG 1 (Ali et al., 2015). Using in silico analysis, the physical position of the *Hg2/hg2* was deduced on Ca1 intergenic region between 15,358,505−15,357,543 Mb in the negative strand and the markers Indel 1 and Indel 2 flanked the *Hg2/hg2* at 9.4 and 14.3 cM, respectively (Ali et al., 2015).

7.6.1.6 Polygenic traits

7.6.1.6.1 Agro-morphological traits

The traits such as yield, plant type, flowers per plant, and pods per plant are complex traits as these are the outcome of interactions of several genes. QTLs for several of such traits have been identified in chickpea. Agronomists desire to shorten crop growth periods without compromising yields as short duration allows to grow crops in short season environments or to accommodate more crops per annum. The crops which flower early are expected to mature early and have more reproductive period; and hence, early flowering is a desirable trait in crop breeding. In crops like Arabidopsis, genes governing early flowering have been identified and well characterized. In chickpea, QTLs governing flowering time (Aryamanesh et al., 2010; Das et al., 2015a; Upadhyaya et al., 2015; Daba et al., 2016; Srivastava et al., 2017) and maturity time (Das et al., 2015a) have been mapped. These QTLs are located on chromosomes 1, 2, 3 (Das et al., 2015a), chromosomes 3 and 8 (Daba et al., 2016), chromosome 5 (Ridge et al., 2017), and chromosome 4 (Srivastava et al., 2017). Daba et al. (2016) delineated the QTL regions to 8.6−23.11 cM (chromosome 3) and 53.88−62.33 cM (chromosome 8). The candidate genes within the flowering time locus on Ca5 were also identified (Ridge et al., 2017). One of the genes in the locus was an orthologue of Arabidopsis *EARLY FLOWERING3* gene (Ridge et al., 2017). The gene in chickpea was named *CaELF3a* (derived from accession ICCV 96,029) and was demonstrated to be strong regulator of flowering time (Ridge et al., 2017). Physical map of QTL regions for flowering time on chromosome 4 was also drawn and the flowering time region was mapped to 907.1 kb and 1.99 Mb and 757.7 kb and 1.39 Mb (Srivastava et al., 2017). Allelic variations in regulatory and coding regions of *efl1* (*early flowering 1*) and *GI* (GIGANTEA) genes were also elucidated (Srivastava et al., 2017). The genes *efl-1*, *efl-2*, and *efl-3* govern flowering time in other crops. In chickpea, these genes are present on different chromosomes as QTLs corresponding to flowering time genes *efl-1* from ICCV 96029 (CaLG04), *efl-3* from BGD 132 (CaLG08), and *efl-4* from ICC 16641 (CaLG06) were mapped on different chromosomes (Mallikarjuna et al., 2017). Gaur et al. (2015) studied allelic relationships of early flowering in several genotypes of chickpea and identified four genes named as *efl-1*, *efl-2*, *efl-31*, and *efl-4* where *efl-4* was a new report for chickpea. Using approach almost similar to these studies, eight potential known/candidate genes showing association with regulation of flowering in chickpea were: *efl1*, *FLD* (*Flowering locus D*), *GI* (*gigantia*), *Myb* (*Myeloblastosis*), *SFH3* (*SEC14-like 3*), *bZIP* (*basic-leucine zipper*), *bHLH* (*basic helix-loop-helix*), and *SBP* (*SQUAMOSA* promoter binding protein) (Upadhyaya et al., 2015). It was also demonstrated that the days to flowering and ABR were negatively correlated (Aryamanesh et al., 2010; Daba et al., 2016) indicating that breeding both for early flowering and blight resistance might be tedious. Further studies are, however, needed to prove universality of this phenomenon and mechanisms underlying negative correlation between these two traits. It is obvious that significant advances in flowering time genetics have been starting from QTLs to physical locations of QTLs on physical maps and identification of genes governing flowering time. Further studies may pin point the role of specific genes governing flowering time in chickpea. Acquisition of such knowledge, however, was not possible without recent advances in biotechnology.

While early flowering reduces crop duration, increased seed weight and pod number enhances productivity. Yield, a complex trait in plants, is governed by several genes and is directly or inversely correlated to several other traits called as yield-related traits. The breeding strategies usually aim to enhance yield in crop plants including chickpea where aim may either be to enhance yield or induce host resistance to diseases/pests that indirectly increase yield by reducing crop losses. Some recent studies focused on identification of QTLs for yield-related traits in chickpea. QTLs (two nos) for pod number were identified on chromosome 4 followed by detection of a gene, *pentatricopeptide repeat* (*PPR*), in the major QTL which showed pod-specific expression (~3.5-fold upregulation in high pod number-containing parental accessions and homozygous individuals of two mapping populations) (Das et al., 2015b). Several QTLs for seed weight are reported and in some studies genes in QTL regions were also identified using in silico approaches. Seed weight was associated with a major QTL called CaqSW1.1 on chromosome 1 (Das et al., 2015c). Physical length of the QTL was 1.37 Mb and the region was further narrowed down to 35 kb interval (Das et al., 2015c). The 35 kb interval contained six genes of which one gene (*photomorphogenic9 [COP9] signalosome complex subunit 8* [*CSN8*] showed seed-specific expression. The gene also showed differential up-/downregulation in low- and high-seed-weight parents and homozygous individuals (Das et al., 2015c). Under rainfed conditions, seed weight was associated with two major genomic regions situated on CaLG01 (1.08 Mb) and CaLG04 (2.7 Mb) which contained four putative genes for seed weight (Singh et al., 2016). Using a combination of high-resolution QTL mapping, marker-based comparative genome mapping, and differential expression profiling, a 126.8 kb region of the genome that harbored seed weight QTL was delineated (Bajaj et al., 2015b). In a subsequent study Bajaj et al. (2016) used a technique called EcoTILLING-Based Association Mapping on a diverse set (192) of desi and kabuli genotypes to identify eight SNP alleles in the eight transcription factor genes regulating seed weight of chickpea. Three major QTLs each for pod number and seed yield were identified on chromosomes 2, 4, and 6 and one of the major QTL was delineated on physical intervals (0.89–2.94 Mb) (Srivastava et al., 2016). Fifteen QTLs (Bajaj et al., 2015b) and 33 genomic regions having 35 QTLs (Kujur et al., 2015) for yield-related traits, i.e., pod number/plant, seed number/plant, and 100 seed weight, were also identified. Further detailed studies for several morphological, phenological, and yield-related traits revealed 45 main-effect QTLs and 973 epistatic QTLs for these traits (Varshney et al., 2014a) or 20 QTLs associated with seed traits (Verma et al., 2015), 21 QTLs associated with plant vigor and canopy conductance of which 13 M-QTLs on CaLG04 were for plant vigor (Sivasakthi et al., 2018). Interestingly, drought tolerance QTL hotspot (see description in a later section) is a plant vigor locus as most of plant vigor QTLs co-mapped with drought tolerance "QTL-hotspot" region on CaLG04 (Sivasakthi et al., 2018). Candidate genes for plant height in chickpea were also identified. Five QTLs were associated with plant height and five candidate genes ("*cytochromec-biosynthesis protein*, *malic oxidoreductase*, *NADH dehydrogenase*, *iron-sulfur protein*, *expressed protein,* and *bZIP transcription factor*") appear to be associated with plant height in chickpea (Kujur et al., 2016). Seed yield in chickpea was associated to six trait-associated genes with SNPs that included quantitative trait nucleotides (QTNs) (Basu et al., 2018). Among these genes, *pentatricopeptide repeat* (*PPR*) gene was found most-promising in chickpea.

Chickpea proteins have good in vitro digestibility and are considered to be the best among pulses. To further knowledge of protein synthesis in chickpea seed, four QTLs that showed marker trait associations, two of which were on LG3 and LG5, were identified (Jadhav et al., 2015). In LG3 marker trait association, 29 genes were identified (Jadhav et al., 2015). Similar to QTLs for ABR, the scientific community has moved from QTL detection to identification of genes for some of the agromorphological traits. These advancements are expected to unravel genic bases of regulation of these traits and exploitation of that knowledge in development of designer chickpea or pulse crops.

7.7 Genetic mapping of abiotic stress tolerance

As described in the beginning, three major abiotic stresses, i.e., drought, cold and heat, occur during chickpea growth period in several parts of the world and are responsible for reducing chickpea grain yields. Chickpea faces drought during all stages of plant growth, and this stress is worldwide in occurrence. Drought at reproductive stage of growth is most devastating as it reduces flower formation and leads to abortion of already developed flowers. Young pods do not bear seeds, whereas seed filling is restricted in older pods. Development of drought-tolerant chickpea is a priority area in chickpea breeding; however, the mechanisms of tolerance to drought are poorly understood. Drought tolerance is a polygenic trait and QTLs governing drought tolerance have been identified (Rehman et al., 2011; Hamwieh et al., 2013; Roorkiwal et al., 2014; Thudi et al., 2014; Varshney et al., 2014a). Varshney et al. (2014a) identified 9 QTL clusters containing QTLs for several drought tolerance traits followed by identification of a region with QTL hotspots for drought tolerance. The QTL hotspot region spanned 29 cM (7.74 Mb on physical map). Jaganathan et al. (2015) refined this "QTL-hotspot" region to ~14 cM (ca. 3 Mb on physical map) within the region described by Varshney et al. (2014a). This region was further enriched with new markers leading to delineation of a ~300 Kb (13,239,546–13,547,009) snapshot of chickpea genome showing 26 candidate genes (QTL-hotspot_a: 15 genes; QTL-hotspot_b: 11 candidate genes) (Kale et al., 2015). Expression of 12 selected genes was studied in roots exposed to drought stress and control conditions in ICC 4958 and ICC 1882, and 7 genes showed significantly higher expression in the ICC 4958 than ICC 1882 (Kale et al., 2015). These studies on drought tolerance collectively highlight the importance of next-generation sequencing technologies in identifying genes from QTL regions. The drought tolerance may be governed by genes associated with plant vigor. 13 QTLs for plant vigor were identified on CaLG04 and these QTLs were co-mapped with those for drought tolerance (Sivasakthi et al., 2018). The knowledge appears to indicate that at least some of the genes for drought tolerance and plant vigor are common and drought tolerance might be the outcome of regulation of plant vigor genes.

QTLs for salinity tolerance were also identified in chickpea. The salinity tolerance was governed by two genomic regions with QTLs that had six and five different salinity tolerance–associated traits (Pushpavalli et al., 2015). The two regions are on CaLG05 (28.6 cM) and CaLG07 (19.4 cM). Among these two regions, one on CaLG05 imparted higher plant vigor, whereas another one (on CaLG07) imparted higher reproductive success. High yield under salinity was also governed by two QTLs which were within these two genomic regions (Pushpavalli et al., 2015). Among the two regions, CaLG05 possessed 31 and the CaLG07, 17 putative genes for salinity stress (Pushpavalli et al., 2015). The number of genes governing salinity stress may be reduced further, once studies on individual genes are initiated. Similar to drought tolerance, the salinity tolerance QTLs were also co-mapped with other agromorphological traits, e.g., a QTL for salinity governs plant vigor and another governs higher reproductive success (Pushpavalli et al., 2015). Cold is another important stress in winter-sown crops of chickpea. Cold tolerance/susceptibility has been studied extensively in chickpea (Sharma and Nayyar 2014, 2016; Kiran et al., 2019); however, there is only one study on identification of QTL for vernalization response (Samineni et al., 2016). The QTL is situated on LG 33 and contributes 55% of phenotypic variation. 84 flowering-related genes were present within the QTL region and a gene that is related to MADS box plays important role in vernalization in cereals. Understanding the mechanisms underlying vernalization response in chickpea is needed for effective exploitation of wild *Cicer* in chickpea improvement.

The correlation between drought and plant vigor (Sivasakthi et al., 2018), that of salinity tolerance with higher plant vigor and higher reproductive success (Pushpavalli et al., 2015), and negative correlation of days to flowering with ABR (Aryamanesh et al., 2010; Daba et al., 2016) indicate that the gene-trait associations in chickpea are far more complex than initially presumed and selection for one trait may influence positively or negatively the other traits. Nonetheless, studies on agromorphological traits and stress tolerance in chickpea have started unraveling the complexity and interlinkages of these two categories of traits. Though there is an indication of interrelationships of stress tolerance to agromorphological traits, the studies on this aspect in chickpea are in the stage of infancy to draw effective conclusions, and further research may reveal the linkages between biotic/abiotic stress tolerance and agromorphological traits.

7.8 Marker-assisted breeding

7.8.1 Introgression of wilt and blight resistance

Markers linked tightly to genes of interest are used to monitor introgression of genes into elite cultivars through a process called marker-assisted foreground selection. Marker-assisted foreground selection is used routinely for development of improved genotypes/cultivars of crops, especially the ones possessing disease resistance. The markers spread over the genome can also be used to assess the recovery of recurrent parent genome in filial or backcross progenies. The selection of desirable progeny plants through markers is called as marker-assisted selection (MAS), whereas breeding of desirable plants with the aid of markers is called as marker-assisted breeding (MAB). Since fusarium wilt is an important disease of chickpea, efforts were made to develop improved cultivars possessing fusarium wilt resistance genes in the background of recurrent parents (Varshney et al., 2014b; Pratap et al., 2017; Mannur et al., 2019). While Varshmey et al. (2014b) introgressed race 1 resistance from a well-characterized wilt resistant genotype WR-315 (Sharma et al., 2004a,b, Sharma et al., 2005, Sharma and Muehlbauer, 2007a,b) to C214, Pratap et al. (2017) introgressed race 2 wilt resistance from cultivar Vijay (resistant to *Foc* races 1, 2, 3, 4 and 5) to a high-yielding desi variety "Pusa 256" and Mannur et al. (2019) incorporated race 4 resistance gene from resistant WR315 into two cultivars (Annigeri 1 and JG 74). All the studies used markers linked to resistance to introgress the traits into the recurrent parent. Varshney et al. (2014b) developed three lines possessing race 1 resistance (BC_3F_4 with 92.7%−95.2% recurrent parent genome) in the background of susceptible but agronomically superior C-214 using marker-assisted foreground and background selection. In case of cultivar Vijay, loci for resistance to *foc2* were already mapped (Gowda et al., 2009) and improved Pusa 256 possessing resistance to fusarium wilt and high yield was developed (Pratap et al., 2017). Similarly high-yielding wilt resistant (race 4) lines "Super Annigeri 1" with 8% yield increase over recurrent parent and "Improved JG 74" with 25.6% and 53.5% yield increase at two locations were developed (Mannur et al., 2019).

Similar to transfer of wilt resistance in chickpea, blight resistance from ILC 3279 was also introgressed into the recurrent parent (C 214) using marker-assisted foreground and background selection (Varshney et al., 2014b). Eight markers (TA194, TR58, TS82, GA16, SCY17, TA130, TA2, and GAA47) linked to two QTLs (ABQTL-I and ABQTL-II) governing resistance to ascochyta blight were used to transfer resistance from ILC 3279 to C 214. Seven progeny lines possessed both the QTLs and have high recovery of recurrent parent genome (81.7%–85.40%) (Varshney et al., 2014b). The progenies of resistant plants can further be advanced and agronomically superior ones can be released as varieties.

7.8.2 Introgression of double podding character

The double podding character was also introduced successfully into the recurrent parent using marker-assisted backcrossing approach wherein closely linked marker TA80 was used for foreground selection (Taran et al., 2013). In addition to double podding, marker-assisted backcrossing was also used to select BC_2F_1 plants of chickpea derived from intraspecific crosses that possess ABR as well as double podding character (Taran et al., 2013). Following this study, markers (CAGM27819 and CAGM27777) that are more tightly linked to double podding gene (*sfl*) in chickpea were also identified (Ali et al., 2016). The gene was mapped onto LG6 and the *sfl* was within an interval of 2 cM between the two markers. The new markers will provide more precision for MAS or MAB than the previously described markers.

7.9 Genomic selection for prediction of agronomic traits

Advances in sequencing technologies and their widespread availability across globe have led to genomic selection (GS) for important traits. In the GS approach, testing (prediction) population is genotyped but is not phenotyped allowing reduction in period of breeding cycle and increasing genetic gain per unit time. The GS predicts values of agronomic traits in lines using genome-wide marker profiling wherein lines are selected prior to phenotyping. The effectiveness of GS in plants vis-à-vis MAS was demonstrated first by Bernardo and Yu (2007). To confirm the robustness of GS models, four yield-related traits (seed yield, 100 seed weight, days to 50% flowering, and days to maturity) were predicted with GS where prediction accuracy varied from 0.138 (seed yield) to 0.912 (100 seed weight) (Roorkiwal et al., 2016). The GS is still in infancy and the technology has not been applied under field conditions so far. Moreover, agronomic superiority is a complex trait, genic interactions leading to development of most of the traits are far from understood, and hence, practical utility of GS in development of new plant varieties seems remote at present. However, rapid advancements in biotechnological tools indicate that it might not be very long when GS becomes a trusted tool for varietal development in crop plants.

7.10 Future perspectives

Chickpea research and development, despite it being the third most important pulse crop of the world, was persistently neglected, prompting the scientific community to hail it as an "orphan legume." Research efforts remained limited even in India which is the largest producer as well as consumer of chickpea. Initial studies on chickpea markers and genomics were carried out in the United States, Germany, and to a smaller extent in other countries, but the efforts were limited. Major boost to marker technologies and genomics came in the beginning of 21st century from India where the national laboratories and International Crop Research Institute on Semi-Arid Tropics (ICRISAT), Hyderabad (a Food and Agricultural Organization institution under "Consultative Group on International Agricultural Research") developed new SSR, EST-SSR, InDel, and SNP markers and used those for trait associations as well as linkage map development. Several thousand markers were added to the repository of already available a few hundred chickpea markers. At the same time, genome of chickpea was sequenced by two different groups, one led by the National Institute of Plant Genome Research, New Delhi, and another led by ICRISAT, Hyderabad. Subsequently, genomes of several genotypes of chickpea were sequenced. The reference genomes paved the way for in silico analysis of genome sequences to find additional markers. Of late, emphasis was on SNP identification and placing those on chickpea reference genomes. Development of markers based on insertion and deletions in DNA sequences has also been reported in chickpea. These markers called as "InDels" can be detected by gel-based assays and hence, may be more accessible to labs with limited facilities and can be ideal marker systems for MAS and MAB, in addition to SSRs. To have same ease of assay, SNPs especially those linked to traits of interest can be converted to allele-specific markers which can also be detected by gel-based assays. A study in this direction has already been initiated by Hiremath et al. (2012) who converted SNPs into allele-specific PCR (KASPar) markers and used those to detect polymorphism in chickpea genotypes. A tool called "KASP-Hedgehog" for generation of allele-specific PCR primers from several types of sequence data has already been developed (Alsamman et al., 2019). Though several thousands of markers have been added for chickpea so far, efforts need to be continued to develop additional markers and place those on chickpea linkage maps. More dense maps have markers linked closely to several genes of interest and such markers can be more useful for MAS or MAB.

Increased development of markers in chickpea has improved marker densities on chickpea maps. Currently, several maps based on SSR, InDel, SNP marker, etc. are available, but only a few have attempted to combine more than one type of markers on the maps. Development of an integrated map, preferably the intraspecific one containing all the presently available markers, would be more useful to identify markers linked closely to the genes of interest and for MAB/MAS. Since wild progenitor of chickpea, i.e., *C. reticulatum*, is a repository of some useful genes for biotic and abiotic stress tolerance, saturated interspecific map will be useful to transfer those genes into the cultigen using MAS. Cloning of chickpea genes governing important traits is a neglected area and needs immediate attention. Cloned genes apart from other uses are important for functional genomics and development of transgenic plants. Beginning in this direction may be made from cloning genes for simpler traits such as fusarium wilt resistance where resistance is monogenic to oligogenic (Sharma and Muehlbauer, 2007a,b) or prostrate growth habit that is governed by one to two genes (Kazan et al., 1993; Ali et al., 2015). Cloned gene sequences can be of immense use in MAB as those can be used to develop gene-derived markers for highly accurate foreground selection. Increased marker densities, especially in QTL regions, are also useful to transfer complete QTLs to the recurrent parents. Majority of the traits in chickpea, e.g., resistance to ascochyta blight, botrytis gray mold, and phytophthora root rot, are governed by QTLs, and transfer of these QTLs to recurrent parents is desirable for improving agronomically desirable cultivars. Efficacy of prediction of ABR in chickpea using markers for QTLs has already been demonstrated in chickpea (Madrid et al., 2013). In addition to foreground selection, markers spread over all the chromosomes are also desirable for background selection. Some advancements have already been made to introgress the fusarium wilt resistance and double podding in chickpea, but the efforts are limited. Rapid advancements in marker and map development in chickpea indicate that marker technology will be used widely to monitor introgression of genes or recovery of the recurrent genome.

Recent advances in biotechnology, especially in sequencing and sequence analysis, allowed delineation of QTL regions on reference genomes and identification of candidate genes in QTL regions. In the absence of reference genomes, WGRS can be used to identify the QTL regions and candidate genes therein. In some cases such as flowering time (Srivastava et al., 2017), ABR (Li et al., 2017; Kumar et al., 2018), and drought tolerance (Kale et al., 2015), etc., candidate genes within QTL regions have been identified. Same strategies can be used to identify candidate genes in QTL regions governing other traits or genes in other QTLs for ABR and drought tolerance. Presence of genes in QTL regions is indicative of their role in a particular trait but not a confirmation. To establish association of genes with particular traits, further studies are needed. Ascertaining the role of a candidate gene in governing a particular trait is a herculean task but may be rewarding in the long run as it will untangle the regulation of gene expression underlying complex traits, and the information may be used to develop designer crops suited to climate change scenarios that may feed the ever-growing world population.

Phenotype of a plant if predicted from the genomic information can reduce the breeding efforts by focusing only on limited individuals. Marker trait associations can be used to predict the phenotype of the plant and the concept is already a reality for characters governed by monogenes such as fusarium wilt resistance in chickpea, blast resistance in rice, and rust resistance in wheat where markers linked closely to the gene of interest are used to predict the phenotype of the plant. A step forward in this direction is "genomic selection for prediction of agronomic traits" wherein the aim is to focus on complex traits governed by several genes and select the individuals based on the genotype that will produce a plant with desirable phenotype. Such selection is expected to avoid phenotypic selection under field conditions and pave the way for speedy development of improved varieties. Currently, such selection for complex traits seems remote, but improved understanding of marker/gene trait associations appears to make it a reality in near future. It may start with selection of a few less complex traits followed by gradual advancement to select highly complex traits such as yield. If recent evidences are some indicative, same QTLs govern more than one trait e.g., drought tolerance and plant vigor QTLs are co-mapped (Sivasakthi et al., 2018), QTLs for salinity also control plant vigor and reproductive success (Pushpavalli et al., 2015); prediction of phenotype based on designer genotype seems to be bit difficult, but with further understanding of marker/gene-trait associations, prediction of complex traits is expected to become a reality.

From slow pace in the beginning, the marker technologies in chickpea have developed rapidly during the last one and a half decade and probably, the crop is no more an "orphan crop." Careful analysis of published information on marker technologies does, however, present a disturbing trend as most of the information is from a few laboratories unlike wheat and rice where research work is widespread. Vast scale exploitation of marker technologies is impossible until more laboratories undertake the research on these aspects and those working on varietal/germplasm development start exploiting these tools routinely for development of improved chickpea.

References

Agrawal, P.K., Katiyar, A.K., 2008. Validation of chickpea-STMS markers and DNA fingerprinting in lentil (*Lens culinaris* subsp. *culinaris*) cultivars of India. Indian J. Genet. Plant Breed. 68 (2), 149–156.

Agarwal, G., Sabbavarapu, M.M., Singh, V.K., Thudi, M., Sheelamary, S., Gaur, P.M., Varshney, R.K., 2015. Identification of a non-redundant set of 202 in silico SSR markers and applicability of a select set in chickpea (*Cicer arietinum* L.). Euphytica 205 (2), 381–394.

Ahmad, F., Slinkard, A.E., 1992. Genetic relationships in the genus *Cicer* L. as revealed by polyacrylamide gel electrophoresis of seed storage proteins. Theor. Appl. Genet. 84 (5–6), 688–692.

Albers, C.A., Lunter, G., MacArthur, D.G., McVean, G., Ouwehand, W.H., Durbin, R., 2011. Dindel: accurate indel calls from short-read data. Genome Res. 21, 961–973. PMID: 20980555 33.

Ali, L., Azam, S., Rubio, J., Kudapa, H., Madrid, E., Varshney, R.K., et al., 2015. Detection of a new QTL/gene for growth habit in chickpea CaLG1 using wide and narrow crosses. Euphytica 204 (2), 473–485.

Ali, L., Deokar, A., Caballo, C., Tar'an, B., Gil, J., Chen, W., et al., 2016. Fine mapping for double podding gene in chickpea. Theor. Appl. Genet. 129 (1), 77–86.

Alsamman, A.M., Abdelsattar, M., Ibrahim, S.D., Hamwieh, A., 2019. KASP-Hedgehog: Efficient and Routine Tool for Allele Specific Primer Design Using Multidisciplinary Data Types. bioRxiv, p. 522045.

Amalraj, A., Taylor, J., Bithell, S., Li, Y., Moore, K., Hobson, K., Sutton, T., 2018. Mapping resistance to Phytophthora root rot identifies independent loci from cultivated (*Cicer arietinum* L.) and wild (*Cicer echinospermum* P.H. Davis) chickpea. Theor. Appl. Genet. 132 (4), 1017–1033.

Amalraj, A., Taylor, J., Bithell, S., Li, Y., Moore, K., Hobson, K., Sutton, T., 2019. Mapping resistance to Phytophthora root rot identifies independent loci from cultivated (*Cicer arietinum* L.) and wild (*Cicer echinospermum* P.H. Davis) chickpea. Theor. Appl. Genet. 132 (4), 1017–1033.

Anbessa, Y., Taran, B., Warkentin, T.D., Tullu, A., Vandenberg, A., 2009. Genetic analyses and conservation of QTL for ascochyta blight resistance in chickpea (*Cicer arietinum* L.). Theor. Appl. Genet. 119 (4), 757–765.

Anupama, K., Kumar, J., Winter, P., 2018. Resistance to race 1 of Fusarium wilt of chickpea (*Cicer arietinum* L.): genetics and mapping of resistance genes in an intraspecific cross. Indian J. Genet. Plant Breed. 78 (4), 527–529.

Anuradha, C., Gaur, P.M., Pande, S., Gali, K.K., Ganesh, M., Kumar, J., Varshney, R.K., 2011. Mapping QTL for resistance to botrytis grey mould in chickpea. Euphytica 182 (1), 1–9.

Aryamanesh, N., Nelson, M.N., Yan, G., Clarke, H.J., Siddique, K.H.M., 2010. Mapping a major gene for growth habit and QTLs for ascochyta blight resistance and flowering time in a population between chickpea and *Cicer reticulatum*. Euphytica 173 (3), 307–319.

Bajaj, D., Das, S., Badoni, S., Kumar, V., Singh, M., Bansal, K.C., et al., 2015a. Genome-wide high-throughput SNP discovery and genotyping for understanding natural (functional) allelic diversity and domestication patterns in wild chickpea. Sci. Rep. 5, 12468.

Bajaj, D., Srivastava, R., Nath, M., Tripathi, S., Bharadwaj, C., Upadhyaya, H.D., et al., 2016. EcoTILLING-based association mapping efficiently delineates functionally relevant natural allelic variants of candidate genes governing agronomic traits in chickpea. Front. Plant Sci. 7, 450.

Bajaj, D., Upadhyaya, H.D., Khan, Y., Das, S., Badoni, S., Shree, T., et al., 2015b. A combinatorial approach of comprehensive QTL-based comparative genome mapping and transcript profiling identified a seed weight-regulating candidate gene in chickpea. Sci. Rep. 5, 9264.

Basu, U., Bajaj, D., Sharma, A., Malik, N., Daware, A., Narnoliya, L., et al., 2018. Genetic dissection of photosynthetic efficiency traits for enhancing seed yield in chickpea. Plant Cell Environ. 42 (1), 158–173.

Benko-Iseppon, A.M., Winter, P., Huettel, B., Staginnus, C., Muehlbauer, F.J., Kahl, G., 2003. Molecular markers closely linked to fusarium resistance genes in chickpea show significant alignments to pathogenesis-related genes located on *Arabidopsis* chromosomes 1 and 5. Theor. Appl. Genet. 107 (2), 379–386.

Bentley, D.R., Balasubramanian, S., Swerdlow, H.P., Smith, G.P., Milton, J., Brown, C.G., et al., 2008. Accurate whole human genome sequencing using reversible terminator chemistry. Nature 456 (7218), 53–59.

Bernardo, R., Yu, J., 2007. Prospects for genome wide selection for quantitative traits in maize. Crop Sci. 47 (3), 1082–1090.

Botstein, D., White, R.L., Skolnick, M., Davis, R.W., 1980. Construction of a genetic linkage map in man using restriction fragment length polymorphisms. Am. J. Hum. Genet. 32 (3), 314.

Buhariwalla, H.K., Jayashree, B., Eshwar, K., Crouch, J.H., 2005. Development of ESTs from chickpea roots and their use in diversity analysis of the *Cicer* genus. BMC Plant Biol. 5 (1), 16.

Caballo, C., Castro, P., Gil, J., Izquierdo, I., Millan, T., Rubio, J., 2018. STMS (sequence tagged microsatellite site) molecular markers as a valuable tool to confirm controlled crosses in chickpea (*Cicer arietinum* L.) breeding programs. Euphytica 214 (12), 231.

Caballo, C., Madrid, E., Gil, J., Chen, W., Rubio, J., Millan, T., 2019. Saturation of genomic region implicated in resistance to *Fusarium oxysporum* f. sp. *ciceris* race 5 in chickpea. Mol. Breed. 39 (2), 16.

Cho, S., Chen, W., Muehlbauer, F., 2004. Pathotype-specific genetic factors in chickpea (*Cicer arietinum* L.) for quantitative resistance to ascochyta blight. Theor. Appl. Genet. 109, 733–739.

Cho, S., Kumar, J., Shultz, J.L., Anupama, K., Tefera, F., Fred, J., Muehlbauer, F.J., 2002. Mapping genes for double podding and other morphological traits in chickpea. Euphytica 128 (2), 285–292.

Choudhary, S., Gaur, R., Gupta, S., Bhatia, S., 2012. EST-derived genic molecular markers: development and utilization for generating an advanced transcript map of chickpea. Theor. Appl. Genet. 124 (8), 1449–1462.

Choudhary, S., Sethy, N.K., Shokeen, B., Bhatia, S., 2009. Development of chickpea EST-SSR markers and analysis of allelic variation across related species. Theor. Appl. Genet. 118, 591–608.

Choumane, W., Winter, P., Baum, M., Kahl, G., 2004. Conservation of microsatellite flanking sequences in different taxa of Leguminosae. Euphytica 138 (3), 239–245.

Choumane, W., Winter, P., Weigand, F., Kahl, G., 2000. Conservation and variability of sequence-tagged microsatellite sites (STMSs) from chickpea (*Cicer aerietinum* L.) within the genus Cicer. Theor. Appl. Genet. 101 (1–2), 269–278.

Cobos, M.J., Fernández, M.J., Rubio, J., Kharrat, M., Moreno, M.T., Gil, J., Millan, T., 2005. A linkage map of chickpea (*Cicer arietinum* L.) based on populations from Kabuli×Desi crosses: location of genes for resistance to fusarium wilt race 0. Theor. Appl. Genet. 110 (7), 1347–1353.

Cobos, M.J., Winter, P., Kharrat, M., Cubero, J.I., Gil, J., Millan, T., Rubio, J., 2009. Genetic analysis of agronomic traits in a wide cross of chickpea. Field Crop. Res. 111 (1–2), 130–136.

Collard, B.C.Y., Pang, E.C.K., Ades, P.K., Taylor, P.W.J., 2003. Preliminary investigation of QTLs associated with seedling resistance to ascochyta blight from *Cicer echinospermum*, a wild relative of chickpea. Theor. Appl. Genet. 107 (4), 719–729.

Daba, K., Deokar, A., Banniza, S., Warkentin, T.D., Tar'an, B., 2016. QTL mapping of early flowering and resistance to ascochyta blight in chickpea. Genome 59 (6), 413–425.

Das, S., Singh, M., Srivastava, R., Bajaj, D., Saxena, M.S., Rana, J.C., et al., 2015b. mQTL-seq delineates functionally relevant candidate gene harbouring a major QTL regulating pod number in chickpea. DNA Res. 23 (1), 53–65.

Das, S., Upadhyaya, H.D., Bajaj, D., Kujur, A., Badoni, S., Kumar, V., et al., 2015c. Deploying QTL-seq for rapid delineation of a potential candidate gene underlying major trait-associated QTL in chickpea. DNA Res. 22 (3), 193–203.

Das, S., Upadhyaya, H.D., Srivastava, R., Bajaj, D., Gowda, C.L.L., Sharma, S., et al., 2015a. Genome-wide insertion-deletion (InDel) marker discovery and genotyping for genomics-assisted breeding applications in chickpea. DNA Res. 22 (5), 377–386.

Datta, S., Kaashyap, M., Kumar, S., 2010. Amplification of chickpea-specific SSR primers in Cajanus species and their validity in diversity analysis. Plant Breed. 129 (3), 334–340.

Dayanandan, S., Bawa, K.S., Kesseli, R., 1997. Conservation of microsatellites among tropical trees (Leguminosae). Am. J. Bot. 84 (12), 1658–1663.

Deokar, A., Sagi, M., Daba, K., Tar'an, B., 2019b. QTL sequencing strategy to map genomic regions associated with resistance to ascochyta blight in chickpea. Plant Biotechnol. 17 (1), 275–288.

Deokar, A., Sagi, M., Tar'an, B., 2019a. Genome-wide SNP discovery for development of high-density genetic map and QTL mapping of ascochyta blight resistance in chickpea (*Cicer arietinum* L.). Theor. Appl. Genet. 132 (6), 1861–1872.

Deokar, A.A., Ramsay, L., Sharpe, A.G., Diapari, M., Sindhu, A., Bett, K., et al., 2014. Genome wide SNP identification in chickpea for use in development of a high density genetic map and improvement of chickpea reference genome assembly. BMC Genomics 15 (1), 708.

Dinari, A., Niazi, A., Afsharifar, A.R., Ramezani, A., 2013. Identification of upregulated genes under cold stress in cold-tolerant chickpea using the cDNA-AFLP approach. PLoS One 8 (1), e52757.

Doddamani, D., Katta, M.A., Khan, A.W., Agarwal, G., Shah, T.M., Varshney, R.K., 2014. CicArMiSatDB: the chickpea microsatellite database. BMC Bioinf. 15 (1), 212.

Flandez-Galvez, H., Ford, R., Pang, E., Taylor, P., 2003. An intraspecific linkage map of chickpea (*Cicer arietinum* L.) genome based on sequence tagged microsatellite site and resistance gene analog markers. Theor. Appl. Genet. 106, 1447–1456.

Garg, R., Patel, R.K., Jhanwar, S., 2011. Gene discovery and tissue-specific transcriptome analysis in chickpea with massively parallel pyrosequencing and web resource development. Plant Physiol. 156, 1661–1678.

Garg, T., Mallikarjuna, B.P., Thudi, M., Samineni, S., Singh, S., Sandhu, J.S., et al., 2018. Identification of QTLs for resistance to *Fusarium wilt* and ascochyta blight in a recombinant inbred population of chickpea (*Cicer arietinum* L.). Euphytica 214 (3), 45.

Gaur, P.M., Stinkard, A.E., 1990b. Inheritance and linkage of isozyme coding genes in chickpea. J. Hered. 81 (6), 455–461.

Gaur, P.M., Slinkard, A.E., 1990a. Genetic control and linkage relations of additional isozyme markers in chickpea. Theor. Appl. Genet. 80 (5), 648–656.

Gaur, R., Azam, S., Jeena, G., Khan, A.W., Choudhary, S., Jain, M., et al., 2012. High-throughput SNP discovery and genotyping for constructing a saturated linkage map of chickpea (*Cicer arietinum* L.). DNA Res. 19 (5), 357–373.

Gaur, R., Jeena, G., Shah, N., Gupta, S., Pradhan, S., Tyagi, A.K., et al., 2015. High density linkage mapping of genomic and transcriptomic SNPs for synteny analysis and anchoring the genome sequence of chickpea. Sci. Rep. 5, 13387.

Gaur, R., Sethy, N.K., Choudhary, S., Shokeen, B., Gupta, V., Bhatia, S., 2011. Advancing the STMS genomic resources for defining new locations on the intraspecific genetic linkage map of chickpea (*Cicer arietinum* L.). BMC Genomics 12, 117.

Gowda, S.J.M., Radhika, P., Kadoo, N.Y., Mhase, L.B., Gupta, V.S., 2009. Molecular mapping of wilt resistance genes in chickpea. Mol. Breed. 24 (2), 177–183.

Gupta, S., Kumar, T., Verma, S., Bharadwaj, C., Bhatia, S., 2015. Development of gene-based markers for use in construction of the chickpea (*Cicer arietinum* L.) genetic linkage map and identification of QTLs associated with seed weight and plant height. Mol. Biol. Rep. 42 (11), 1571–1580.

Gupta, S., Nawaz, K., Parween, S., Roy, R., Sahu, K., Kumar, P.A., et al., 2016. Draft genome sequence of *Cicer reticulatum* L., the wild progenitor of chickpea provides a resource for agronomic trait improvement. DNA Res. 24 (1), 1–10.

Gutierrez, M.V., Patto, M.V., Huguet, T., Cubero, J.I., Moreno, M.T., Torres, A.M., 2005. Cross-species amplification of *Medicago truncatula* microsatellites across three major pulse crops. Theor. Appl. Genet. 110 (7), 1210–1217.

Halila, I., Cobos, M.J., Rubio, J., Millán, T., Kharrat, M., Marrakchi, M., Gil, J., 2009. Tagging and mapping a second resistance gene for *Fusarium wilt* race 0 in chickpea. Eur. J. Plant Pathol. 124, 87–92.

Hamwieh, A., Imtiaz, M., Malhotra, R.S., 2013. Multi-environment QTL analyses for drought-related traits in a recombinant inbred population of chickpea (*Cicer arientinum* L.). Theor. Appl. Genet. 126, 1025–1038.

Hiremath, P.J., Kumar, A., Penmetsa, R.V., Farmer, A., Schlueter, J.A., Chamarthi, S.K., et al., 2012. Large-scale development of cost-effective SNP marker assays for diversity assessment and genetic mapping in chickpea and comparative mapping in legumes. Plant Biotechnol. 10 (6), 716–732.

Huttel, B., Winter, P., Weising, K., Choumane, W., Weigand, F., Kahl, G., 1999. Sequence-tagged microsatellite site markers for chickpea (*Cicer arietinum* L.). Genome 42 (2), 210–217.

International Human Genome Sequencing Consortium, 2001. Initial sequencing and analysis of the human genome. Nature 409, 860–921.

Iruela, M., Rubio, J., Barro, F., Cubero, J.I., Millan, T., Gil, J., 2006. Detection of two quantitative trait loci for resistance to ascochyta blight in an intraspecific cross of chickpea (*Cicer arietinum* L.): development of SCAR markers associated with resistance. Theor. Appl. Genet. 112, 278–287.

Jadhav, A.A., Rayate, S.J., Mhase, L.B., Thudi, M., Chitikineni, A., Harer, P.N., et al., 2015. Marker-trait association study for protein content in chickpea (*Cicer arietinum* L.). J. Genet. 94 (2), 279–286.

Jaganathan, D., Thudi, M., Kale, S., Azam, S., Roorkiwal, M., Gaur, P.M., et al., 2015. Genotyping-by-sequencing based intra-specific genetic map refines a "QTL-hotspot" region for drought tolerance in chickpea. Mol. Genet. Genom. 290 (2), 559–571.

Jain, A., Roorkiwal, M., Kale, S., Garg, V., Yadala, R., Varshney, R.K., 2019. InDel markers: an extended marker resource for molecular breeding in chickpea. PLoS One 14 (3), e0213999.

Jain, M., Misra, G., Patel, R.K., Priya, P., Jhanwar, S., Khan, A.W., et al., 2013. A draft genome sequence of the pulse crop chickpea (*Cicer arietinum* L.). Plant J. 74 (5), 715–729.

Jayashree, B., Buhariwalla, H.K., Shinde, S., Crouch, J.H., 2005. A legume genomics resource: the chickpea root expressed sequence tag database. Electron. J. Biotechnol. 8 (2), 8–13.

Jendoubi, W., Bouhadida, M., Millan, T., Kharrat, M., Gil, J., Rubio, J., Madrid, E., 2016. Identification of the target region including the foc0 1/foc0 1 gene and development of near isogenic lines for resistance to *Fusarium Wilt* race 0 in chickpea. Euphytica 210 (1), 119–133.

Jhanwar, S., Priya, P., Garg, R., Parida, S.K., Tyagi, A.K., Jain, M., 2012. Transcriptome sequencing of wild chickpea as a rich resource for marker development. Plant Biotechnol. 10 (6), 690–702.

Jingade, P., Bhosale, L.V., Sanjayrao, J.A., Rajanna, R., Jain, M., Ravikumar, R.L., 2014. Characterization of microsatellite markers, their transferability to orphan legumes and use in determination of genetic diversity among chickpea (*Cicer arietinum* L.) cultivars. J. Crop Sci. Biotechnol. 17 (3), 191–199.

Jingade, P., Ravikumar, R.L., 2015. Development of molecular map and identification of QTLs linked to *Fusarium wilt* resistance in chickpea. J. Genet. 94 (4), 723–729.

Kale, S.M., Jaganathan, D., Ruperao, P., Chen, C., Punna, R., Kudapa, H., et al., 2015. Prioritization of candidate genes in "QTL-hotspot" region for drought tolerance in chickpea (*Cicer arietinum* L.). Sci. Rep. 5, 15296.

Kazan, K.M.F.J., Muehlbauer, F.J., Weeden, N.E., Ladizinsky, G., 1993. Inheritance and linkage relationships of morphological and isozyme loci in chickpea (*Cicer arietinum* L.). Theor. Appl. Genet. 86 (4), 417–426.

Khajuria, Y.P., Saxena, M.S., Gaur, R., Chattopadhyay, D., Jain, M., Parida, S.K., Bhatia, S., 2015. Development and integration of genome-wide polymorphic microsatellite markers onto a reference linkage map for constructing a high-density genetic map of chickpea. PLoS One 10 (5), e0125583.

Kiran, A., Kumar, S., Nayyar, H., Sharma, K.D., 2019. Low temperature induced aberrations in male and female reproductive organ development cause flower abortion in Chickpea. Plant Cell Environ. 42, 2075–2089.

Krawczak, M., 1999. Informativity assessment for biallelic single nucleotide polymorphisms. Electrophoresis 20, 1676–1681.

Kujur, A., Bajaj, D., Saxena, M.S., Tripathi, S., Upadhyaya, H.D., Gowda, C.L.L., et al., 2013. Functionally relevant microsatellite markers from chickpea transcription factor genes for efficient genotyping applications and trait association mapping. DNA Res. 20 (4), 355–374.

Kujur, A., Bajaj, D., Upadhyaya, H.D., Das, S., Ranjan, R., Shree, T., et al., 2015. Employing genome-wide SNP discovery and genotyping strategy to extrapolate the natural allelic diversity and domestication patterns in chickpea. Front. Plant Sci. 6, 162.

Kujur, A., Upadhyaya, H.D., Bajaj, D., Gowda, C.L.L., Sharma, S., Tyagi, A.K., Parida, S.K., 2016. Identification of candidate genes and natural allelic variants for QTLs governing plant height in chickpea. Sci. Rep. 6, 27968.

Kumar, K., Purayannur, S., Kaladhar, V.C., Parida, S.K., Verma, P.K., 2018. mQTL-seq and classical mapping implicates the role of an AT-HOOK MOTIF CONTAINING NUCLEAR LOCALIZED (AHL) family gene in ascochyta blight resistance of chickpea. Plant Cell Environ 41 (9), 2128–2140.

Labdi, M., Malhotra, R.S., Benzohra, I.E., Imtiaz, M., 2013. Inheritance of resistance to *Ascochyta rabiei* in 15 chickpea germplasm accessions. Plant Breed. 132, 197–199.

Li, Y., Ruperao, P., Batley, J., Edwards, D., Davidson, J., Hobson, K., Sutton, T., 2017. Genome analysis identified novel candidate genes for ascochyta blight resistance in chickpea using whole genome re-sequencing data. Front. Plant Sci. 8, 359.

Lichtenzveig, J., Bonfil, D.J., Zhang, H.B., Shtienberg, D., Abbo, S., 2006. Mapping quantitative trait loci in chickpea associated with time to flowering and resistance to *Didymella rabiei* the causal agent of ascochyta blight. Theor. Appl. Genet. 113 (7), 1357–1369.

Madrid, E., Chen, W., Rajesh, P.N., Castro, P., Millán, T., Gil, J., 2013. Allele-specific amplification for the detection of ascochyta blight resistance in chickpea. Euphytica 189 (2), 183–190.

Madrid, E., Rajesh, P.N., Rubio, J., Gil, J., Millan, T., Chen, W., 2012. Characterization and genetic analysis of an EIN4-like sequence (CaETR-1) located in QTLAR1 implicated in ascochyta blight resistance in chickpea. Plant Cell Rep. 31, 1033–1042.

Mallikarjuna, B.P., Samineni, S., Thudi, M., Sajja, S.B., Khan, A.W., Patil, A., et al., 2017. Molecular mapping of flowering time major genes and QTLs in chickpea (*Cicer arietinum* L.). Front. Plant Sci. 8, 1140.

Mannur, D.M., Babbar, A., Thudi, M., Sabbavarapu, M.M., Roorkiwal, M., Sharanabasappa, B.Y., et al., 2019. Super Annigeri 1 and improved JG 74: two Fusarium wilt-resistant introgression lines developed using marker-assisted backcrossing approach in chickpea (*Cicer arietinum* L.). Mol. Breed. 39 (1), 2.

Mayer, M.S., Tullu, A., Simon, C.J., Kumar, J., Kaiser, W.J., Kraft, J.M., Muehlbauer, F.J., 1997. Development of a DNA marker for fusarium wilt resistance in chickpea. Crop Sci. 37 (5), 1625–1629.

Millan, T., Clarke, H.J., Siddique, K.H., Buhariwalla, H.K., Gaur, P.M., Kumar, J., et al., 2006. Chickpea molecular breeding: new tools and concepts. Euphytica 147 (1–2), 81–103.

Millan, T., Winter, P., Jungling, R., 2010. A consensus genetic map of chickpea (*Cicer arietinum* L.) based on 10 mapping populations. Euphytica 175, 175–189.

Mishra, R.K., Gangadhar, B.H., Nookaraju, A., Kumar, S., Park, S.W., 2012. Development of EST-derived SSR markers in pea (*Pisum sativum*) and their potential utility for genetic mapping and transferability. Plant Breed. 131 (1), 118–124.

Mullis, K., Faloona, F., Scharf, S., Saiki, R.K., Horn, G.T., Erlich, H., 1986. Specific enzymatic amplification of DNA in vitro: the polymerase chain reaction. In: Cold Spring Harbor symposia on quantitative biology, vol. 51. Cold Spring Harbor Laboratory Press, pp. 263–273.

Nayak, S.N., Zhu, H., Varghese, N., Datta, S., Choi, H., Horres, R., et al., 2010. Integration of novel SSR and gene-based SNP marker loci in the chickpea genetic map and establishment of new anchor points with *Medicago truncatula* genome. Theor. Appl. Genet. 120, 1415–1441.

Pandian, A., Ford, R., Taylor, P.W., 2000. Transferability of sequence tagged microsatellite site (STMS) primers across four major pulses. Plant Mol. Biol. Report. 18 (4), 395-395.

Parween, S., Nawaz, K., Roy, R., Pole, A.K., Suresh, B.V., Misra, G., et al., 2015. An advanced draft genome assembly of a desi type chickpea (*Cicer arietinum* L.). Sci. Rep. 5, 12806.

Peakall, R., Gilmore, S., Keys, W., Morgante, M., Rafalski, A., 1998. Cross-species amplification of soybean (*Glycine max*) simple sequence repeats (SSRs) within the genus and other legume genera: implications for the transferability of SSRs in plants. Mol. Biol. Evol. 15 (10), 1275–1287.

Pfaff, T., Kahl, G., 2003. Mapping of gene-specific markers on the genetic map of chickpea (*Cicer arietinum* L.). Mol. Genet. Genom. 269 (2), 243–251.

Pratap, A., Chaturvedi, S.K., Tomar, R., Rajan, N., Malviya, N., Thudi, M., et al., 2017. Marker-assisted introgression of resistance to *Fusarium wilt* race 2 in Pusa 256, an elite cultivar of desi chickpea. Mol. Genet. Genom. 292 (6), 1237–1245.

Pushpavalli, R., Krishnamurthy, L., Thudi, M., Gaur, P.M., Rao, M.V., Siddique, K.H., et al., 2015. Two key genomic regions harbour QTLs for salinity tolerance in ICCV 2×JG 11 derived chickpea (*Cicer arietinum* L.) recombinant inbred lines. BMC Plant Biol. 15 (1), 124.

Qadir, S.A., Datta, S., Singh, N.P., Kumar, S., 2007. Development of highly polymorphic SSR markers for chickpea (*Cicer arietinum* L.) and their use in parental polymorphism. Indian J. Genet. Plant Breed. 67 (4), 329–333.

Radhika, P., Gowda, S.J.M., Kadoo, N.Y., Mhase, L.B., Jamadagni, B.M., Sainani, M.N., et al., 2007. Development of an integrated intraspecific map of chickpea (*Cicer arietinum* L.) using two recombinant inbred line populations. Theor. Appl. Genet. 115, 209–216.

Rajesh, P.N., Muehlbauer, F.J., 2008. Discovery and detection of single nucleotide polymorphism (SNP) in coding and genomic sequences in chickpea (*Cicer arietinum* L.). Euphytica 162 (2), 291–300.

Rajesh, P.N., Tekeoglu, M., Gupta, V.S., Ranjekar, P.K., Muehlbauer, F.J., 2002. Molecular mapping and characterization of an RGA locus RGAPtokin1-2 171 in chickpea. Euphytica 128 (3), 427–433.

Rajesh, P.N., Coyne, C., Meksem, K., Sharma, K.D., Gupta, V., Muehlbauer, F.J., 2004. Construction of a HindIII bacterial artificial chromosome library and its use in identification of clones associated with disease resistance in chickpea. Theor. Appl. Genet. 108 (4), 663–669.

Ratnaparkhe, M.B., Santra, D.K., Tullu, A., Muehlbauer, F.J., 1998a. Inheritance of inter-simple-sequence-repeat polymorphisms and linkage with a *Fusarium wilt* resistance gene in chickpea. Theor. Appl. Genet. 96 (3–4), 348–353.

Ratnaparkhe, M.B., Tekeoglu, M., Muehlbauer, F.J., 1998b. Inter-simple-sequence-repeat (ISSR) polymorphisms are useful for finding markers associated with disease resistance gene clusters. Theor. Appl. Genet. 97 (4), 515–519.

Rehman, A.U., et al., 2011. Mapping QTL associated with traits affecting grain yield in chickpea (*Cicer arietinum* L.) under terminal drought stress. Crop Sci. 51, 450–463.

Reuter, J.A., Spacek, D.V., Snyder, M.P., 2015. High-throughput sequencing technologies. Mol. Cell 58 (4), 586–597.

Ridge, S., Deokar, A., Lee, R., Daba, K., Macknight, R.C., Weller, J.L., Tar'an, B., 2017. The chickpea early flowering 1 (Efl1) locus is an ortholog of Arabidopsis ELF3. Plant Physiol. 175 (2), 802–815.

Roorkiwal, M., Nayak, S.N., Thudi, M., Upadhyaya, H.D., Brunel, D., Mournet, P., et al., 2014. Allele diversity for abiotic stress responsive candidate genes in chickpea reference set using gene based SNP markers. Front. Plant Sci. 5, 248. https://doi.org/10.3389/fpls.2014.00248.

Roorkiwal, M., Jain, A., Kale, S.M., Doddamani, D., Chitikineni, A., Thudi, M., Varshney, R.K., 2018. Development and evaluation of high-density Axiom® Cicer SNP array for high-resolution genetic mapping and breeding applications in chickpea. Plant Biotechnol. J. 16 (4), 890–901.

Roorkiwal, M., Rathore, A., Das, R.R., Singh, M.K., Jain, A., Srinivasan, S., et al., 2016. Genome-enabled prediction models for yield related traits in chickpea. Front. Plant Sci. 7, 1666.

Roorkiwal, M., Sawargaonkar, S.L., Chitikineni, A., Thudi, M., Saxena, R.K., Upadhyaya, H.D., et al., 2013. Single nucleotide polymorphism genotyping for breeding and genetics applications in chickpea and pigeonpea using the BeadXpress platform. Plant Genome 6, 2.

Rubio, J., Hajj-Moussa, E., Kharrat, M., Moreno, M.T., Millan, T., Gil, J., 2003. Two genes and linked RAPD markers involved in resistance to *Fusarium oxysporum f.* sp. *ciceris* race 0 in chickpea. Plant Breed. 122 (2), 188–191.

Ruperao, P., Chan, C.K.K., Azam, S., Karafiatova, M., Hayashi, S., Cızkova, J., Saxena, R.K., et al., 2014. A chromosomal genomics approach to assess and validate the desi and kabuli draft chickpea genome assemblies. Plant Biotechnol. J. 12, 778–786.

Sabbavarapu, M.M., Sharma, M., Chamarthi, S.K., Swapna, N., Rathore, A., Thudi, M., et al., 2013. Molecular mapping of QTLs for resistance to *Fusarium wilt* (race 1) and Ascochyta blight in chickpea (*Cicer arietinum* L.). Euphytica 193 (1), 121–133.

Samineni, S., Kamatam, S., Thudi, M., Varshney, R.K., Gaur, P.M., 2016. Vernalization response in chickpea is controlled by a major QTL. Euphytica 207 (2), 453–461.

Santra, D.K., Tekeoglu, M., Ratnaparkhe, M., Kaiser, W.J., Muehlbauer, F.J., 2000. Identification and mapping of QTLs conferring resistance to ascochyta blight in chickpea. Crop Sci. 40 (6), 1606–1612.

Sethy, N.K., Choudhary, S., Shokeen, B., Bhatia, S., 2006b. Identification of microsatellite markers from *Cicer reticulatum*: molecular variation and phylogenetic analysis. Theor. Appl. Genet. 112 (2), 347–357.

Sethy, N.K., Shokeen, B., Edwards, K.J., Bhatia, S., 2006a. Development of microsatellite markers and analysis of intraspecific genetic variability in chickpea (*Cicer arietinum* L.). Theor. Appl. Genet. 112 (8), 1416–1428.

Sharma, K.D., Muehlbauer, F.J., 2007a. Genetic mapping of *Fusarium oxysporum* f. sp. *ciceris* race-specific resistance genes in chickpea (*Cicer arietinum* L.). In: Abstract of the International Food Legume Research Conference—IV. Indian Agricultural Research Institute, New Delhi, India, pp. 18–22.

Sharma, K.D., Nayyar, H., 2016. Regulatory networks in pollen development under cold stress. Front. Plant Sci. 7, 402.

Sharma, K.D., Chen, W., Muehlbauer, F.J., 2004a. A consensus set of differential lines for identifying races of *Fusarium oxysporum* f. sp. *ciceris*. Int Pigeonpea Chickpea Newsl 11, 34–36.

Sharma, K.D., Chen, W., Muehlbauer, F.J., 2005. Genetics of chickpea resistance to five races of *Fusarium wilt* and a concise set of race differentials for *Fusarium oxysporum* f. sp. *ciceris*. Plant Dis. 89 (4), 385–390.

Sharma, K.D., Muehlbauer, F.J., 2007b. *Fusarium wilt* of chickpea: physiological specialization, genetics of resistance and resistance gene tagging. Euphytica 157 (1–2), 1–14.

Sharma, K.D., Nayyar, H., 2014. Cold stress alters transcription in meiotic anthers of cold tolerant chickpea (*Cicer arietinum* L.). BMC Res 7 (1), 717.

Sharma, K.D., Winter, P., Kahl, G., Muehlbauer, F.J., 2004b. Molecular mapping of *Fusarium oxysporum* f. sp. *ciceris* race 3 resistance gene in chickpea. Theor. Appl. Genet. 108 (7), 1243–1248.

Sharma, P.C., Huttel, B., Winter, P., Kahl, G., Gardner, R.C., Weising, K., 1995b. The potential of microsatellites for hybridization-and polymerase chain reaction-based DNA fingerprinting of chickpea (*Cicer arietinum* L.) and related species. Electrophoresis 16 (1), 1755–1761.

Sharma, P.C., Winter, P., Bunger, T., Huttel, B., Weigand, F., Weising, K., Kahl, G., 1995a. Abundance and polymorphism of di-, tri-and tetra-nucleotide tandem repeats in chickpea (*Cicer arietinum* L.). Theor. Appl. Genet. 90 (1), 90–96.

Simon, C.J., Muehlbauer, F.J., 1997. Construction of a chickpea linkage map and its comparison with maps of pea and lentil. J. Hered. 88 (2), 115–119.

Singh, V.K., Khan, A.W., Jaganathan, D., Thudi, M., Roorkiwal, M., Takagi, H., et al., 2016. QTL-seq for rapid identification of candidate genes for 100-seed weight and root/total plant dry weight ratio under rainfed conditions in chickpea. Plant Biotechnol. J. 14 (11), 2110–2119.

Sivasakthi, K., Thudi, M., Tharanya, M., Kale, S.M., Kholová, J., Halime, M.H., et al., 2018. Plant vigour QTLs co-map with an earlier reported QTL hotspot for drought tolerance while water saving QTLs map in other regions of the chickpea genome. BMC Plant Biol. 18 (1), 29.

Srivastava, R., Singh, M., Bajaj, D., Parida, S.K., 2016. A high-resolution InDel (insertion–deletion) markers-anchored consensus genetic map identifies major QTLs governing pod number and seed yield in chickpea. Front. Plant Sci. 7, 1362.

Srivastava, R., Upadhyaya, H.D., Kumar, R., Daware, A., Basu, U., Shimray, P.W., et al., 2017. A multiple QTL-Seq strategy delineates potential genomic loci governing flowering time in chickpea. Front. Plant Sci. 8, 1105.

Stephens, A., Lombardi, M., Cogan, N.O.I., Forster, J.W., Hobson, K., Materne, M., Kaur, S., 2014. Genetic marker discovery, intraspecific linkage map construction and quantitative trait locus analysis of ascochyta blight resistance in chickpea (*Cicer arietinum* L.). Mol. Breed. 33 (2), 297–313.

Tar'an, B., Warkentin, T.D., Tullu, A., Vandenberg, A., 2007. Genetic mapping of ascochyta blight resistance in chickpea (*Cicer arietinum* L.) using a simple sequence repeat linkage map. Genome 50 (1), 26–34.

Taran, B., Warkentin, T.D., Vandenberg, A., 2013. Fast track genetic improvement of ascochyta blight resistance and double podding in chickpea by marker-assisted backcrossing. Theor. Appl. Genet. 126 (6), 1639–1647.

Tekeoglu, M., Rajesh, P.N., Muehlbauer, F.J., 2002. Integration of sequence tagged microsatellite sites to the chickpea genetic map. Theor. Appl. Genet. 105, 847–854.

Tekeoglu, M., Tullu, A., Kaiser, W.J., Muehlbauer, F.J., 2000. Inheritance and linkage of two genes that confer resistance to *Fusarium wilt* in chickpea. Crop Sci. 40 (5), 1247–1251.

Temesgen, B., Brown, G.R., Harry, D.E., Kinlaw, C.S., Sewell, M.M., Neale, D.B., 2001. Genetic mapping of expressed sequence tag polymorphism (ESTP) markers in loblolly pine (*Pinus taeda* L.). Theor. Appl. Genet. 102 (5), 664–675.

Thiel, T., Michalek, W., Varshney, R., Graner, A., 2003. Exploiting EST databases for the development and characterization of gene-derived SSR-markers in barley (*Hordeum vulgare* L.). Theor. Appl. Genet. 106 (3), 411–422.

Thudi, M., Upadhyaya, H.D., Rathore, A., Gaur, P.M., Krishnamurthy, L., Roorkiwal, M., et al., 2014. Genetic dissection of drought and heat tolerance in chickpea through genome-wide and candidate gene-based association mapping approaches. PLoS One 9, e96758.

Thudi, M., Bohra, A., Nayak, S.N., Varghese, N., Shah, T.M., Penmetsa, R.V., et al., 2011. Novel SSR markers from BAC-end sequences, DArT arrays and a comprehensive genetic map with 1,291 marker loci for chickpea (*Cicer arietinum* L.). PLoS One 6 (11), e27275.

Tullu, A., Kaiser, W.J., Kraft, J.M., Muehlbauer, F.J., 1999. A second gene for resistance to race 4 of fusarium wilt in chickpea and linkage with a RAPD marker. Euphytica 109 (1), 43–50.

Tullu, A., Muehlbauer, F.J., Simon, C.J., Mayer, M.S., Kumar, J., Kaiser, W.J., Kraft, J.M., 1998. Inheritance and linkage of a gene for resistance to race 4 of fusarium wilt and RAPD markers in chickpea. Euphytica 102 (2), 227–232.

Udupa, S.M., Baum, M., 2003. Genetic dissection of pathotype-specific resistance to aschochyta blight disease in chickpea (*Cicer arietinum* L.) using microsatellite markers. Theor. Appl. Genet. 106, 1196–1202.

Ulrich, C.M., Bigler, J., Velicer, C.M., Greene, E.A., Farin, F.M., Potter, J.D., 2000. Searching expressed sequence tag databases: discovery and confirmation of a common polymorphism in the thymidylate synthase gene. Cancer Epidemiol. Biomark. Prev. 9 (12), 1381–1385.

Upadhyaya, H.D., Bajaj, D., Das, S., Saxena, M.S., Badoni, S., Kumar, V., et al., 2015. A genome-scale integrated approach aids in genetic dissection of complex flowering time trait in chickpea. Plant Mol. Biol. 89 (4–5), 403–420.

Upadhyaya, H.D., Haware, M.P., Kumar, J., Smithson, J.B., 1983a. Resistance to wilt in chickpea. I. Inheritance of late-wilting in response to race 1. Euphytica 32 (2), 447–452.

Upadhyaya, H.D., Smithson, J.B., Haware, M.P., Kumar, J., 1983b. Resistance to wilt in chickpea. II. Further evidence for two genes for resistance to race 1. Euphytica 32 (3), 749–755.

Varshney, R.K., et al., 2014a. Genetic dissection of drought tolerance in chickpea (*Cicer arietinum* L.). Theor. Appl. Genet. 127, 445–462.

Varshney, R.K., Hiremath, P.J., Lekha, P., et al., 2009. A comprehensive resource of drought and salinity responsive ESTs for gene discovery and marker development in chickpea (*Cicer arietinum* L.). BMC Genomics 15, 523.

Varshney, R.K., Mohan, S.M., Gaur, P.M., Chamarthi, S.K., Singh, V.K., Srinivasan, S., et al., 2014b. Marker-assisted backcrossing to introgress resistance to Fusarium wilt race 1 and Ascochyta blight in C 214, an elite cultivar of chickpea. Plant Genome 7, 1–11.

Varshney, R.K., Song, C., Saxena, R.K., Azam, S., Yu, S., Sharpe, A.G., et al., 2013. Draft genome sequence of chickpea (*Cicer arietinum* L.) provides a resource for trait improvement. Nat. Biotechnol. 31 (3), 240.

Verma, S., Gupta, S., Bandhiwal, N., Kumar, T., Bharadwaj, C., Bhatia, S., 2015. High-density linkage map construction and mapping of seed trait QTLs in chickpea (*Cicer arietinum* L.) using Genotyping-by-Sequencing (GBS). Sci. Rep. 5, 17512.

Weising, K., Kaemmer, D., Weigand, F., Epplen, J.T., Kahl, G., 1992. Oligonucleotide fingerprinting reveals various probe-dependent levels of informativeness in chickpea (*Cicer arietinum* L.). Genome 35 (3), 436–442.

Weising, K., Winter, P., Hüttel, B., Kahl, G., 1997. Microsatellite markers for molecular breeding. J. Crop Prod. 1 (1), 113–143.

Williams, J.G., Kubelik, A.R., Livak, K.J., Rafalski, J.A., Tingey, S.V., 1990. DNA polymorphisms amplified by arbitrary primers are useful as genetic markers. Nucleic Acids Res. 18 (22), 6531–6535.

Winter, P., Benko-Iseppon, A.M., Huttel, B., Ratnaparkhe, M., Tullu, A., Sonnante, G., et al., 2000. A linkage map of chickpea (*Cicer arietinum* L.) genome based on recombinant inbred lines from a *C. arietinum* × *C. reticulatum* cross: localization of resistance genes for *Fusarium wilt* races 4 and 5. Theor. Appl. Genet. 101, 1155–1163.

Winter, P., Pfaff, T., Udupa, S.M., Huttel, B., Sharma, P.C., Sahi, S., et al., 1999. Characterization and mapping of sequence-tagged microsatellite sites in the chickpea (*Cicer arietinum* L.) genome. Mol. Genet. Genom. 262, 90–101.

Xing, C., Schumacher, F.R., Xing, G., Lu, Q., Wang, T., Elston, R.C., 2005. Comparison of microsatellites, single-nucleotide polymorphisms (SNPs) and composite markers derived from SNPs in linkage analysis. BMC Genet. 6, S29.

Yu, J.K., Paik, H., Choi, J.P., Han, J.H., Choe, J.K., Hur, C.G., 2010. Functional domain marker (FDM): an in silico demonstration in Solanaceae using simple sequence repeats (SSRs). Plant Mol. Biol. Report. 28, 352–356.

CHAPTER

Genetic transformation

8

Alok Das[1], Shallu Thakur[1], Alok Shukla[1], Prateek Singh[1], Jamal Ansari[1], N.P. Singh[2]

[1]*Division of Plant Biotechnology, ICAR-Indian Institute of Pulses Research, Kanpur, Uttar Pradesh, India;* [2]*ICAR-Indian Institute of Pulses Research, Kanpur, Uttar Pradesh, India*

8.1 Introduction

Chickpea (*Cicer arietinum* L.) is an important grain legume, which has worldwide acceptance as a major source of dietary protein. Besides high protein content (20%−25%), it is a good source of carbohydrate (48.2%−67.6%), starch (41%−50%), fat (6%), and nutritionally important minerals. Chickpeas are grown in more than 50 countries spread over Asia, Africa, Europe, Australia, North America, and South America and are second most important food legume after common bean (*Phaseolus vulgarius* L.). Based on seed morphology, cultivated chickpeas are of two types: *desi* type with small seeds (100 seed weight up to 30 g), thick seed coat with dark brown color and mostly grown in semiarid tropical region, and *kabuli* type with larger seeds (100 seed weight up to 50 g), thin seed coat of white to cream color and usually grown in temperate climate. More than 90% of the chickpea production is consumed locally, demonstrating its importance to the community where it is grown. The Indian subcontinent is the principal chickpea producing and consuming region contributing to almost 65% of the world's production basket. Chickpea also plays an important role in diversifying cereal-based cropping system due to its ability to fix atmospheric nitrogen and breaking disease cycles.

8.2 Genetic base of chickpea

Chickpea is an annual, self-pollinated diploid crop (2n = 16) of the family Fabaceae with genome size of 740 Mb (Arumuganathan and Earle, 1991). Its origin of cultivation could be traced back to 7500 years ago in south eastern Turkey adjoining Syria and later extended to central Asia. *Cicer* species occurs from sea level to over 5000 m above mean sea level. The chickpea gene pool comprises of three classes: GP1 (a and b), GP2, and GP3. Species of GP1a (*Cicer arietinum* L. and *Cicer reticulatum* L.) are closely related to GP1b (*Cicer echinospermum*), as compared to GP2 (*Cicer bijugum*, *C. judaicum* and *C. pinnatifidum*) and GP3 (other *Cicer* species) (Kupicha, 1977). The wild species *Cicer reticulatum* is regarded as the wild progenitor of the cultivated chickpea (*Cicer arietinum* L.).

Chickpea: Crop Wild Relatives for Enhancing Genetic Gains. https://doi.org/10.1016/B978-0-12-818299-4.00008-7
Copyright © 2020 Elsevier Inc. All rights reserved.

Unlike other crops, the wild progenitor of chickpea is reported to have narrow distribution, and during domestication, a series of evolutionary bottlenecks resulted in a narrow genetic base among cultivated chickpea varieties. This in turn attenuates the efforts of chickpea improvement program based on conventional breeding to a large extend. Genetic diversity (measured as the average pairwise nucleotide diversity within population ($\theta\pi$), Watterson's estimator of segregating sites (θw), and Tajima's D) of *kabuli* was found lesser compared to *desi* type. These may be due to the fact that *kabuli* chickpeas are defined primarily on the basis of traits derived after domestication including large and light-colored seeds, so *kabuli* varieties have likely undergone a more recent, secondary bottleneck (Varshney et al., 2013). In another study including wild and cultivated species, the genetic distance between the *desi* and *kabuli* cultivar groups (distance coefficient 0.35) was least as compared to that of *desi* versus *wild* (0.42) and *kabuli* versus *wild* (0.45) (Gupta et al., 2016).

8.3 Constraints to chickpea productivity

Chickpea production is affected by biotic stresses like root diseases like Fusarium wilt (*Fusarium oxysporum* f. sp. *ciceris*), collar rot (*Sclerotium rolfsii*), dry root rot (*Rhizoctonia bataticola*); foliar diseases like Ascochyta blight (*Ascochyta rabiei*), Botrytis gray mold (*Botrytis cinerea*); insect pests like gram pod borer (*Helicoverpa armigera*), black aphid (*Apis craccivora*), beet armyworm (*Spodoptera exigua*), termites (*Microtermes* spp. and *Odontotermes* spp.), cutworms (*Agrotis* spp.), bruchids (*Callosobruchus* spp.), and parasitic weeds (*Bromus* spp., *Cuscuta* spp. and *Orobanche* spp.). Abiotic stresses like drought, salinity, and heat stress coupled with unpredictable climate change patterns too take a heavy toll on production.

Over past few decades, breeding efforts have been restricted to the limited introduction of diverse germplasm available in the repositories. Moreover, chickpeas are cultivated mostly semiarid environments and on soils of marginal agricultural quality and combined with its vulnerability to various biotic and abiotic stresses, have restricted yields to <1.0 ton/ha, which is considerably below the yield potential. Genetic improvement, either by traditional or molecular methods, has been hampered by the limited genomic resources coupled with narrow genetic diversity in the elite gene pool.

8.4 Genetic engineering as complement to plant breeding

Modern crop breeding techniques and genomics-based approaches have been useful to create improved chickpea varieties that are resistant to diseases (*Ascochyta* and *Fusarium*) and tolerant to drought. Further, chemical/cultural control measures to a certain extent help in controlling crop losses due to diseases and insect pests. However, most intractable stresses like insect pest (gram pod borer, aphids, bruchids), weeds, drought, salinity, and low methionine content in the seeds are challenging because breeding efforts for these traits are limited, due to cross-incompatibility and lack of resistant sources in the available germplasm (Acharjee and Sarmah, 2013). Genetic engineering can provide a potential alternative to address the issue by employing genes available across kingdoms. Genetically engineered crops are currently grown in 189.8 mha in over 24 countries across the world.

8.4.1 *Agrobacterium tumefaciens*—mediated genetic transformation

Agrobacterium tumefaciens is a gram-negative bacterium that uses horizontal gene transfer to create tumors in plants. *Agrobacterium tumefaciens* has been extensively used as major agent for the production of transgenic plants in a wide variety of plant species (Hooykaas, 1989). These bacteria are well-recognized plant pathogens which induce the diseases: crown gall (incited by *A. tumefaciens*) and hairy root (incited by related *Agrobacterium rhizogenes*) in many dicotyledons. Both diseases are caused by the transfer and stable integration of part transfer DNA (T-DNA) of a large Ti or Ri plasmid from the bacterium into the genome of recipient plant cells. In plasmids from wild-type strains of *Agrobacterium*, expression of oncogenicity genes on the T-DNA normally modifies the physiology of plant cells to undergo tumor formation. However, removal of such genes results in disarmed Ti or Ri plasmids, which can be used to introduce foreign genes into plant cells without affecting their endogenous growth regulator balance. Any foreign gene(s) or DNA sequences inserted between the borders (left border and right border) of T-DNA get integrated into recipient plant genomes. T-DNA transfer from *Agrobacterium* to plant cell requires mainly three genetic elements. The first of these elements are the T-DNA border that consists of 25 bp direct repeats, the second one is *vir* (*virulence*) regulons. *virA* and *virG* form a two-component regulatory system responsible for transcriptional inactivation of the *vir* regulon in *Agrobacterium*, and the third element is chromosomal genes containing chromosomal *virulence* (*chv*) (Bourras et al., 2015). Binary vectors are currently being used for genetic transformation comprising of two plasmid vectors: one consisting of T-DNA (binary plasmid) and the other consists of *vir* genes (helper plasmid) (Hoekema et al., 1983). In few cases, two independent T-DNAs are present in a binary plasmid for cotransformation studies or production of marker-free transgenics.

Genetic transformation of chickpea has been successfully reported employing *Agrobacterium tumefaciens*/*Agrobacterium rhizogenes* mostly in *desi* chickpea genotypes. Selection of genotype is dependent mainly on amenability to in vitro regeneration including multiple shoot induction, health and length of the shoots, ability to form roots, etc. The *desi* chickpea cultivar, ICCV 37, has been widely used for genetic transformation studies, other promising genotypes include ICCV 1, ICCV 6, PG 1, PG 12, Chafa, Turkey, CM 72, C-236, DCP 92-3, PUSA 256, C 235, ICCV 89314, Semsen, PG 186, and PUSA 372 (Kar et al., 1996, 1997; Krishnamurthy et al., 2000; Sanyal et al., 2003; Anbazhagan et al., 2014; Datta et al., 2015; Das et al., 2017). Recently, a *desi* chickpea variety DCP 92-3 developed by Indian Institute of Pulses Research, Kanpur, has been widely used for shoot organogenesis from embryonic axis or parts thereof for genetic transformation studies primarily based on faster regeneration time, generation of multiple shoots, health of shoots, and suitability of kanamycin selection. In vitro plant regeneration in chickpea has been reported through organogenesis from shoot meristems (Sharma et al., 1979; Kartha et al., 1981), immature cotyledons (Shri and Davis, 1992), and through embryogenesis from immature cotyledons (Sagare et al., 1993) and leaflet callus (Barna and Wakhlu, 1993, 1994; Kumar et al., 1995).

An explant is a part of the plant by which a whole plant can be produced through plant tissue culture technique. In chickpea, among the explants, the embryonic axis and axillary meristem explants are highly responsive over other tissues for the regeneration of multiple shoot buds. In addition, other explants reported in chickpea are hypocotyl (Husnain et al., 2000), cotyledonary node (Sanyal et al., 2003; Mehrotra et al., 2011), plumule slices (Senthil et al., 2004), cotyledon with half embryonic axis (Acharjee et al., 2010; Das and Sharma, 2005), whole seeds (Akbulut et al., 2008), and axillary meristem explants (Datta et al., 2015; Das et al., 2017) (Table 8.1).

A number of *Agrobacterium* strains containing nononcogenic (disarmed) *vir* helper plasmids have been developed, including LBA4404, GV3101, AGL0, EHA 101, EHA 105, etc. Most of the reports utilize indirect method of transformation (*Agrobacterium tumefaciens* mediated) in chickpea by using different *Agrobacterium* strains such as LBA 4404 (Fontana et al., 1993; Kar et al., 1996; Sanyal et al., 2003; Senthil et al., 2004; Indurker et al., 2010; etc.), C58C1 (Krishnamurthy et al., 2000; Akbulut et al., 2008; Anbazhagan et al., 2014), EHA 101 (Krishnamurthy et al., 2000; Akbulut et al., 2008, Tewari-Singh et al., 2004), EHA 105 (Bhattacharjee et al., 2010; Das et al., 2017), AGL1 (Senthil et al., 2004; Sarmah et al., 2004; Acharjee et al., 2010; Bhattacharjee et al., 2010), and LBG 66 (Polowick et al., 2004). Among the all reports, it has been observed that LBA 4404 is a hypervirulent strain and is more effectively used in the transformation of chickpea with high transformation frequency (Table 8.2).

8.4.2 Gene gun–mediated genetic transformation

Gene gun was developed to successfully overcome transformation barriers allowing the transformation of species that were previously difficult or impossible to transform. Various cell organelles like chloroplast of *Chlamydomonas* (Boynton and Gillham, 1993), plastids, leaves (Denchev et al., 1997), roots (Warkentin et al., 1992), petals (Griesbach and Klein, 1993), pollen (Twell et al., 1989), endosperm (Knudsen and Muller, 1991), callus (Wan et al., 1995), meristems (Cao et al., 1990), and embryos immature inflorescences (Casas et al., 1997) can be specifically genetically transformed using gene gun. Gene gun employs use of gene coated microcarrier particles (0.5–1.5 μm spheres) of gold or tungsten, which uses high-velocity microprojectiles stream that delivers the coated gene of interest into the

Table 8.1 Reports on genetic transformation in chickpea.

Sl. No.	Genotype/cultivar	Explants	Method/strain/vector	Transformed gene	Marker gene	References
1	Local ecotypes	Embryos	ATMT/LBA 4404/pBI121	*uid A*, *npt II*	nptII	Fontana et al. (1993)
2	ICCV 1, ICCV 6, Desi genotypes	Embryonic axis	ATMT/LBA 4404/pBI121	*uid A*, *npt II*	nptII	Kar et al. (1996)
3	ICCV 1, ICCV 6	Embryonic axis	Gene gun/pCry1A(c)BDS1 and pKAN	*cry1Ac*, *nptII*	nptII	Kar et al. (1997)
4	Red chickpea, Canitez 87, MB-10	Shoot primordial of mature embryo (from 24 h seedlings)	ARMT/LBA 9402/Binary vector with NPTII and GUS	*uid A*, *npt II*	nptII	Altinkut et al. (1997)
5	PG 1, PG 12, Chafa, Turkey	Embryonic axis	ATMT/C58C1/GV2260/ p35SGUSINT; EHA 101/ pIBGUS	*uid A*, *npt II*, *PAT*	nptII, PAT	Krishnamurthy et al. (2000)
6	6153, CM 72	Hypocotyl	Gene gun/pBI121	*GUS*, *nptII*	nptII	Husnain et al. (2000)
7	C-236, BG-256, Pusa 362, Pusa 372	Embryonic axis (EAs) and cotyledonary node (CN)	Gene gun and ATMT/LBA 4404, GV 2260/pBI121, pCAMBIA1303, *P*35S-GUS INT	*uid A*, *nptII*, *gfp*, *hpt*	nptII, hyg	Sanyal et al. (2003)
8	*Desi*: H 208, ICCL 87322, K 850, Annigeri *Kabuli*: ICCV 5	Plumule slices	ATMT/C58, LBA 4404, AGL 1/pGIN1	PGIP, *uidA*, *bar*	PPT	Senthil et al. (2004)
9	Semsen	Cotyledon with half embryonic axis	ATMT/AGL 1/pRM50	*α-ai*, *nptII*	nptII	Sarmah et al. (2004)
10	P-362, P-1042, P-1043	Decapitated embryo with cotyledon	ATMT/EHA 101/pIBGUS/Gene gun/pBI121-dsAK	*uidA*, *bar*, *AK*, *nptII*	nptII, bar, LT	Tewari-Singh et al. (2004)

Continued

Table 8.1 Reports on genetic transformation in chickpea.—*cont'd*

Sl. No.	Genotype/cultivar	Explants	Method/strain/vector	Transformed gene	Marker gene	References
11	*Kabuli*: CDC Yuma	Embryo slices	ATMT/EHA 105/LBG 66	*uidA, nptII*	nptII	Polowick et al. (2004)
12	C 235, BG-256, Pusa 362, Pusa 372	Cotyledonary node (CN)	ATMT/LBA 4404/pGUS-INT, pRAT	*uidA, nptII, cry1Ac*	nptII	Sanyal et al. (2005)
13	K 850	Embryo axis	ATMT/LBA 4404/pAPSKαAI1	*α-ai, uidA, nptII*	nptII	Ignacimuthu and Prakash (2006)
14	Chaffa, PG 12, ICCC 37, ICCC 32	Stem, epicotyl, and embryonic axis	Gene gun/pHS102, pHS101, pBI121	*uidA, cry1Ac, nptII*	nptII	Indurker et al. (2007)
15	ICC 10943, ICC 10386	Decapitated embryo	ATMT/SAAT/LBA 4404/pCAMBIA 1305.2	*GUS Plus, hptII*	hptII	Pathak and Hamzah (2008)
16	Gokce, Er, Akcin, Uzunlu, Kusmen	Embryo axis (shoot tip and cotyledonary region)	IPT/ATMT/AI/C58C1, EHA 105, KYRT1/pTJK136	*uidA, npt II,*	nptII	Akbulut et al. (2008)
17	ICCV 89314	Dissected cotyledon with half embryo	ATMT/AGL-1/pCAMBIA35SASAL/rolCASAL	*ASAL, nptII*	nptII	Chakraborti et al. (2009)
18	C 235, Annegeri, K 850	Apical dome of shoot apex	ATMT/AGL0, LBA 4404, EHA 105/pCGP1258	*uidA, bar*	PPT	Singh et al. (2009)
19	C 235	Embryo axis with cotyledons	ATMT/AI/GV3101/pNOV2819	*pmi*	pmi	Patil et al. (2009)
20	C 235	Axillary meristem	ATMT/C58/pBI *P5CSF129A*	*uidA::nptII P5CSF129A*	nptII	Bhatnagar-Mathur et al. (2009)
21	ICCV 89314, Semsen	Cotyledon with half embryonic axis	ATMT/AGL 1/TTD/pBK101	*cry2Aa, nptII*	nptII	Acharjee et al. (2010)

22	Chaffa, PG 12, ICCC 37, ICCC 32	Mature zygotic embryonic axis, stem, and epicotyl	ATMT/LBA 4404/pHS101, pHS102, pBI121	*nptII*, *uidA*, *cry1Ac*	nptII	Indurker et al. (2010)
23	Pusa 256, KWR-108, Pusa 1003, local line	Embryonic axes, cotyledons, and cotyledonary nodes	ATMT/LBA 4404, EHA 105, AGL 1/pCAMBIA1301	*uidA*, *hpt*	hptII	Bhattacharjee et al. (2010)
24	P-362	Cotyledonary node	ATMT/LBA 4404/pBin200, pRD400, pCEC.Ac, pRD401, pCEC.Ab.Ac	*cry1Ac*, *cry1Ab*, *nptII*	nptII	Mehrotra et al. (2011)
25	KAK-2	Embryonic axes	ATMT/EHA 105/pBin-Bt8	*cry 1X*, *nptII*	nptII	Ashrani et al. (2011)
26	C 235	Axillary meristem	ATMT/C58-C1/pBI29ApNot	*DREB1A*, *nptII*	nptII	Anbazhagan et al. (2014)
27	DCP 92-3	Embryonic axes	ATMT/pCAMBIA1301	*cry1Ab/Ac*, *hpt*	hpt	Ganguly et al. (2014)
28	C 235	Embryonic axes and embryonic slices	ATMT/EHA 105/pBIN1Aa3	*cry1Aa3*, *nptII*	nptII	Khatodia et al. (2014)
29	ICCV89314	Single cotyledons with half embryos	ATMT/AGL-1/pCAMBIA2301	*uidA*, *nptII*, *Cry1Ac*	nptII	Chakroborty et al. (2016)
30	DCP 92-3	Axillary meristem explants	ATMT/EHA 105/pBinAR	*cry1Aabc*, *nptII*	nptII	Das et al. (2017)
31	Annigeri, C 235, CPS1, JG 62, K850, Vijay, WR-315	Cotyledonary node and hypocotyl	ARMT/K599/pCAMBIA1302	gfp, hyg, AtTT2	hyg	Aggarwal et al. (2018)

AI, *agroinfiltration;* AK, *aspartate kinase;* ARMT, Agrobacterium rhizogenes–*mediated transformation;* ASAL, Allium sativum *leaf agglutinin;* ATMT, Agrobacterium tumefaciens–*mediated transformation;* AtTT2, Arabidopsis thaliana *transparent testa 2;* bar, *basta (bialaphos) resistance;* cry1A3, *crystal 1A3 from* Bacillus thuringiensis*;* cry1Ab/Ac, *fusion of crystal 1Ab and crystal 1Ac from* Bacillus thuringiensis*;* cry1Aabc, *fusion of domain I from crystal 1Aa, domain II from crystal 1Ab and domain III from crystal 1Ac from* Bacillus thuringiensis*;* cry1Ac, crystal *1Ac from* Bacillus thuringiensis*;* cry1AX, *crystal 1AX from* Bacillus thuringiensis*;* cry2Aa, crystal *2Aa from* Bacillus thuringiensis*;* DREB1A, *dehydration-responsive element-binding protein 1A;* gfp, *green fluorescent protein;* GUS Plus, *GUS gene with* intron *from castor bean catalase gene and rice glycine-rich protein (GRP) signal peptide for* in planta *secretion;* hpt, *hygromycin phosphotransferase;* hyg, *hygromycin B;* IPT, in planta *transformation;* LT, *lysine and threonine;* nptII, *neomycin phosphotransferase II;* P5CSF129 A, Δ^1 –pyrroline-5-carboxylate synthetase F129A*;* PAT, *phosphinothricin acetyltransferase;* PGIP, *polygalacturonase-inhibiting protein genes from raspberry/kiwifruit;* pmi, *phosphomannose isomerase;* PPT, *phosphinothricin;* SAAT, *sonication-assisted* Agrobacterium tumefaciens–*mediated transformation;* TTD, *twin T-DNA vector;* uid A *or* GUS, *β-glucuronidase gene;* uidA::nptII, *fused β-glucuronidase and neomycin phosphotransferase II;* α-ai, *alpha-amylase inhibitor.*

Table 8.2 List of medium composition, selection agent, and transformation frequency of reported transgenic chickpea.

Sl. No	Explant preparation media	Shooting media	Rooting media	Selection agent concentration	Frequency/ Efficiency of transformation	References
1	MS media	MS + 1.0 mg/L kinetin	MS + 0.5 mg/L IAA + 0.5 mg/L kinetin, 1% sucrose (w/v)	50 mg/L (NPT)	4.0%	Fontana et al. (1993)
2	MS macrosalts (1X) + MS microsalts (4X) + B5 vitamins + 3.0 mg/L BAP + 0.04 mg/L NAA (SR medium)	MS macrosalts (1X) + MS microsalts (4X) + B5 vitamins + 3.0 mg/L BAP + 0.04 mg/L NAA	MS salts + MS vitamins + 0.05 mg/L IBA	25 mg/L, 50 mg/L and 100 mg/L (NPT)	1.40%–1.96%	Kar et al. (1996)
3	SR medium	MS macrosalts (1X) + MS microsalts (4X) + B5 vitamins + 3.0 mg/L BAP + 0.04 mg/L NAA	MS salts + MS vitamins + 0.05 mg/L IBA	25 mg/L and 50 mg/L (NPT)	Cotransformation frequency (45.8%)	Kar et al. (1997)
4	Callus-derived plantlets (CDPss): MS + 0.5 mg/L 2,4 D + 0.1 mg/L NAA Seedlings: Asceptically grown for 15 days	MS + 0.2 mg/L IAA + 0.5 mg/L kinetin + 0.2 mg/L BAP	MS + 1.0 mg/L NAA	50 mg/L (NPT)	CDP: 5.37% –10.4% Seedlings: 3.17% –4.44%	Altinkut et al. (1997)
5	MS + 0.5 mg/L BAP	MS + 0.5 mg/L BAP	In vitro grafting	100 mg/L (NPT) 10 mg/L (PPT)	<0.2%–<1.5%	Krishnamurthy et al. (2000)
6	Callusing: MS + 3 μM NAA + 3 μM BAP	MS + 3 μM 2,4 D + 1 μM kinetin	MS + 0.5 mg/L IBA	NR	NR	Husnain et al. (2000)
7	–	MS + 1.0 mg/L BAP + 0.05 mg/L IBA	MS + 0.5 mg/L IBA	100 mg/L (NPT) 20 mg/L (HYG)	0.6%–0.8%	Sanyal et al. (2003)

8	MS salts + B5 vitamins + 10 μM TDZ	MS salts + B5 vitamins + 2.5 μM TDZ	MS salts + B5 vitamins + 2.5 mM NAA, half strength MS media and grafting	1.0 mg/L and 2.5 mg/L (PPT)	5.6%–13.3%	Senthil et al. (2004)
9	–	MS + 0.5 mg/L BAP + 0.5 mg/L kinetin + 0.05 mg/L NAA + 10 mM MES (R1); MS + 0.5 mg/L BAP + 0.5 mg/L kinetin (R2); MS + 0.1 mg/L BAP + 0.1 mg/L kinetin (R3)	MS + 1 mg/L IBA + 10 mM MES and grafting	200 mg/L (NPT)	0.56%	Sarmah et al. (2004)
10	MS salts + B5 vitamins + 10 μM TDZ (ATMT) MS + B5 + 5 μM TDZ (GG)	MS salts + B5 vitamins + 10 μM TDZ; MS salts + B5 vitamins + 2 μM TDZ; MS salts + B5 vitamins + 2.3 μM kinetin	MS salts + B5 vitamins + 4.9 μM IBA and grafting	250 mg/L (NPT) 5 mg/L (PPT) 5 mM (LT)	0.7%–1.2% (PPT) 0.5%–1.3% (NPT) 1.0% (LT)	Tewari-Singh et al. (2004)
11	–	B5 + 3 mg/L BAP; MS + 1 mg/L BAP	B5 salts + B5 vitamins + 1 μM NAA	50 mg/L, 75 mg/L and 150 mg/L (NPT)	1.3%–3.1%	Polowick et al. (2004)
12	MS + 1.0 mg/L BAP	MS + 1.0 mg/L BAP + 0.05 mg/L IBA	MS + reduced nitrogen (0.084 g/L NH_4NO_3 and 0.096 g/L KNO_3) + 0.5 mg/L IBA and micrografting	100 mg/L (NPT) and 150–200 mg/L (NPT)	0.32%–1.12%	Sanyal et al. (2005)
13	MS salts + MS Iron + B5 vitamins + 3 mM MES	MS salts + B5 vitamins + 1.0 mg/L BAP; MS salts + MS Iron + B5 vitamins + 3 mM MES + 0.1–5.0 mg/L BAP	MS salts + B5 vitamins + 1 mg/L IBA	100 mg/L (NPT)	0.3%	Ignacimuthu and Prakash (2006)

Continued

Table 8.2 List of medium composition, selection agent, and transformation frequency of reported transgenic chickpea.—*cont'd*

Sl. No	Explant preparation media	Shooting media	Rooting media	Selection agent concentration	Frequency/ Efficiency of transformation	References
14	MS + 0.5 mg/L BAP + 0.1 mg/L kinetin	MS + 0.5 mg/L BAP + 0.1 mg/L kinetin	MS + 10 mM IBA (30 s)	50 mg/L and 75 mg/L (NPT)	Stem: 9.0 ± 0.58% Embryonic axis: 6.0 ± 1.15% Epicotyl: 16.0 ± 0.33%	Indurker et al. (2007)
15	–	MS salts + B5 vitamins + 2.3 μM 2.4D + 2.2 μM BAP + 1.2 μM IAA; MS salts + B5 vitamins + 2.2 μM BAP + 2.3 μM kinetin + 1.2 μM IAA	MS salts + B5 vitamins + 0.9 μM NAA + 1.2 μM IAA + 0.9 μM kinetin	50 mg/L (HYG)	ATMT: 9%–11% SAAT: 24%–26%	Pathak and Hamzah (2008)
16	MS basal	*In planta*	*In planta*	*In planta*	–	Akbulut et al. (2008)
17	MS salts + B5 vitamins + 1.5 mg/L BAP + 0.04 mg/L NAA	MS salts + B5 vitamins + 1.5 mg/L BAP + 0.04 mg/L NAA; MS salts + B5 vitamins + 0.2 mg/L IAA	Grafting	100 mg/L (NPT)	0.066% ± 0.003%	Chakraborti et al. (2009)
18	MS salts + B5 vitamins + 2.0 mg/L BAP + 0.1 mg/L NAA	MS salts + B5 vitamins + 2.0 mg/L BAP + 0.1 mg/L NAA	Rooting (1/4 MS + 2 mg/L NAA) and grafting	10 mg/L (PPT)	1.29%–3.33%	Singh et al. (2009)
19	–	MS salts + vitamins + 0.5 mg/L TDZ; MS salts + vitamins + 0.5 mg/L zeatin	½ MS salts + vitamins + 1 mg/L IBA	15–20 g/L (MANN)	Approximately, 3.0%	Patil et al. (2009)
20	MS + 4 μM TDZ + 10 μM 2iP + 2 μM kinetin	MS + 4 μM TDZ + 10 μM 2iP + 2 μM kinetin; MS + 5 μM 2iP + 2 μM	MS + 9.4 mM KNO_3 + 5 μM IBA	25 mg/L, 50 mg/L and 100 mg/L (NPT)	70%	Bhatnagar-Mathur et al. (2009)

		kinetin; MS + 2 μM GA_3				
21	–	MS + 0.5 mg/L BAP + 0.5 mg/L kinetin + 0.05 mg/L NAA + 10 mM MES (R1); MS + 0.5 mg/L BAP + 0.5 mg/L kinetin (R2); MS + 0.1 mg/L BAP + 0.1 mg/L kinetin (R3)	Grafting	200 mg/L (NPT)	Cotransformation frequency (87%)	Acharjee et al. (2010)
22	MS + 0.5 mg/L BAP + 0.1 mg/L kinetin	MS + 0.5 mg/L BAP + 0.1 mg/L kinetin	MS + 0.5 mg/L IBA	50 mg/L and 75 mg/L (NPT)	EA: 15.08% Epicotyl: 15.92% Stem: 8.77%	Indurker et al. (2010)
23a	Embryonic axis: MS medium	MS salts + B5 vitamins + 2 mg/L BAP + 0.05 mg/L NAA; MS salts + B5 vitamins + 1.1 mg/L TDZ; MS salts + B5 vitamins + 2 mg/L BAP + 0.05 mg/L NAA; MS salts + B5 vitamins + 1.1 mg/L TDZ	½ MS + ½ B5 vitamins + 0.1 mg/L IBA; ½ MS + 0.001 mg/L IBA	10 mg/L, 20 mg/L and 30 mg/L (HYG)	7.56 ± 0.54 −13.68 ± 0.56	Bhattacharjee et al. (2010)
23b	Callus from cotyledons and cotyledonary nodes: MS salts + B5 vitamins + 2 mg/L BAP + 1.0 mg/L NAA; MS salts + B5 vitamins + 1 mg/L zeatin + 0.2 mg/L IAA;	MS salts + B5 vitamins + 1 mg/L zeatin + 0.2 mg/L IAA	½ MS + ½ B5 vitamins + 0.1 mg/L IBA; ½ MS + 0.001 mg/L IBA		Cotyledon-derived callus: 11.67 ± 0.98 −19.35 ± 0.43 Cotyledonary node −derived callus: 14.68 ± 0.23 −25.56 ± 0.78	
24	MS salts + B5 vitamins + 2 mg/L BAP	MS salts + 1.0 mg/L BAP + 0.05 mg/L IBA	MS salts + reduced nitrogen (0.084 g/L NH_4NO_3 and 0.096 g/L KNO_3) + 0.5 mg/L IBA and micrografting	50 mg/L (NPT)	1.69%–2.77%	Mehrotra et al. (2011)

Continued

Table 8.2 List of medium composition, selection agent, and transformation frequency of reported transgenic chickpea.—*cont'd*

Sl. No	Explant preparation media	Shooting media	Rooting media	Selection agent concentration	Frequency/ Efficiency of transformation	References
25	–	*In planta*	*In planta*	*In planta*	15.63%	Ashrani et al. (2011)
26	MS + 4 μM TDZ + 10 μM 2iP + 2 μM kinetin	MS + 4 μM TDZ + 10 μM 2iP + 2 μM kinetin; MS + 5 μM 2iP + 2 μM kinetin; MS + 2 μM GA3	MS + 9.4 mM KNO_3 + 5 μM IBA (hydrophonics system)	100 mg/L (NPT)	–	Anbazhagan et al. (2014)
27	–	MS + 0.5 mg/L BAP + 0.5 mg/L kinetin + 0.05 mg/L NAA + 10 mM MES (R1); MS + 0.5 mg/L BAP + 0.5 mg/L kinetin (R2); MS + 0.1 mg/L BAP + 0.1 mg/L kinetin (R3)	MS + 1 mg/L IBA + 10 mM MES and grafting	50 mg/L (HYG)	–	Ganguly et al. (2014)
28	MS salts + B5 vitamins + 1.5 mg/L BAP + 0.04 mg/L NAA	MS salts + B5 vitamins + 1.5 mg/L BAP + 0.04 mg/L NAA; MS salts + B5 vitamins + 0.25 mg/L IAA	½ MS salts + B5 vitamins + 1 mg/L IBA	100 mg/L (NPT)	0.8%–1.72%	Chakroborty et al. (2016)

29	MS salts + B5 vitamins + 2.0 mg/L 2iP + 0.43 mg/L kinetin + 0.88 mg/L TDZ	MS salts + B5 vitamins + 1.0 mg/L 2iP + 0.43 mg/L kinetin MS salts + B5 vitamins + 1.0 mg/L GA_3	Grafting	100 mg/L (NPT)	0.076%	Das et al. (2017)
30	MS salts	–	–	20 mg L (HYG)	38.66 ± 3.05 (CN) 73.33 ± 1.15 (hypocotyl)	Aggarwal et al. (2018)

2,4 D, *2,4 dichlorophenoxyacetic acid;* 2iP, *(2-isopentenyl) adenine;* B5, *Gamborg medium (1968);* BAP, *6-benzylaminopurine;* GA3, *gibberellic acid 3;* HYG, *hygromycin B;* IAA, *indole-3-acetic acid;* IBA, *indole-3-butyric acid;* Kinetin, *N-furfuryladenine;* LT, *lysine and threonine;* MANN, *mannose;* MS, *Murashige and Skoog medium (1962);* NAA, *α-Naphthalene-3 acetic acid;* NPT, *Neomycin phospho transferase;* PPT, *phosphinothricin;* TDZ, *thidiazuron.*

target cell. The process is affected by an electric discharge or a pressurized helium pulse resulting into its transient expression that can led to its stable incorporation into host chromosomes. A large number of targeted cells can be bombarded in a single delivery. Efficiency of transformation via. gene gun depends on amount of macromolecules; distance to target; ratio, dimension, and material of carrier particles; method opted for coating macromolecules; bombardment helium pressure; type of targeted cells and tissues; and density of carrier particles and macromolecules.

Genetic transformation of chickpea has been successfully reported employing *gene gun* or *particle gun* method (Kar et al., 1997; Husnain et al., 2000; Sanyal et al., 2003; Tewari-Singh et al., 2004; Indurker et al., 2007) (Table 8.1). The choice of explants is the most important parameter governing transformation frequency. Cotyledonary node, decapitated embryo with cotyledon, plant stem, and epicotyl are choice explants for various gene gun–based transformation reports in chickpea. Bombardment of DNA-coated tungsten particle into calli leading to somatic embryogenesis and subsequently development of transgenic chickpea was also reported (Husnain et al., 2000). In another report, bombardment with gold and tungsten microcarrier onto cotyledonary node responded better compared to shoot tip, embryonic axes, and leaf explants. Further, gold particles have deeper penetration into the cell layer than the tungsten particles that did not pass beyond the epidermal layer for direct transformation of chickpea (Sanyal et al., 2003). Indurker et al. (2007) reported the most suitable burst pressure is 900 psi using gold particles as a microcarrier (transformation frequency 18%), while tungsten microcarriers resulted in 15% frequency (Table 8.2).

8.5 Selectable/scorable markers

Selectable markers are required in genetic transformation for the identification of transformed cells harboring gene of interest inside the plant cell (Walden, 1989; Klein et al., 1989). Selectable genes comprise selectable markers which are used for genetic transformation, allowing plant cells to grow under conditions that prevent the growth of untransformed tissue. These marker genes often exhibit dominant phenotype, usually of microbial origin, and often placed under the control of promoter elements, originating from plant, bacterial, or viral elements for strong constitutive eukaryotic expression.

Antibiotic selection marker mostly used in genetic transformation of chickpea is neomycin phosphotransferase II (*nptII*) sourced from *Tn*5 transposon of *Escherichia coli*. The *nptII* gene confers transgenic plants resistance to antibiotics like kanamycin monosulphate, G418, neomycin, etc. Kanamycin normally inhibits protein synthesis in plants by binding to ribosomes resulting in bleaching of plant. However, in the transgenic plants, the encoded enzyme, NPTII, catalyzes the phosphorylation of kanamycin, once it enters into plant cells. The phosphorylated kanamycin is inactive and, therefore, unable to shutdown the protein synthesis machinery in plants. This enables transgenic plants to grow in the presence of these antibiotics. Further, there is no endogenous NPTII activity in chickpea (Das et al., unpublished), and hence kanamycin can normally inhibit the growth of nontransgenic chickpea plants. Further, the use of *nptII* gene as a selectable marker does not pose any risk to human or animal health or environment (European Food Safety Authority, GMO Panel retrieved from https://www.efsa.europa.eu), making it a selectable marker of choice. Other selectable markers used for transgenic chickpea selection are *hygromycin phosphotransferase* (*hpt* or *aphIV*) isolated from *E. coli* conferring resistance against antibiotics, hygromycin B (Sanyal et al., 2003; Pathak and Hamzah, 2008; Bhattacharjee et al., 2010; Ganguly et al., 2014; Aggarwal et al., 2018); bialaphos resistance (*bar*) or phosphinothricin acetyltransferase (PAT) from *Streptomyces* spp. encoding acetyltransferase that inactivates L-phosphinothricin by acetylating free amino group of L-PPT active component of herbicide glufosinate, PPT, or basta (Krishnamurthy et al., 2000; Senthil et al., 2004; Tewari-Singh et al., 2004; Singh et al., 2009); desensitized *aspartate kinase* (AK) from *E. coli* for selection in lysine plus threonine (LT) medium, that otherwise would inhibit AK enzyme leading to starvation of methionine (Tewari-Singh et al., 2004); and *phosphomannose isomerase* (*pmi*) from *E. coli* that can utilize mannose as carbon source instead of sucrose (Patil et al., 2009).

Scorable selectable marker used extensively for standardization of genetic transformation of chickpea employed β-glucuronidaseA (*gus* A/*uid* A) from *E. coli*. This enzyme when incubated with some specific colorless or nonfluorescent substrate can transform them into colored or fluorescent products. Common substrate for GUS histochemical staining is 5-bromo-4-chloro-3-indolyl glucuronide (X-Gluc), the product of the reaction is clear blue color. Other common substrates are p-nitrophenyl β-D-glucuronide for the spectrophotometric assay and 4-methylumbelliferyl-beta-D-glucuronide (MUG) for the fluorimetrical assay (Jefferson et al., 1987). GUS *Plus* is the synthetic version of GUS based on sequence from *Staphylococcus* spp. that produces a GUS protein detectable at up to tenfold higher sensitivity compared to *E. coli* GUS gene (Jefferson et al., 2003). GUS *Plus* gene with intron from castor bean catalase gene is available in CAMBIA binary vector (https://cambia.org/). The intron inserted into the GUS Plus coding sequence prevents expression in bacterial cells, thus restricting activity to plant cells that had incorporated and expressed the transgene. Green fluorescent protein (GFP) is another scorable marker first isolated from jellyfish *Aequorea victoria*, which exhibits bright green fluorescence when exposed to light in the blue to ultraviolet (UV) range under fluorescent microscope.

8.6 Root transformation

Chickpeas are also amenable to *Agrobacterium rhizogenes*—mediated transformation. Root transformation using *Agrobacterium rhizogenes* has emerged as an important alternative for reverse genetic studies in chickpea, principally for quick *overexpression* and *knock-out* of genes in roots. The regenerated roots are expected to be nonchimeric as they are reported to originate from single cell lineage and can be maintained for substantial time in controlled conditions, providing sufficient time for detailed functional studies. Composite chickpea plant comprising of transgenic root (stock) and *wild-type* shoot (scion) generated by micrografting procedures can serve as an ideal system for root biology studies, functional studies in association with other organisms and importantly, and root-pathogen interactome studies.

Reports of root-transformation employing *Agrobacterium rhizogenes* strain LBA9402 in three Turkish chickpea cultivars Canitez87, red chickpea, and MB10 using calli-derived plantlets and 15 day old seedlings resulted in hairy roots with transformation frequency (%) ranging from 5.3 to 10.4 (for plantlets) and 3.17–4.44 (for seedlings) (Altinkut et al., 1997). In another study, basal cotyledonary region of three Turkish chickpea cultivars Gokce, Akcin91, and Izmir92 were infected with *Agrobacterium rhizogenes* strain 15834, and hairy roots were obtained in the variable frequency range in light and dark conditions (Khawar and Ozcan, 2004). Recently, a genotype-independent *Agrobacterium rhizogenes* (strain K 599)–mediated root transformation method was reported employing seven different chickpea cultivars *viz.* K 850, Annigeri, C 235, CPS 1, JG 62, Vijay, and WR-315 with variable transformation frequency. Reported method transforms chickpea roots at early developmental stage with high efficiency (up to 74%). Functional analysis of chickpea hairy roots by overexpressing *Arabidopsis TRANSPARENT TESTA 2* (*AtTT2*) gene involved in proanthocyanidins biosynthesis resulted in decreased colonization of fungal pathogen, *Fusarium oxysporum* (Aggarwal et al., 2018).

In future, composite chickpea plants shall be helpful in elucidating biology of dreaded soilborne diseases like wilt (*Fusarium oxysporum* f. sp. *ciceri*), dry root rot (*Rhizoctonia bataticola*), various nematodes, weedy root parasites (Broomrapes), and also microbes associated with symbiotic nitrogen fixation in chickpea and large spectrum of abiotic stresses.

8.7 Conclusion

Significant progress has been made toward genetic transformation of chickpea, once considered recalcitrant to transformation. Genetic engineering of chickpea is very essential considering the narrow genetic base and absence of useful traits in chickpea gene pool. Till date, confined field trials of transgenic chickpea for insect-resistant trait are reported; however, the efforts require constant support to bring forth viable transgenics, addressing all regulatory issues. Efforts should be directed toward development of large number of transgenic lines, evaluate them in open field conditions, and identify elite event(s) with high trait efficacy. Efforts should also be initiated for targeted integration of transgene in genome using genome editing techniques (CRISPR/Cas9 systems).

References

Acharjee, S., Sarmah, B.K., Kumar, P.A., Olsen, K., Mohan, R., et al., 2010. Transgenic chickpea (*Cicer arietinum* L.) expressing a sequence modified *cry2Aa* gene. Plant Sci. 178 (3), 333–339.

Acharjee, S., Sarmah, B.K., 2013. Biotechnologically generating "super chickpea" for food and nutritional security. Plant Sci. 207, 108–116.

Aggarwal, P.R., Nag, P., Choudhary, P., Chakraborty, N., Chakroborty, S., 2018. Genotype-independent *Agrobacterium rhizogenes*-mediated root transformation of chickpea: a rapid and efficient method for reverse genetics studies. Plant Methods 14, 55.

Akbulut, M., Yucel, M., Oktem, H.A., 2008. Analysis and organisation of DNA delivery in chickpea (*Cicer arietinum* L.) seedlings by *Agrobacterium tumefaciens*. Afr. J. Biotechnol. 7, 1011–1017.

Altinkut, A., Bajrovic, K., Gozukirmizi, N., 1997. High percentage of regeneration and transformation in chickpea. Acta Hortic. 447, 319–320.

Anbazhagan, K., Bhatnagar-Mathur, P., Vadez, V., Dumbala, S.R., Kishor, P.B.K., Sharma, K.K., 2014. *DREB1A* over expression in transgenic chickpea alters key traits influencing plant water budget across water regimes. Plant Cell Rep. https://doi.org/10.1007/s00299-014-1699-z.

Arumuganathan, K., Earle, E., 1991. Nuclear DNA content of some important plant species. Plant Mol. Biol. Report. 9, 208–218.

Ashrani, B.M., Ganeshaiah, K.N., Raja, A., Kumar, V., Makarla, U., 2011. Transformation of chickpea lines with cry1X using *in planta* transformation and characterisation of putative transformants T_1 lines for molecular and biochemical characterization. J. Plant Breed Crop Sci. 3 (16), 413–423.

Barna, K.S., Wakhlu, A.K., 1993. Somatic embryogenesis and plant regeneration from callus cultures of chickpea (*Cicer arietinum* L.). Plant Cell Rep. 12, 521–524.

Barna, K.S., Wakhlu, A.K., 1994. Whole plant regeneration of *Cicer arietinum* from callus cultures *via* organogenesis. Plant Cell Rep. 13 (9), 510–513.

Bhatnagar-Mathur, P., Vadez, V., Devi, M.J., Lavanya, M., Vani, G., Sharma, K.K., 2009. Genetic engineering of chickpea (*Cicer arietinum* L.) with the *P5CSF129A* gene for osmoregulation with implications on drought tolerance. Mol. Breed. 23, 591–606.

Bhattacharjee, B., Mohan, M., Nair, S., 2010. Transformation of chickpea: effect of genotype, explant, *Agrobacterium*-strain and composition of culture medium. Biol. Plant. 54 (1), 21–32.

Bourras, S., Rouxel, T., Meyer, M., 2015. *Agrobacterium tumefaciens* gene transfer: how a plant pathogen hacks the nuclei of plant and non plant organisms. Phytopathology 105, 1288–1301.

Boynton, J.E., Gillham, N.W., 1993. Chloroplast transformation in *Chlamydomonas*. Methods Enzymol. 217, 510–536.

Cao, J., Wang, Y.C., Klein, T.M., Sanford, J., Wu, R., 1990. Transformation of rice and maize using the biolistic process. UCLA Symp. Mol. Cell. Biol. 129, 21–33.

Casas, A.M., Kononowicz, A.K., Zhang, L., Tomes, D.T., Bressan, R.A., Hasegawa, P.M., 1997. Transgenic sorghum plants obtained after microprojectile bombardment of immature inflorescences. In Vitro Cell. Dev. Biol. 33, 92–100.

Chakraborti, D., Sarkar, A., Mondal, H.S., Das, S., 2009. Tissue specific expression of protein insecticidal, *Allium sativum leaf agglutinin* (ASAL) in important pulse crop, chickpea

(*Cicer arietinum* L.) to resist the phloem feeding *Aphis craccivora*. Transgenic Res. 18, 529–544.

Chakroborty, J., Sen, S., Ghosh, P., Sengupta, A., Basu, D., Das, S., 2016. Homologous promoter derived constitutive and chloroplast targeted expression of synthetic *cry1Ac* in transgenic chickpea confers resistance against *Helicoverpa armigera*. Plant Cell Tissue Organ Cult. 125 (3), 521–535.

Das, A., Datta, S., Thakur, S., Shukla, A., Ansari, J., Sujayanand, G.K., Chaturvedi, S.K., Kumar, P.A., Singh, N.P., 2017. Expression of a chimeric gene encoding insecticidal crystal protein *Cry1Aabc* of *Bacillus thuringiensis* in chickpea (*Cicer arietinum* L.) confers resistance to gram pod borer (*Helicoverpa armigera* Hubner). Front. Plant Sci. 8, 1423.

Das, A., Sharma, B.K., 2005. Genetic transformation of chickpea (*Cicer arietinum* L.). Res. J. Contemp. Concerns 1, 1–6.

Datta, S., Das, A., Shukla, A., Kumar, M., Ansari, M.J., 2015. Effect of cytokinins and exogenous amino acids on *in vitro* plant regeneration and *Agrobacterium* mediated genetic transformation of chickpea (*Cicer arietinum* L.). J. Food Legumes 28 (4), 281–285.

Denchev, P.D., Songstad, D.D., McDaniel, J.K., Conger, B.V., 1997. Transgenic orchard grass (*Dactylis glomerata*) plants by direct embryogenesis from microprojecticle bombarded leaf. Plant Cell Rep. 16, 813–819.

Fontana, G.S., Santini, L., Caretto, S., Frugis, G., Mariotti, D., 1993. Genetic transformation in the grain legume *Cicer arietinum* L. (chickpea). Plant Cell Rep. 12 (4), 194–198.

Ganguly, M., Molla, K.A., Karmakar, S., Datta, K., Datta, S.K., 2014. Development of pod borer-resistant transgenic chickpea using a pod specific and a constitutive promoter-driven fused *cry1Ab/Ac* gene. Theor. Appl. Genet. 127, 2555–2565.

Griesbach, R.J., Klein, T.M., 1993. *In situ* complementation of a flower color mutant in *Doritis pulcherrima* (orchidaceae). Lindleyana 8, 223–226.

Gupta, S., Nawaz, K., Parween, S., Roy, R., Sahu, K., Pole, A.K., Khandal, H., Srivastava, R., Parida, S.K., Chattopadhyay, D., 2016. Draft genome sequence of *Cicer reticulatum* L., the wild progenitor of chickpea provides a resource for agronomic trait improvement. DNA Res. 24 (1), 1–10.

Hoekema, A., Hirsch, P.R., Hooykaas, P.J.J., Schilperoot, R.A., 1983. A binary plant vector strategy based on separation of *vir* and T-region of the *Agrobacterium tumefaciens* Ti plasmid. Nature 303, 179–180.

Hooykaas, P.J.J., 1989. Transformation of plant cells *via Agrobacterium*. Plant Mol. Biol. 13, 327–336.

Husnain, T., Fatima, T., Islam, R., Riazuddin, S., 2000. Plant regeneration and expression of beta-glucorinidase gene in hypocotyls tissue of chickpea (*Cicer arietinum* L.). Pak. J. Biol. Sci. 3 (5), 842–845.

Ignacimuthu, S., Prakash, S., 2006. *Agrobacterium*-mediated transformation of chickpea with α-amylase inhibitor gene for insect resistance. J. Biol. Sci. 31 (3), 339–345.

Indurker, S., Misra, H.S., Eapen, S., 2007. Genetic transformation of chickpea (*Cicer arietinum* L.) with insecticidal crystal protein gene using particle gun bombardment. Plant Cell Rep. 26, 755–763.

Indurker, S., Mishra, H.S., Eapen, S., 2010. *Agrobacterium*-mediated transformation in chickpea (*Cicer arietinum* L.) with an insecticidal protein gene: optimization of different factors. Physiol. Mol. Biol. Plants 16 (3), 273–284.

Jefferson, R.A., Kavanagh, T.A., Bevan, M.W., 1987. GUS fusions: β-glucuronidase as a sensitive and versatile gene fusion marker in higher plants. EMBO J. 6, 3901–3907.

Jefferson, R.A., Harcourt, R.L., Kilian, A., Wilson, K.J., Keese, P.K., 2003. Microbial β-glucuronidase Genes, Gene Products and Uses There of. US Patent 6, pp. 391–547.

Khawar, K.M., Ozcan, S., 2004. Hairy root transformation in Turkish chickpea (*Cicer arietinum* L.) cultivars. Biotechnol. Biotechnol. Equip. 18 (3), 51–54.

Kar, S., Basu, D., Das, S., Ramakrishnan, N.A., Mukherjee, P., et al., 1997. Expression of *cry1Ac* gene of *Bacillus thuringiensis* in transgenic chickpea plants inhibits development of pod borer (*Heliothis armigera*) larvae. Transgenic Res. 6, 177–185.

Kar, S., Johnson, T.M., Nayak, P., Sen, S.K., 1996. Efficient transgenic plant regeneration through *Agrobacterium*-mediated transformation of chickpea (*Cicer arietinum* L.). Plant Cell Rep. 16, 32–37.

Kartha, K.K., Pahal, K., Leung, N.L., Mroginski, L.A., 1981. Plant regeneration from meristems of grain legumes: soybean, cowpea, peanut, chickpea and bean. Can. J. Bot. 59 (9), 1671–1679.

Khatodia, S., Kharb, P., Batra, P., Kumar, P.A., Chowdhury, V.K., 2014. Molecular characterization of Bt-chickpea (*Cicer arietinum* L.) plants carrying *cry1Aa3* gene. Int. J. Curr. Microbiol. Appl. Sci. 3 (8), 636–642.

Klein, T.M., Kornstein, L., Sanford, J.C., Fromm, M.E., 1989. Genetic transformation of maize cell by particle bombardment. Plant Physiol. 91, 440–444.

Knudsen, S., Muller, M., 1991. Transformation of the developing barley endosperm by particle bombardment. Planta 185, 330–336.

Krishnamurthy, K.V., Suhasini, K., Sagare, A.P., Meixner, M., Kathen, A.D., Pickardt, T., Schieder, O., 2000. *Agrobacterium* mediated transformation of chickpea (*Cicer arietinum* L.) embryo axes. Plant Cell Rep. 19, 235–240.

Kumar, V.D., Kirti, P.B., Sachan, J.K.S., Chopra, V.L., 1995. Picloram induced somatic embryogenesis in chickpea (*Cicer arietinum* L.). Plant Sci. 109, 207–213.

Kupicha, F.K., 1977. The delimitation of the tribe Vicieae and the relationship of Cicer L. Bot. J. Linn. Soc. 74, 131–162.

Mehrotra, M., Singh, A.K., Sanyal, I., Altosaar, I., Amla, D.V., 2011. Pyramiding of modified *cry1Ab* and *cry1Ac* genes of *Bacillus thuringiensis* in transgenic chickpea (*Cicer arietinum* L.) for improved resistance to pod borer insect *Helicoverpa armigera*. Euphytica 182, 87–102.

Pathak, M.R., Hamzah, R.Y., 2008. An effective method of sonication-assisted *Agrobacterium* mediated transformation of chickpeas. Plant Cell Tissue Organ Cult. 93, 65–71.

Patil, G., Dearker, A., Jain, P.K., Thengane, R.J., Srinivasan, R., 2009. Development of a phoshomannose isomerase-based *Agrobacterium*-mediated transformation system for chickpea (*Cicer arietinum* L.). Plant Cell Rep. 28, 1669–1676.

Polowick, P.L., Baliski, D.S., Mahon, J.D., 2004. *Agrobacterium* tumefaciens-mediated transformation of chickpea (*Cicer arietinum* L.): gene integration, expression and inheritance. Plant Cell Rep. 23, 485–491.

Sagare, A.P., Suhasini, K., Krishnamurthy, K.V., 1993. Plant regeneration *via* somatic embryogenesis in chickpea (*Cicer arietinum* L.). Plant Cell Rep. 12, 652–655.

Sanyal, I., Singh, A.K., Amla, D.V., 2003. *Agrobacterium tumefaciens*-mediated transformation of chickpea (*Cicer arietinum* L.) using mature embryonic axes and cotyledonary nodes. Indian J. Biotechnol. 2, 524–532.

Sanyal, I., Singh, A.K., Kaushik, M., Amla, D.V., 2005. *Agrobacterium* mediated transformation of chickpea (*Cicer arietinum* L.) with *Bacillus thuringiensis cry1Ac* gene for resistance against pod borer insect *Helicoverpa armigera*. Plant Sci. 168 (4), 1135–1146.

Sarmah, B.K., Moore, A., Tate, W., Molvig, L., Morton, R.L., Rees, D.P., Chiaieste, P., Chrispeels, M.J., Tabe, L.M., Higgins, T.J.V., 2004. Transgenic chickpea seeds expressing high levels of a bean α-amylase inhibitor. Mol. Breed. 14, 73–82.

Senthil, G., Williamson, B., Dinkins, R.D., Ramsay, G., 2004. An efficient transformation system for chickpea (*Cicer arietinum* L.). Plant Cell Rep. 23, 297–303.

Singh, R., Singh, N.P., Datta, S., Yadav, I.S., Singh, A.P., 2009. Agrobacterium mediated transformation of chickpea using shoot meristem. Indian J. Biotechnol. 8, 78–84.

Sharma, D.R., Kumari, R., Chowdhury, J.B., 1979. Plant regeneration in *Cicer* species through tissue culture. Indian J. Exp. Biol. 17, 607–609.

Shri, P.V., Davis, T.M., 1992. Zeatin-induced shoot regeneration from immature chickpea (*Cicer arietinum* L.) cotyledons. Plant Cell Tissue Organ Cult. 28, 45–51.

Tewari-Singh, N., Sen, J., Kiesecker, H., Reddy, V.S., Jacobsen, H.J., Mukherjee, S.G., 2004. Use of a herbicide or lysine plus threonine for non antibiotic selection of transgenic chickpea. Plant Cell Rep. 22, 576–583.

Twell, D., Klein, T.M., Fromm, M.E., McCormick, S., 1989. Transient expression of chimeric genes delivered into pollen by microprojectile bombardment. Plant Physiol. 91, 1270–1274.

Varshney, R.K., Song, C., Saxena, R.K., Azam, S., Yu, S., Sharpe, A.G., et al., 2013. Draft genome sequence of chickpea (*Cicer arietinum* L.) provides a resource for trait improvement. Nat. Biotechnol. 31, 240–246.

Walden, R., 1989. Genetic Transformation in Plants. Prentice Hall, Englewood Cliffs, NJ.

Wan, Y., Widholm, J.M., Lemaux, P.G., 1995. Type I callus as a bombardment target for generating fertile transgenic maize (*Zea mays* L.). Planta 196, 7–14.

Warkentin, T.D., Jordan, M.C., Hobbs, S.L.A., 1992. Effect of promoter-leader sequences on transient reporter gene expression in particle bombarded pea (*Pisum sativum* L.) tissues. Plant Sci. 87, 171–177.

CHAPTER

Chickpea economy in India

9

Subhash Sharma, Samriti, Ravinder Sharma

Department of Social Sciences, Dr YS Parmar University of Horticulture and Forestry, Solan, Himachal Pradesh, India

9.1 Introduction

Food security has long been the mainstay of public policy in India. Only in the 1970s did the country begin to come out of the shadow of severe food scarcity with its history of acute famines. Even in the mid-1960s, its import dependency concerning cereals was as high as 16%, which coupled with the country's small export base also had serious balance-of-payment repercussions. Consequently, India's agricultural policy has always aimed to maximize agricultural production and achieve self-sufficiency, particularly in cereals. Several technological, infrastructural, and institutional changes were implemented to increase agricultural production. Even with those advances, however, rapid growth in the country's population meant its land and other resources remained under great pressure to meet food needs. Therefore, high-yielding crops such as rice and wheat, which could lead to increased food production, received greater attention. The late 1960s and 1970s witnessed a spectacular growth in the production of these crops with the introduction of Green Revolution technologies 1 The spread of these technologies did not confer equal benefits across crops or regions of the country. Vast areas in the eastern and western regions remained untouched by the technologies. An additional challenge was that with the adoption of technology and new access to irrigation, wheat and rice recorded significant growth at the cost of displacing other crops, particularly coarse cereals, pulses, and oilseeds.

Pulses historically have been one of the most important constituents of the Indian cropping and consumption patterns. Pulses, long considered "the poor man's meat" because of their protein profile, occupy a unique place in India. India ranks as the country with the world's largest number of malnourished people. Because of this, coupled with the country's high incidence of vegetarianism, the future of pulses is of special significance to India's large poor population. Pulses are also among the crops that have been adversely affected by the dominance of cereals over the past several decades, so it is clear that agricultural policy is implicated in both the nutritional challenges and, as we hope to show, in potential solutions. Pulses are consumed equally by India's rich and poor as it is one of the less expensive sources of protein (Mohanty and Satyasai, 2015). Currently, only 11% of India's protein

Chickpea: Crop Wild Relatives for Enhancing Genetic Gains. https://doi.org/10.1016/B978-0-12-818299-4.00009-9
Copyright © 2020 Elsevier Inc. All rights reserved.

needs are met by pulses. The balance is either met through other sources or not met at all. India is still a home to about 24% of undernourished people in the world. It is a fact that Indians today are consuming far less protein than they used to do so in the past. Their importance as a source of protein for masses in India is well recognized, and, therefore, their production and availability assume special significance for the nutritional security of the people. Pulses are the major source of protein among all cultivated crops and important constituent of Indian diet after cereals. Pulses can be grown in all type of soils and leguminous in nature (except arhar), fix atmospheric nitrogen into the soil, and thus contribute significantly to environmental and economic sustainability. India is the world's largest producer and consumer of pulses. Major pulses grown in India include chickpea or Bengal gram, pigeon pea or red gram, lentil, black matpe, mung bean or green gram, lablab bean, moth bean, horse gram, pea, grass pea or khesari (*Lathyrus sativus*), cowpea (*Vigna unguiculata*), and broad bean or faba bean (*Vicia faba*). Popular pulses in India are chickpea, pigeon pea, green gram, black matpe, and lentil. Pulses are mostly grown in two seasons: (1) the warmer, rainy season or kharif (June—October) and (2) the cool, dry season or rabi (October—April) (Gowda et al., 2013). In India, pulses can be produced with a minimum use of resources, and hence, it becomes less costly even than animal protein. Pulses can be cultivated as an intercrop and also as mixed crop. Pulses are mostly cultivated under rainfed conditions and do not require intensive irrigation facility, and this is the reason why pulses are grown in areas left after satisfying the demand for cereals/cash crops. Even in such conditions, pulses give better returns.

The recent rise in prices has led to an increase in demand for legumes worldwide through both income and population growth. The increasing demand for livestock feed in developing countries is a significant change in the demand structure. The potential of pulses to help address future global food security, nutrition, and environmental sustainability needs has been acknowledged through the UN declaration of the 2016 International Year of Pulses. Pulses are among the most extensively used foods in the world. They provide protein (20 g per 100 g) and fiber and are a great source of vitamins and minerals such as iron, zinc, and magnesium. Pulses should be eaten as part of a healthy diet. Pulses have a high nutritional value. They are low in saturated fats and sodium (Table 9.1).

Chickpea or chana is an important pulse crop from the point of view of food basket for growing population in the country. Scientifically is named *Cicer arietinum* belonged to Leguminosae family. Chickpea is a highly nutritious pulse crop and placed third in the list of the food legumes that are cultivated throughout the world. It contains 25% proteins and 60% carbohydrates. Jukanti et al. (2012) provide the nutrition benefits of chickpea specifically and present it as a good source of carbohydrates and protein, with protein quality in particular better than other pulses. Chickpea has significant amounts of all the essential amino acids except the sulfur-containing types, which can be complemented by adding cereals to daily diet as cereals are rich in sulfur-containing amino acids. Jukanti et al. (2012) list chickpea as an affordable source of protein, carbohydrates, minerals and vitamins, dietary fiber, folate, β-carotene, and health-promoting fatty acids.

Table 9.1 Nutritional level of various pulses.

Name of food stuff	Gram	Urad	Mung	Kulthi	Lentil	Pea	Tur	Moth	Khesari	Cowpea
Protein (%)	20	24	25	22	25	22	22	25	31	23
Vit. A (I.U.)	316	64	83	119	450	31	220	16	200	60
Vit. C	3	–	–	1	–	–	–	2	–	–
Vit. K	0.29	0.19	–	–	0.25	–	–	–	–	–
Thiamine	0.3	0.41	0.72	0.42	0.45	0.47	0.45	0.45	0.39	0.5
Riboflavin	0.51	0.37	0.15	0.2	0.49	0.21	0.51	0.09	0.41	0.48
Nicotinic-acid	2.1	2	2.4	1.5	1.5	3.5	2.6	1.5	2.2	1.3
Biotin (g/100 g)	10	7.5	–	–	13.2	–	7.6	–	7.5	202
Choline	194	206	–	–	299	–	183	–	–	–
Folic acid (g/100 g)	125	144	–	–	107	–	83	–	100	–
Inositol	240	90	–	–	130	–	100	–	140	–
Pantothenic acid	1.3	3.5	–	–	1.6	–	1.5	–	2.6	–
Total number of vitamins/minerals	12	11	5	6	11	5	10	6	9	6

Source: Indian Council of Medical Research (ICMR)-National Institute of Nutrition (NIN), Hyderabad (2012).

There are mainly two types of chickpea produced, i.e., Desi and Kabuli. The Desi-type chickpea contributes to around 80% and the Kabuli type around 20% of the total chickpea production. India is the largest producer of chickpea of around 6 million tons followed by Pakistan, Turkey, and Iran and contributes to around 70% of the world's total chickpea production. The major producing countries of Desi chickpea are India, Pakistan, Myanmar, Australia, and Bangladesh. Countries like Turkey, Iran, Ethiopia, Mexico, Syria, Spain, Canada, United States, Algeria, Ethiopia, Malawi, Sudan, Tanzania, Tunisia, Spain, and Portugal are the major producers of Kabuli chickpea. Chickpea producing states in India are Madhya Pradesh, Uttar Pradesh, Rajasthan, Maharashtra, and Andhra Pradesh. Madhya Pradesh has major share, i.e., around 40% in the India's chickpea production. India also exports chickpea to different countries namely the United States, United Kingdom, Saudi Arabia, UAE, and Sri Lanka. The major trading centers of chickpea in India are Indore (Madhya Pradesh), Bhopal (Madhya Pradesh), Vidisha (Madhya Pradesh), Jalgaon (Maharashtra), Latur (Maharashtra), Mumbai (Maharashtra), Akola (Maharashtra), Jaipur (Rajasthan), Bikaner (Rajasthan), Kota (Rajasthan), Jodhpur (Rajasthan), Sri Ganganagar (Rajasthan), Hanumangarh (Rajasthan), Delhi, Chennai, Kanpur, Hapur, Hyderabad, Vijayawada, Gulbarga, Sirsa, Jalandhar, Ludhiana, and Sangrur.

9.2 Global pulse economy: an overview

Pulses are an important food crops. Since 1960s, production of pulses has made steady advances all across the world. Pulses are grown both in developed and developing countries. In developed countries, pulses are grown on industrial scale, whereas in developing countries, these are grown on subsistence level and considered as staple food. Asia accounts for 73% contribution of world pulse production followed by America (5%), Europe (4%), and Africa (4%). In Asia, India is the largest producer and consumer of pulses in the world contributing around half of the total global production (Fig. 9.1).

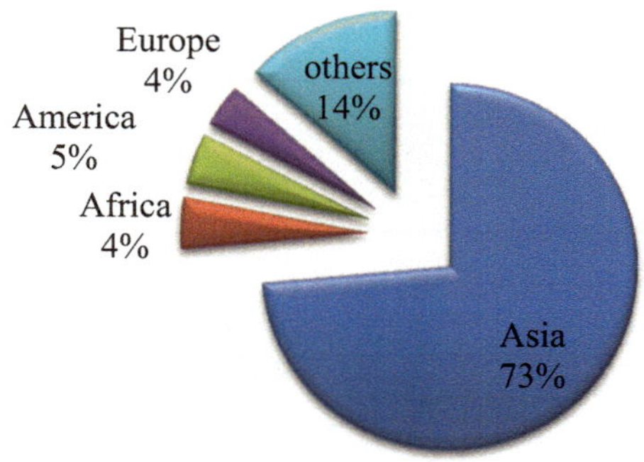

FIGURE 9.1

Region-wise world production of pulses in 2018.

9.3 Global scenario of chickpea

India rank first both in area and production of chickpea in the world as it contributes 70.57% to the total world area and 69.21% to the total world production followed by Australia as shown in Table 9.2. The highest productivity of chickpea is observed in Ethiopia (2038 kg/ha) followed by Mexico (1918 kg/ha) and Australia (1875 kg/ha). Ethiopia just contributes 2.92% and 1.55% in production and area of the world, but productivity is very high as compared to India. Productivity of India was found to be 1063 kg/ha. So, in productivity, India ranks sixth in the world, this is a point of concern. It can be suggested from the data that more focus should be given on the improvement of the chickpea varieties (Fig. 9.2).

9.4 Indian pulse scenario

It is clear from Table 9.3 that the highest area under the major pulses was recorded in 2017–18, i.e., about 300 lakh ha. Maximum production (252.3 lakh tonnes) and yield (841 kg/ha) of major pulses were recorded on the same year. There is tremendous increase in the area under the pulses after 2014. The production (252.3 lakh tonnes) and productivity (841 kg/ha) of pulses have also sharply increased during 2017–18 from the level of 110.75 lakh tonnes and 544 kg/ha during 2000–01. This is the result of the various govt. policies and scientific interventions.

The highest production (252.30 lakh tonnes) and yield (841 kg/ha) of major pulses were recorded in 2017–18. However, the maximum growth rate in area of

Table 9.2 World area, production, and productivity of chickpea in 2017 –18.

Country	Area (lakh ha)	% Contribution	Production (lakh tonnes)	% Contribution	Yield (kg/ha)
India	105.61	70.57	112.29	69.21	1063
Australia	10.69	7.14	20.04	12.35	1875
Pakistan	9.71	6.49	3.3	2.03	340
Iran	5.66	3.78	2.71	1.67	480
Russian federation	4.57	3.05	4.18	2.58	916
Turkey	3.92	2.62	4.7	2.90	1197
Myanmar	3.76	2.51	5.27	3.25	1402
USA	2.43	1.62	3.13	1.93	1291
Ethiopia	2.32	1.55	4.74	2.92	2038
Mexico	0.99	0.66	1.89	1.16	1918
World	149.66		162.25		1252

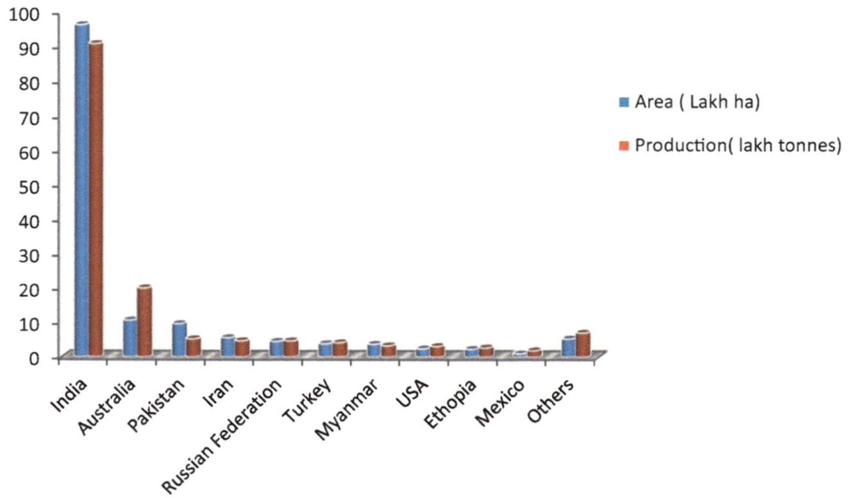

FIGURE 9.2

Major chickpea producing countries in 2017–18.

Table 9.3 Area, production, and productivity of pulses in India.

Year	Area (lakh ha)	Production (lakh tonnes)	Yield (kg/ha)
2000–01	203.48	110.75	544
2001–02	220.08	133.68	607
2002–03	204.96	111.25	543
2003–04	234.58	149.05	635
2004–05	227.63	131.03	577
2005–06	223.91	133.84	598
2006–07	231.92	141.98	612
2007–08	236.33	147.62	625
2008–09	220.94	145.66	659
2009–10	232.82	146.62	630
2010–11	864.02	182.41	691
2011–12	844.62	170.89	699
2012–13	232.57	183.42	789
2013–14	252.18	192.55	764
2014–15	235.53	171.52	728
2015–16	249.11	163.48	656
2016–17	292.77	213.30	765
2017–18	299.90	252.30	841

Source: Statistical Year Book, 2018. Ministry of Statistics and Program Implementation. Government of India.

pulses was recorded between the period from 2015–16 to 2016–17 (17.53%). Whereas the maximum production growth rate (33.98%) and maximum yield growth rate (16.94%) of pulses were recorded between 2002–03 and 2003–04 (Table 9.4).

Table 9.4 Annual growth in area, production, and productivity of major pulses of India.

Year	Area (lakh ha)	Growth rate (%)	Production (lakh tonnes)	Growth rate (%)	Yield (kg/ha)	Growth rate (%)
2000–01	203.48	–	110.75	–	544.00	–
2001–02	220.08	8.16	133.68	20.70	607.00	11.58
2002–03	204.96	–6.87	111.25	–16.78	543.00	–10.54
2003–04	234.58	14.45	149.05	33.98	635.00	16.94
2004–05	227.63	–2.96	131.03	–12.09	577.00	–9.13
2005–06	223.91	–1.63	133.84	2.14	598.00	3.64
2006–07	231.92	3.58	141.98	6.08	612.00	2.34
2007–08	236.33	1.90	147.62	3.97	625.00	2.12
2008–09	220.94	–6.51	145.66	–1.33	659.00	5.44
2009–10	232.82	5.38	146.62	0.66	630.00	–4.40
2010–11	864.02	271.11	182.41	24.41	691.00	9.68
2011–12	844.62	–2.25	170.89	–6.32	699.00	1.16
2012–13	232.57	–72.46	183.42	7.33	789.00	12.88
2013–14	252.18	8.43	192.55	4.98	764.00	–3.17
2014–15	235.53	–6.60	171.52	–10.92	728.00	–4.71
2015–16	249.11	5.77	163.48	–4.69	656.00	–9.89
2016–17	292.77	17.53	213.30	30.47	765.00	16.62
2017–18	299.90	2.44	252.30	18.28	841.00	9.93

Source: Statistical Year Book, 2018. Ministry of Statistics and Program Implementation. Government of India.

Table 9.5 Per capita availability of pulses in India.

Year	g per capita per day	kg per capita per year
2000	31.78	11.60
2001	29.86	10.90
2002	35.34	12.90
2003	29.04	10.60
2004	35.89	13.10
2005	31.51	11.50
2006	32.33	11.80
2007	35.34	12.90
2008	41.92	15.30
2009	36.99	13.50
2010	35.34	12.90
2011	43.01	15.70
2012	41.64	15.20
2013	43.29	15.80
2014	46.30	16.90
2015	43.84	16.00
2016	43.01	15.70
2017	52.88	19.30
2018	55.89	20.40

Source: Department of Economics and Statistics. Ministry of Agriculture, Cooperation and Farmers Welfare, Government of India, 2017–18.

9.5 Per capita availability of pulses in India

Per capita availability of pulses was increased from 31.78 g per day in the year 2000 to 56 g/day in the year 2018. While WHO recommends per capita consumption at 80 g/day for India, resulting wide gap in pulses availability due to increasing population of country (Table 9.5).

9.5.1 Indian scenario of chickpea

The highest area under chickpea was recorded in 2017–18 i.e., 105.61 lakh ha. Maximum production (112.29 lakh tonnes) and yield (1063 kg/ha) of major pulses was recorded on the same year. The area production and productivity (1063 kg/ha) of pulses has sharply increased during 2017–18 from the level of 51.85 lakh tonnes and 744 kg/ha during 2000–01 (Table 9.6).

9.6 Growth in chickpea area, production, and productivity

The highest production (112.29 lakh tonnes) and yield (1063 kg/ha) of major pulses were recorded in 2017–18. However, the maximum growth rate in area (23.74%)

Table 9.6 Area, production, and productivity of chickpea in India.

Year	Area (lakh ha)	Production (lakh tonnes)	Yield (kg/ha)
2000–01	51.85	38.55	744
2001–02	64.16	54.73	853
2002–03	59.06	42.37	717
2003–04	70.48	57.18	811
2004–05	67.15	54.69	815
2005–06	69.26	56.00	808
2006–07	74.94	63.34	845
2007–08	75.44	57.49	762
2008–09	78.93	70.6	895
2009–10	81.69	74.76	915
2010–11	91.86	82.21	895
2011–12	82.99	77.02	928
2012–13	85.22	88.32	1036
2013–14	99.27	95.26	960
2014–15	82.51	73.32	889
2015–16	83.99	70.58	840
2016–17	95.39	90.75	951
2017–18	105.61	112.29	1063

Source: Statistical Year Book, 2018. Ministry of Statistics and Program Implementation. Government of India.

and production (41.97%) of chickpea were recorded between the period from 2000–01 and 2001–02. Whereas the maximum yield growth rate (17.45%) of chickpea was recorded between 2007–08 and 2008–09 (Table 9.7).

9.7 Contribution of chickpea area to the total pulses area in India

Per cent share of chickpea to the total pulses in the country in term of area was 25.48 in 2000–01, which showed the increasing trend, and it is found that the chickpea occupies 35.22% of the total area under pulses in 2017–18 as it is one of the major pulses grown in India (Table 9.8).

9.8 Availability status of pulses

India is the largest producer of pulses in the world. However, India is also the largest consumer of pulses in the world due to its increasing population. To meet out the domestic demand, India imports around 18% of its pulses requirements. The yield

Table 9.7 Annual growth in area, production, and productivity of chickpea in India.

Year	Area (lakh ha)	Growth rate (%)	Production	Growth rate (%)	Yield	Growth rate (%)
2000−01	51.85	–	38.55	–	744.00	–
2001−02	64.16	23.74	54.73	41.97	853.00	14.65
2002−03	59.06	−7.95	42.37	−22.58	717.00	−15.94
2003−04	70.48	19.34	57.18	34.95	811.00	13.11
2004−05	67.15	−4.72	54.69	−4.35	815.00	0.49
2005−06	69.26	3.14	56.00	2.40	808.00	−0.86
2006−07	74.94	8.20	63.34	13.11	845.00	4.58
2007−08	75.44	0.67	57.49	−9.24	762.00	−9.82
2008−09	78.93	4.63	70.60	22.80	895.00	17.45
2009−10	81.69	3.50	74.76	5.89	915.00	2.23
2010−11	91.86	12.45	82.21	9.97	895.00	−2.19
2011−12	82.99	−9.66	77.02	−6.31	928.00	3.69
2012−13	85.22	2.69	88.32	14.67	1036.00	11.64
2013−14	99.27	16.49	95.26	7.86	960.00	−7.34
2014−15	82.51	−16.88	73.32	−23.03	889.00	−7.40
2015−16	83.99	1.79	70.58	−3.74	840.00	−5.51
2016−17	95.39	13.57	90.75	28.58	951.00	13.21
2017−18	105.61	10.71	112.29	23.74	1063.00	11.78

Source: Statistical Year Book, 2018. Ministry of Statistics and Program Implementation. Government of India.

of pulses in India is quite low at 1063 kg/ha in 2017−18, which might be due to policy neglect. The post−green revolution era saw a sharp decline in per capita production and availability of pulses with record 112.82 lakh tonnes imports of pulses in 2016−17. Availability status of major pulses based on the domestic production,

Table 9.8 Share of chickpea in area under major pulses in India from 2000 −01 to 2017−18.

Year	Area under major pulses (lakh ha)	Area under chickpea (lakh ha)	% Contribution To total area
2000 −01	203.48	51.85	25.48
2001 −02	220.08	64.16	29.15
2002 −03	204.96	59.06	28.82
2003 −04	234.58	70.48	30.05
2004 −05	227.63	67.15	29.50
2005 −06	223.91	69.26	30.93
2006 −07	231.92	74.94	32.31
2007 −08	236.33	75.44	31.92
2008 −09	220.94	78.93	35.72
2009 −10	232.82	81.69	35.09
2010 −11	864.02	91.86	10.63
2011 −12	844.62	82.99	9.83
2012 −13	232.57	85.22	36.64
2013 −14	252.18	99.27	39.36
2014 −15	235.53	82.51	35.03
2015 −16	249.11	83.99	33.72
2016 −17	292.77	95.39	32.58
2017 −18	299.9	105.61	35.22

Source: Statistical Year Book, 2018. Ministry of Statistics and Program Implementation. Government of India.

Table 9.9 Availability status of pulses in India (lakh tonnes).

Year	Production	Import	Export	Total availability
2000–01	110.75	3.50	2.44	111.81
2001–02	133.68	22.18	1.61	154.25
2002–03	111.25	19.92	1.48	129.69
2003–04	149.05	17.23	1.54	164.74
2004–05	131.03	13.39	2.71	141.71
2005–06	133.84	16.96	4.47	146.33
2006–07	141.98	22.56	2.47	162.07
2007–08	147.62	28.35	1.64	174.33
2008–09	145.66	24.74	1.36	169.04
2009–10	146.62	35.10	1.00	180.72
2010–11	182.41	26.99	2.08	207.32
2011–12	170.89	33.65	1.74	202.8
2012–13	183.42	38.39	2.02	219.79
2013–14	192.55	30.49	3.43	219.61
2014–15	171.52	45.85	2.22	215.15
2015–16	163.48	57.98	2.56	218.90
2016–17	213.30	112.82	1.37	324.75
2017–18	252.30	56.10	2.00	306.40

Source: Directorate General of Commercial Intelligence and Statistics (DGCI&S), Kolkata, 2018.

import, and export is summarized under Table 9.9. It is evident from the table that total availability of pulses has followed the increasing trend after 2004 (Fig. 9.3).

9.9 Availability status of chickpea in India

Availability status of chickpea based on the domestic production, import, and export is summarized in Table 9.10. Due to increase in the population, requirement for chickpea also increased, and hence yield of chickpea in the country is quite low, import quantity increased from 4.19 lakh tonnes during 2014–05 to 10.81 lakh tonnes in 2016–17 to fulfill the demand of the population and thus reduced the export quantity from 3.46 lakh tonnes to 1.37 lakh tonnes in 2016–17.

9.9.1 Different state's contribution in area and production of total pulses in India

In India, area and production of total pulses in 2017–18 have been greater than 300 lakh hectares and 252 lakh tonnes, respectively. Out of total area, Madhya Pradesh singly contributes 25% earning a prime status in pulse production with 32.15% contribution to the total production followed by Rajasthan and Maharashtra with 13.42 and 13.09% contribution to the total production (Table 9.11). More than

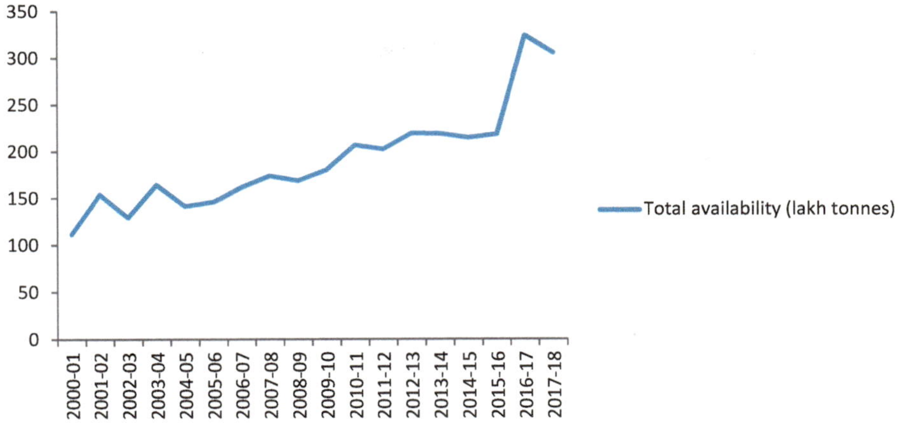

FIGURE 9.3

Total availability of pulses in India.

Table 9.10 Availability status of chickpea in India (lakh tonnes).

Year	Domestic production	Import	Export	Total availability
2013–14	95.26	2.76	3.46	94.56
2014–15	73.32	4.19	2.22	75.29
2015–16	70.58	10.31	2.56	78.33
2016–17	93.78	10.81	1.37	103.22
2017–18	112.3			

Source: Directorate General of Commercial Intelligence and Statistics (DGCI&S), Kolkata, 2018.

90% of total production has been contributed by 10 states, i.e., MP, Rajasthan, Maharashtra, Karnataka, UP, AP, Gujarat, TN, Jharkhand, and Chhattisgarh, respectively, as shown in Figs. 9.4 and 9.5.

9.9.2 Different states' contribution in area and production of chickpea

It is found from the table that in 2017–18, chickpea was cultivated in about 106 lakh ha and recorded a production of greater than 112 lakh tonnes (Fig. 9.7). Madhya Pradesh has contributed 34% of the total chickpea area and 41% of the total chickpea production in the country (Fig. 9.6). Therefore, Madhya Pradesh (MP) ranking first both in area and production of chickpea in the country is followed by Maharashtra (19% and 16%) and Rajasthan (15% and 17%) as shown in figure (Table 9.12).

Table 9.11 States' contribution in area and production of total pulses in 2017–18.

States	Area(lakh ha)	% Contribution	States	Production(lakh tonnes)	% Contribution
Madhya Pradesh	74.80	24.94	Madhya Pradesh	81.12	32.15
Rajasthan	53.29	17.77	Rajasthan	33.86	13.42
Maharashtra	43.52	14.51	Maharashtra	33.04	13.09
Karnataka	30.21	10.07	Uttar Pradesh	22.08	8.75
Uttar Pradesh	22.68	7.56	Karnataka	18.55	7.35
Andhra Pradesh	14.07	4.69	Andhra Pradesh	12.24	4.85
Gujarat	9.10	3.03	Gujarat	9.26	3.67
Tamil Nadu	8.65	2.88	Jharkhand	8.45	3.35
Jharkhand	7.93	2.64	Tamil Nadu	5.50	2.18
Chhattisgarh	7.84	2.61	Chhattisgarh	5.43	2.15
Others	27.84	9.28	Others	22.81	9.04
Total	299.93		Total	252.34	

Source: Department of Economics and Statistics. Ministry of Agriculture, Cooperation and Farmers Welfare, Government of India, 2017–18.

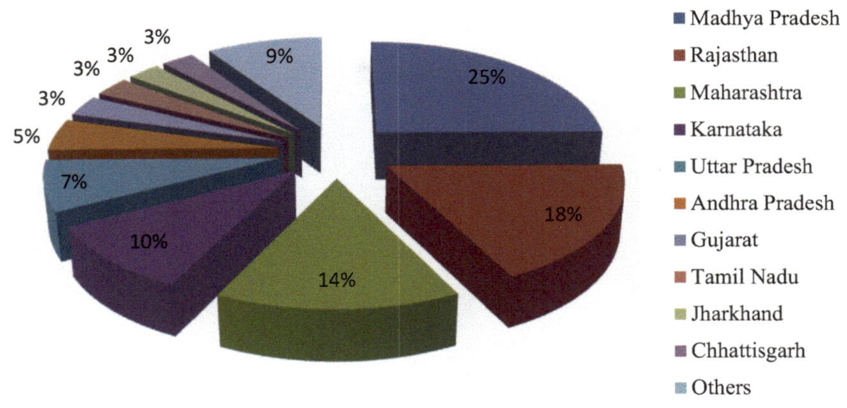

FIGURE 9.4

Different States' contribution in area of pulses in India in 2017–18.

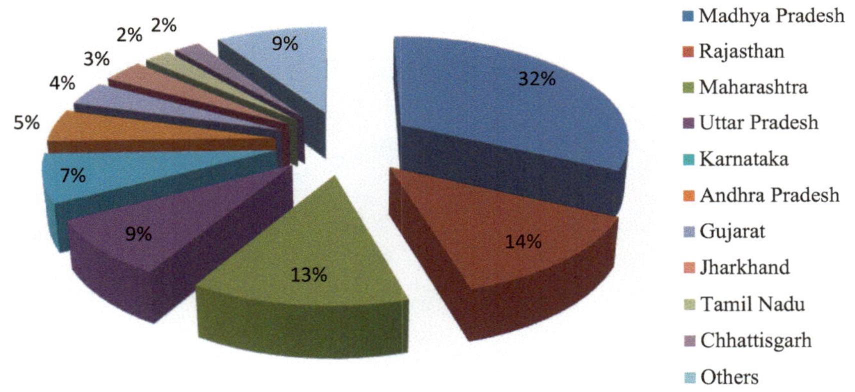

FIGURE 9.5

Different states' contribution in production of pulses in India in 2017–18.

9.10 Export and import of chickpea

Australia is the major exporter of chickpea in the world. The chickpea export of the country was 9.26 lakh tonnes followed by Russia (2.43 lakh tonnes), India (2.12 lakh tonnes), Mexico (1.49 lakh tonnes), and Canada (0.81 lakh tonnes) (Table 9.13).

The table revealed that India is the major importer of chickpea due to its large population size. India imports 5.90 lakh tonnes chickpea followed by Bangladesh (2.08 lakh tonnes) and Algeria (0.72 lakh tonnes) (Table 9.14).

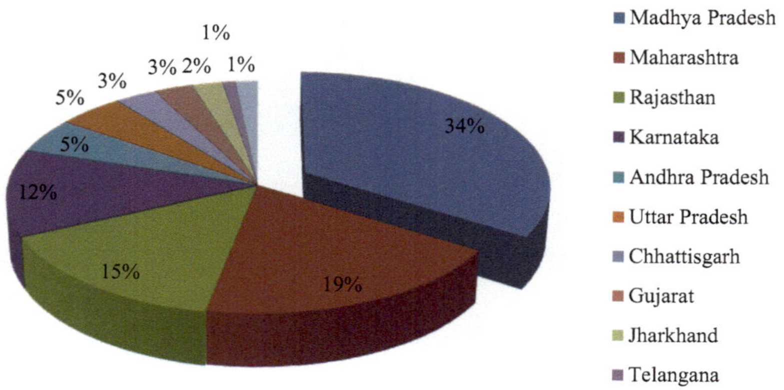

FIGURE 9.6

States contribution in area of chickpea in 2017–18.

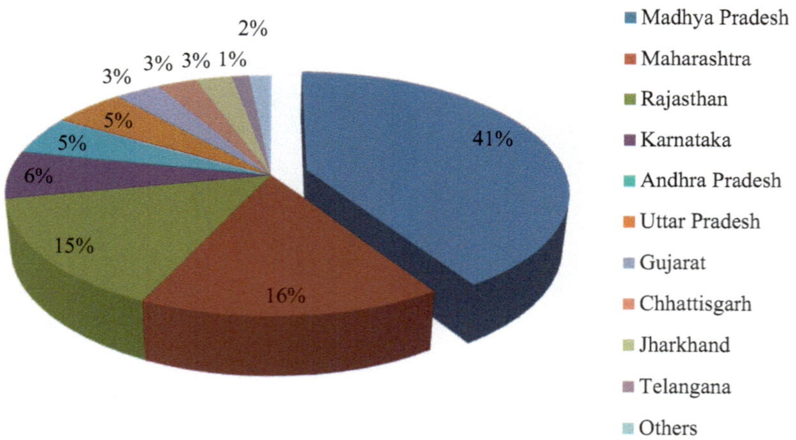

FIGURE 9.7

States contribution in production of chickpea in 2017–18.

9.11 Trade destination of chickpea

Major export destinations of India for chickpea were Pakistan (21.60%) followed by Algeria (11.6%). India imports maximum share of chickpea from Australia (85.10%) followed by Russia (4.7%) and Tanzania (3.8%) (Table 9.15 and 9.16).

Table 9.12 Different states' contribution in area and production of chickpea in India in 2017–18 (area—lakh ha, production—lakh tonnes).

States	Area	% Contribution	States	Production	% Contribution
Madhya Pradesh	35.90	33.99	Madhya Pradesh	45.95	40.92
Maharashtra	20.00	18.94	Maharashtra	17.84	15.89
Rajasthan	15.72	14.88	Rajasthan	16.70	14.87
Karnataka	12.65	11.98	Karnataka	7.21	6.42
Andhra Pradesh	5.20	4.92	Andhra Pradesh	5.85	5.21
Uttar Pradesh	5.01	4.74	Uttar Pradesh	5.79	5.16
Chhattisgarh	3.18	3.01	Gujarat	3.67	3.27
Gujarat	2.93	2.77	Chhattisgarh	3.16	2.81
Jharkhand	2.33	2.21	Jharkhand	2.93	2.61
Telangana	0.97	0.92	Telangana	1.38	1.23
Others	1.72	1.63	Others	1.81	1.61
Total	105.61			112.29	

Source: Department of Economics and Statistics. Ministry of Agriculture, Cooperation and Farmers Welfare, Government of India, 2017–18.

Table 9.13 Major chickpea exporting countries from 2010 to 2018.

Sr. No.	Country	Exports (lakh tonnes)
1	Australia	9.26
2	Russian federation	2.43
3	India	2.12
4	Mexico	1.49
5	Canada	0.81
6	Argentina	0.77
7	USA	0.70
8	Ethiopia	0.59
9	Turkey	0.21

Source: Food and Agriculture Organization (FAO), 2019. FAOSTAT Statistical Database of the United Nation Food and Agriculture Organization (FAO) Statistical Division. Rome.

Table 9.14 Major chickpea importing countries from 2010 to 2018.

Sr. No.	Country	Imports (tonnes)
1	India	590,656.80
2	Bangladesh	208,308.40
3	Algeria	71,675.00
4	Sri Lanka	24,869.60
5	Egypt	26,834.20
6	Iran	18,131.80
7	Iraq	8485.80
8	France	5856.60
9	Greece	3909.40

Source: Food and Agriculture Organization (FAO), 2019. FAOSTAT Statistical Database of the United Nation Food and Agriculture Organization (FAO) Statistical Division. Rome.

Table 9.15 India's trade destinations of chickpea in 2017–18.

Sr. No.	Export destinations	Import sources
1	Pakistan (21.60%)	Australia (85.1%)
2	Algeria (11.60%)	Russia (4.7%)
3	UAE (10.60%)	Tanzania (3.8%)
4	Saudi Arabia (9.50%)	USA (1.4%)
5	Sri Lanka (7.30%)	Canada (0.91%)

Figures in parenthesis indicates percentage share of global import/export.
Source: Ministry of Commerce and Industry. Government of India.

Table 9.16 India pulses imports from top trading partners (metric tons).

Country/year	2013	2014	2015	2016	2017
Canada	1,382,456	2,113,160	2,510,605	2,257,035	2,250,744
Australia	575,365	339,893	543,484	870,631	1,880,232
Myanmar	898,406	991,390	803,760	634,355	678,056
Russia	231,084	198,655	432,623	490,895	540,267
Ukraine	22,794	61,286	66,270	173,037	387,498
Romania	100	0	0	43.14	177,182
Mozambique	72,464	84,476	101,590	143,754	156,438

9.12 Import duty on important pulses

Due to the augmentation in the pulse prices, the Central Government had prohibited the export of pulses in 2006. With the country's pulse production reaching a record high of 224 lakh tonnes in 2016–17 as against 163 lakh tonnes in the previous year, the government, in September 2017, lifted the ban on export of chickpea, black gram, and green gram to help farmers to get remunerative prices. chickpea and lentils are also allowed to be exported in limited quantities provided, the export contracts are registered with APEDA, exports are made through customs electronic data interchange ports, and APEDA certifies for the produce to be organic. The export of pulses has been permitted in restricted quantities to the Republic of Maldives from 2014–15 to 2016–17. On the other hand, the import of pulses is free without any quantitative restrictions. Import duty on Gram was raised from zero to 30% w.e.f. December 21, 2017. It was further raised to 40% w.e.f. February 06, 2018. It was further raised to 60% w.e.f. March 01, 2018 and from 60% to 70% on 20th June 2018. The bound, statutory, and applied duties on some of the important pulses are given in the table (Table 9.17).

Table 9.17 Import duty on important pulses in India.

	Tariff schedule		
Commodity	Bound duty	Statuary duty	Applied duty
Peas	50%	50%	50%
Chickpeas	100%	30%	60%
Mung/Urad	100%	30%	Nil
Lentil	100%	30%	30%
Pigeon pea	100%	30%	10%

Source: Directorate General of Foreign Trade (DGFT), 2018.

9.13 Cost of cultivation of chickpea

The cost of cultivation of chickpea is Rs. 23,899 per ha. The total variable cost accounts for 55.46% of the total cost of cultivation. Seed cost accounts for 13.81% of the total cost of cultivation followed by hired labor (13.79%), machine labor (7.88%), plant protection material (5.30%), fertilizers (5.06%), irrigation charges (4.02%), and family labor (3.71%). The managerial cost accounts for about 9.09% of the total cost of cultivation (Table 9.18).

From Table 9.19, it can be observed that cost of production of chickpea has increased from Rs. 1530.30 per quintal to Rs. 2911.56 per quintal from 2009–10 to 2014–15, i.e., almost double in last 6 years. The cost A_2 + FL and cost C_2 have also increased more than double in last 6 years, which is a major concern to the farmers, scientists, and policy planners.

9.14 MSP for major pulses

The minimum support prices (MSPs) are announced by the Government of India for certain crops on the basis of the recommendations of the Commission for

Table 9.18 Break-up cost of cultivation of chickpea in India (Rs/ha).

Particulars	Chickpea
Variable cost	
Hired human labor	3295.20 (13.79)
Family labor	886.00 (3.71)
Machine labor	1882.40 (7.88)
Seed	3305.60 (13.83)
Fertilizers	1210.40 (5.06)
Plant protection material	1267.20 (5.30)
Irrigation charges	960.00 (4.02)
Interest on working capital	448.24 (1.88)
Total variable cost	13255.04 (55.46)
Fixed cost	
Land revenue	15.80 (0.07)
Rental value on owned land	7100.00 (29.71)
Interest on fixed capital	559.00 (2.34)
Depreciation	796.53 (3.33)
Total fixed cost	8471.33 (35.45)
Managerial cost	2172.63 (9.09)
Total cost	23,899.00

Figures in parentheses are percentages to the total.
Source: CACP, 2019. Price Policy for Rabi Crops: The Marketing Season 2018–19. Commission for Agricultural Costs and Prices, New Delhi.

Table 9.19 Economics of chickpea in India.

Year	Cost of cultivation (Rs/ha)		Cost of production (Rs/Quintal)
	Cost A_2 + FL	Cost C_2	
2009–10	10,978.24	18,468.11	1530.30
2010–11	11,447.59	18,657.75	1728.81
2011–12	16,280.13	29,948.11	2434.90
2012–13	19,629.37	31,634.35	2714.67
2013–14	19,538.39	28,298.52	2946.02
2014–15	21,376.7	33,952.40	2911.56

Source: Statistical Year Book, 2018. Ministry of Statistics and Program Implementation. Government of India.

Agricultural Costs and Prices (CACP). MSP is price fixed by Government of India to protect the producer—farmers against excessive fall in price during bumper production years. MSP for major pulses has been shown in the table, which shows the increasing trend from 2010–11 to 2018–19 (Table 9.20).

9.15 Prices of chickpea in major domestic markets

In this table, we tried to check the prices of chickpea in important domestic markets of India on September 20, 2019. It is inferred from the table the wholesale prices are more than that of cost of production and MSP, which is a good sign for the farmers and the government (Table 9.21).

9.16 Market segmentation in chickpea market

The Indian chickpea market is highly segmented and the demand depends on end use. Four main segments were identified: (1) the direct food use market, (2) the split (dal) market, (3) the chickpea flour market, and (4) the roasted and puffed chickpea market.

Table 9.20 Minimum support price of major pulses (Rs/Quintal).

Year/pulses	Tur	Gram	Mung	Urad	Lentil
2010–11	3000	2100	3170	2900	2250
2011–12	3200	2800	3500	3300	2800
2012–13	3850	3000	4400	4300	2900
2013–14	4300	3100	4500	4300	2950
2014–15	4350	3175	4600	4350	3075
2015–16	4625	3500	4850	4625	3400
2016–17	5050	4000	5225	5000	3950
2017–18	5450	4400	5575	5400	4250
2018–19	5675	4620	6975	5600	4475

Source: CACP, 2019. Price Policy for Rabi Crops: The Marketing Season 2018–19. Commission for Agricultural Costs and Prices, New Delhi.

Table 9.21 Price movement for chickpea in major domestic markets on September 09, 2019.

Market	Minimum price (Rs/ Quintal)	Maximum price (Rs/ Quintal)	Modal price (Rs/ Quintal)
Delhi	3610	3850	3730
Akola	3650	4275	4175
Indore	3700	4710	4100
Bikaner	3560	3854	3708

Source: CACP, 2019. Price Policy for Rabi Crops: The Marketing Season 2018–19. Commission for Agricultural Costs and Prices, New Delhi.

9.16.1 The direct food use market

Chickpea is used predominantly for human consumption in India. Kabuli chickpea is mostly consumed as whole seed. It is either consumed separately or combined with other dishes. The market participants interviewed pointed out that imported kabuli chickpea is highly differentiated based on size into three main groupings: 7–8 mm, 8–9 mm, and 9 mm and above. Australia's kabuli chickpea is not graded because it is uniform in size. However, it is smaller than the most preferred size in the market.

There are certain quality traits of kabuli chickpea that are mostly preferred by consumers. These are light yellow/cream color, large and uniform size, good taste, and thin seed coat. Seed size, color, and good taste were thought to be the most preferred quality attributes followed by thin seed coat. It was noted that the Mexican kabuli chickpea is the number one choice, followed by Australia, Iran, and Turkey in that order. The Mexican kabuli chickpea is distinguished by its larger size, attracting a premium of about $10 -$100 per tonne. The Indian kabuli chickpea is of a smaller size and of lower quality than imported varieties.

In terms of desi chickpea, consumers are thought to prefer bright or golden yellow color, soft and light seed, which tastes good. Imported chickpeas are not considered suitable for direct food use because of the bitter taste. In the recent past, there has been an increase in demand for desi chickpea imported from Myanmar due to similar soil and climatic conditions that make the desi chickpea from Myanmar similar in taste to domestically produced Indian varieties.

9.16.2 The split (dal) chickpea market

Desi chickpea is mostly consumed in split (dal) form in India. The market surveys indicate that consumers prefer a desi type that has a high recovery rate, is of uniform shape and large size, and free from foreign matter. Millers assess the quality of desi chickpea by visually inspecting small samples from the lot. After dehulling, the dal is graded by size. A large-sized dal may attract a price premium of about $100–$150 per tonne. Australia dominates the desi chickpea market for making dal. About 80%–90% of its export is used for making dal. Australia's desi chickpea is considered superior to other imported and domestically produced desi-type chickpea for making dal because of its high recovery rate, which is estimated to be about 3%–5% higher than the others. It has the added advantage of being free of foreign matter and has a uniform size. Over the years, Myanmar's desi chickpea exports to India have grown and become increasingly competitive with Australia. In addition to its proximity advantage, Myanmar's desi chickpea is produced in similar climatic conditions to that of India. Therefore, the chickpea produced has a similar taste to that produced in India.

9.16.3 The chickpea flour market

The use of whole chickpea for making flour depends largely on the type of plant used for processing. It was observed that, for flour making, the size of split chickpea is not important. The most important factor identified in the interview is the recovery rate. Flour millers often use desi-type chickpea of slightly lower quality than that used for making dal. Locally produced desi chickpea are mostly used for making flour, although millers sometimes use imported ones. The color of dal is not a needed quality trait for flour making.

9.16.4 The roasted and puffed chickpea market

There are two types of roasted chickpea market in India: the puffed chickpea (phutana) market and the roasted split chickpea (roasted dal) market. Domestically produced desi-type chickpeas are often used because imported ones have a bitter taste, except for those from Tanzania and Myanmar. Desi chickpeas from Myanmar and Tanzania are used for making roasted dal because the taste is similar to that of locally produced varieties. The preferred quality traits of roasted chickpea are large size and uniform shape. Although Gulabi (a local desi-type chickpea variety) is small compared to other desi-types, it is often used for making roasted chickpea because of its good taste, puffing quality, and round shape. Gulabi chickpea fetches the highest price in the roasted chickpea market because of its superior qualities.

9.17 Marketing channel of chickpea in India

The marketing system of chickpeas in India is schematically depicted in Fig. 9.8. Chickpeas are sold in primary or secondary wholesale markets directly by the producer. Importers who are licensed to import chickpeas into India sell directly to a broker, commission agent, middlemen, or secondary wholesaler, miller, or processor. The bulk of the chickpeas from brokers and commission agents are sold to the primary wholesaler, who in turn sells to millers and processors of dal or to secondary wholesalers. Small quantity of whole chickpeas moves from the primary market to the secondary market and reaches the consumer via the retailer. The dal produced by the millers is sold to large mills or to secondary wholesalers. The flour

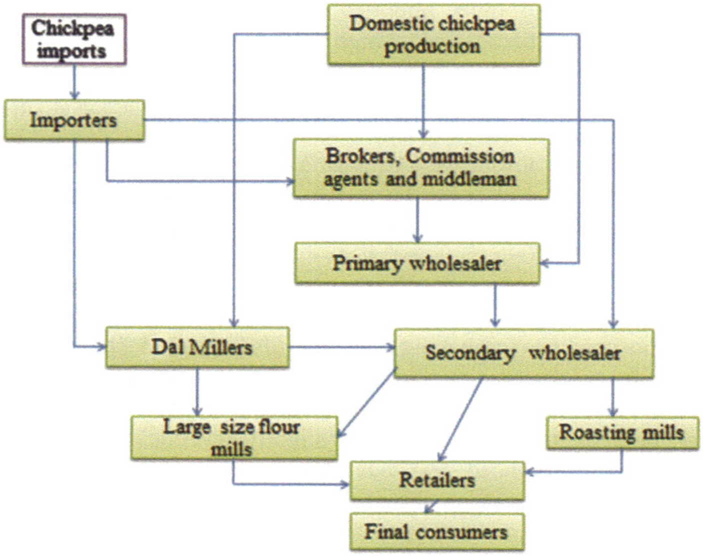

FIGURE 9.8

Marketing channels of chickpea in India.

moves to consumers via retail markets. A proportion of the dal from dal millers and primary wholesalers goes to secondary wholesalers, which is then sold to consumers as dal through the retailer. Whole chickpeas from secondary wholesalers are sold to frying mills. Puffed or roasted chickpeas move to consumers via retail markets.

9.18 Government intervention for chickpea in India

In 2017, India put in place both higher applied tariff rates and quantitative restrictions on certain pulses in an effort to address falling prices resulting from a supply glut in the domestic market. These policies effectively shut off one of the major global pulse markets from exporters, including the United States, Canada, Tanzania, and Myanmar. The import restrictions sent domestic prices for pigeon peas in Tanzania and Myanmar plummeting and threw the pulses industry into chaos, impacting small farmer incomes in both countries. These import controls have carried through to 2018 with additional import restrictions being put in place as well. The issue of pulses is high on the minds of governments and farmers, both inside and outside India right now, with numerous policy actions geared toward increasing Indian domestic pulse prices. These policies are in part being driven by an effort to win over the support of India's largest voting population—small, rural farmers who are protesting declining farm incomes and distress sales when domestic prices fall below MSP—in the face of current and upcoming state and national elections in

India. Starting in March 2017, in order to get a handle on the domestic oversupply, the Government of India (GoI) began implementing policies to restrict imports. On July 14, 2017, the ban on chickpeas futures contracts was lifted in order to help increase prices through price discovery and hedging, but reintroduction of future contracts did not result in increased prices. On December 21, 2017, the Ministry of Finance (MoF) announced a 30% tariff on imports of the pigeon peas (Tur HTS 0713.60.00), chickpeas (HTS 0713.20.00), and lentils (HTS 0713.40.00). 21 On February 6, 2018, the chickpea tariff was further raised to 40%. Further, on March 1, 2018, the import duty on chickpeas of Desi variety (small and dark chickpeas or black gram, the main variety of chickpeas grown in India) was raised from 40% to 60%, but excluded Kabuli chana (large chickpeas) wherein the import duty remains unchanged at 40%. In April 2018, the Government of India (GoI) revised the import policy for yellow peas, restricting imports to 100,000 MT for a period of 3 months, from April 1 to June 30, 2018, and on July 2, 2018, the import restrictions were further extended until September 30, 2018, to curb cheaper imports and boost domestic prices. In April 2018, India's Directorate General of Foreign Trade introduced a 7% (of the Free On Board value of the product) export subsidy on chickpeas (HTS 0713.20.00) under the Merchandise Exports from India Scheme (MEIS) for a 3-month period. In 2015, the Government of India (GoI) decided to set up a buffer stock of pulses with the National Agriculture Cooperative Marketing Federation of India (NAFED), Small Farmers Agri-Business Consortium (SFAC), and Food Corporation of India (FCI)—all procuring domestic product—and state-run Metals and Minerals Trading Corporation of India (MMTC) and SFAC—both procuring imports, under the authority of the Price Stabilization Fund (PSF). On March 28, 2017, the Government of India (GoI) Press Information Bureau announced an increase in the buffer stock for pulses to 2 MMT under the PSF.

9.19 Conclusion

India ranks first in terms of area and production but lags behind several countries in terms of productivity of chickpea. In India, major states producing chickpea are mainly (Madhya Pradesh, Rajasthan, Maharashtra, and Uttar Pradesh) as a rainfed crop. Because of the high demand and less availability of pulses, India is the major importer of pulses and chickpea. Over the past years, cost for cultivation and cost of production of chickpea is following an increasing trend. The MSP for chickpea has increased in the past years. The main concern with the chickpea in India is its lowest productivity as compared to the other countries.

References

Agbola, F.W., Bent, J.M., Kelley, T.G., Rao, P.P., 2000. Factors influencing the demand for chickpea in India: implications for marketing and promotion in the Indian chickpea market. In: 44th Australian Agricultural and Resource Economics Society Conference, Sydney, Australia.

Agricultural Statistics at a Glance 2017. Directorate of Economics and Statistics. Department of Agriculture, Cooperation & Farmers Welfare, Ministry of Agriculture, Government of India, New Delhi.

Annual Report 2016–17. Directorates of Pulses Development, Government of India, Bhopal.

Annual Report 2016–17. Department of Agriculture, Corporation and Farmers' Welfare, Ministry of Agriculture and Farmers Welfare, Government of India.

Annual Report 2017–18. Directorates of Pulses Development, Government of India, Bhopal.

CACP, 2019. Price Policy for Rabi Crops: The Marketing Season 2018–19. Commission for Agricultural Costs and Prices, New Delhi.

Department of Economics and Statistics, Ministry of Agriculture, Cooperation and Farmers Welfare, Government of India.

Food and Agriculture Organization of United Nations (FAO), FAOSTAT. http://apps.fao.org/default.html.

Food and Agriculture Organization (FAO), (2019). FAOSTAT Statistical Database of the United Nation Food and Agriculture Organization (FAO) Statistical Division. Rome.

Gowda, C.L.L., Srinivasan, S., Gaur, P.M., Saxena, K.B., 2013. Enhancing the productivity and production of pulses in India. In: Shetty, P.K., Ayyappan, S., Swaminathan, M.S. (Eds.), Climate Change and Sustainable Food Security. National Institute of Advanced Studies, Bangalore, pp. 145–159. New Delhi: Indian Institute of Pulses Research (IIPR).

Indian Council of Medical Research, 2012. New Delhi.

Jukanti, A.K., Gaur, P.M., Gowda, C.L.L., Chibbar, R.N., 2012. Nutritional quality and health benefits of chickpea (*Cicer arietinum* L.): a review. Br. J. Nutr. 108, S12–S26.

Kumar, S., Singh, R., Sharma, V., Choubey, A.K., Shah, S., 2014. Yield gap and economics of chick pea (*Cicer arietinum* L.) cultivation in low hills of Himachal Pradesh. Himachal J. Agric. Res. 40, 45–49.

Maurya, O.P., Kumar, H., 2018. Growth of chickpea production in India. J. Pharmacogn. Phytochem. 7, 1175–1177.

Merga, B., Haji, J., 2019. Economic importance of chickpea: production, value, and world trade. Cogent Food Agric. 5, 1–12.

Mohanty, S., Satyasai, K.J., 2015. Feeling the pulse, Indian pulses sector. NABARD Rural Pulse Issue X (July–August), 1–4.

Singh, P., Shahi, B., Singh, K.M., 2017. Trends of Pulses Production, Consumption and Import in India: Current Scenario and Strategies. Available from: https://mpra.ub.uni-muenchen.de/81589/MPRA Paper No. 81589.

Statistical Year Book, 2018. Ministry of Statistics and Program Implementation. Government of India.

Tiwari, A.K., Shivhare, A.K., 2017. Pulses in India: Retrospect and Prospects. Government of India, Ministry of Agri. & Farmers Welfare (DAC&FW), Directorate of Pulses Development, Vindhyachal Bhawan, Bhopal, MP.

Index

'*Note*: Page numbers followed by "f" indicate figures and "t" indicates tables.'

A

Abiotic stresses, 3–5, 67
Abiotic stress tolerance, 188–190
Advanced backcross quantitative trait loci (AB-QTL), 132, 151–152
Agrobacterium tumefaciens, 207–208
Agro-morphological traits, 186–188
Agronomic traits, 191
Albinism, 115–116
All India Coordinated Pulse Improvement Program (AICPIP), 46
Ascochyta blight resistance (ABR), 65, 149, 167
Australian Temperate Field Crops Collection (ATFCC), 41–42

B

Bacterial artificial chromosome (BAC), 134–135, 169
Biotic stresses, 3–5
  transfer of resistance to, 77–79
Blight resistance, 190–191
Botrytis gray mold (BGM), 3–5, 66
Bruchids, 8, 66–67

C

Chickpea
  abiotic stresses, 3–5
  biotic stresses, 3–5
  botrytis gray mold, 3–5
  characterization, 10
  chromosomal manipulations, 11
  Cicer species, 4t
  collection, 10
  conservation, 10
  conventional cytogenetic manipulations, 11
  crop wild relatives (CWRs), 1–2
  crossability groups, 31–32
  cultivation, 23
  distribution, 10, 21–23
  economy, 13–14
  embryology, 112t
    early-heart, 107
    globular, 107
    heart, 107
    late-heart, 107
    parameters, 106–107
    path coefficient analysis, 107
    pollination, 106–107
    regression analysis, 107
    torpedo, 107
    zygotic embryo, 110
  embryo rescue, 11
  embryo rescue protocols, 116
    L6 media, 116–118
    MS media, 118–120
  evaluation, 2–6
  flour market, 247
  gene pools, 10, 23–31, 98
    primary gene pool, 27–28
    secondary gene pool, 28–31
    tertiary gene pool, 31
  gene pyramiding, 11–12
  genetic resources, 10, 133–134
  genetic transformation, 13
  germplasm characterization, 2–6
  germplasm conservation, 2
  global production, 95–96
  hybridization, 95–96
  interspecific hybridization, 31–32
  kabuli-type chickpea, 19–21
  maintenance, 10
  marker-assisted selection (MAS) approach, 5–6
  marker-trait associations, 12–13
  molecular markers, 12–13
  multiple character breeding, 11–12
  origin, 10, 21–23
  production-related problems
    bruchids, 8
    cold tolerance, 8
    drought tolerance, 8–9
    fusarium wilt, 6–7
    heat tolerance, 9–10
    pod borer, 7–8
  simple sequence repeat (SSR) markers, 5–6
  somatic embryogenesis, 111–113
  taxonomy, 23, 29t–30t
Chlamydomonas, 208–218
Chromosomal rearrangement, 28–31
Chromosomes, 69
Chromosome segment substitution lines (CSSLs), 149
Cicer arietinum, 226
Cicer arietinum sequence-tagged microsatellite site (CaSTMS) markers, 169

Cloned gene sequences, 192–193
Cold tolerance, 8
Conventional breeding, 27–28
Conventional chickpea breeding strategies, 68–69
Conventional cytogenetic manipulations, 11
  bridge crosses, 75
  chickpea improvement, 69
  conventional chickpea breeding strategies, 68–69
  crop improvement, 74–75
  cytogenetics, 69
  direct hybridization, 74
  embryo culture, 75
  future strategy, 79–81
  genetic variability, 71–72
  high-yielding genotypes., 63
  hybridization, 63, 70–71
    chromosome doubling prior to, 74
    interspecific hybridization, 74–75
  marker-assisted selection, 70
  mutagenesis, 71–72
  narrow genetic bases, 67–68
  plant transformation, 74–75
  pre- and postfertilization crossing barrier, 74–75
  production-related problems, 64–67
    abiotic stresses, 67
    Ascochyta blight, 65
    botrytis gray mold, 66
    bruchids, 66–67
    dry root rot, 66
    fusarium wilt, 65–66
    pod borer, 66
    powdery mildew, 66
  somatic hybridization, 74–75
  wide hybridization
    biotic stresses, transfer of resistance to, 77–79
    new germplasm/varieties, 75–77
  wild *Cicer* species, 72–73
Convergent breeding, 95–96
Crop wild relatives (CWRs), 1–2
Cytogenetics, 122
  chromosomal manipulations, 125–126
  chromosome morphology and DNA content, 122–124
  visualization, 125–126

D

Direct food use market, 246
Direct hybridization, 74
Disease resistance, genetic mapping of, 181–188
  ABR, 183–185
  agro-morphological traits, 185
  monogenic/digenic traits, 185–186
  polygenic traits, 186–188
  QTLs, 183–185
  wilt resistance, 181–183, 182t
Diversity arrays technology (DArT), 173
DNA-based markers, 169–172
  expressed sequence tags (ESTs), 171–172
  genic markers. *See* Genic markers
  nongenic markers. *See* Nongenic markers
  simple sequence repeat markers, 169–171
  transcription factor gene–derived microsatellite markers, 172
Domestication, 68
Double podding character, 191
Drought tolerance, 8–9
Dry root rot, 66

E

Economy
  availability status, 233–237, 236t, 238t
  chickpea flour market, 247
  chickpea, global scenario, 229
  chickpea, India, 232, 233t
  Cicer arietinum, 226
  contribution, 233, 235t, 237
  cultivation cost, 244, 244t
  direct food use market, 246
  export, 239
  food security, 225
  global pulse economy, 228, 228f
  government intervention, 248–249
  import, 239
  import duty, 243, 243t
  India, 229–231, 230t–231t
  marketing channel, 247–248
  market segmentation, 245–247
  minimum support prices (MSPs), 244–245
  per capita availability, 232, 232t
  prices, 245
  productivity, 232–233, 234t
  puffed chickpea market, 247
  roasted chickpea market, 247
  split (dal) chickpea market, 246–247
  trade destination, 240, 242t
Embryo culture, 75
Embryo rescue, 11
Embryo rescue/chromosomal manipulations, 95–96, 121
  cytogenetics and distant hybridization, 122–126
  gene pools and wide hybridization, 98, 103t
    chickpea embryology, 106–111
    Cicer, 99
    culturing media, 102–104

embryo culture, 104–105, 105t
embryo rescue protocols, 116–120
intervention, 104
literatures and reports, 99, 100t–101t
mature embryo culture, 102–104
seedlings, hybrid embryos, 115–116
somatic embryogenesis, 111–113
sterile isolation and in vitro culturing, 102
genus *Cicer*, 97–98
origin, 96–97
taxonomy, 96–97
Export, 239
Expressed sequence tags (ESTs), 135
Ex situ conservation, 41–42, 43t

F

Fertile hybrids, 76
Functional maps, 138
Fusarium wilt, 6–7, 65–66

G

GBS. *See* Genotyping-by-sequencing (GBS)
GEBVs. *See* Genomic-estimated breeding values (GEBVs)
Gene gun–mediated genetic transformation, 208–218
Gene pools, 10, 23–31, 39–40
primary gene pool, 27–28
secondary gene pool, 28–31
tertiary gene pool, 31
Gene pyramiding, 11–12
Genetic base, 205–206
Genetic disharmony, 31–32
Genetic diversity assessment, 49–51
Genetic maps, 138, 139t–142t
Genetic resources
advanced backcross QTL analysis–based breeding, 151–152
ascochyta blight (AB), 149
Australian Temperate Field Crops Collection (ATFCC), 41–42
bacterial artificial chromosome (BAC) clones, 134–135
challenges, 53
characterization, 49–51
chickpea, 133–134
chromosome segment substitution lines (CSSLs), 149
core and mini-core, 51–53
definition, 134–154
duplicates, 45
evaluation, 45–53
expressed sequence tags (ESTs), 135
ex situ conservation, 41–42, 43t
functional maps, 138
gene pools, 39–40
genetic and physical maps, 138, 139t–142t
genetic diversity assessment, 49–51
genome resequencing, 143
genome sequencing, 143
genomic-estimated breeding values (GEBVs), 153
genomics-assisted breeding, 145
MAGIC lines, 151
genomic selection (GS), 153–154
genotyping by sequencing (GBS), 136
germplasm characterization, 45–53, 46f
germplasm conservation, 40–45
ICRISAT, 41–42, 43t, 45–46
isozymes, 134
larger linkage disequilibrium (LD), 153–154
linkage disequilibrium (LD), 153
marker-assisted backcrossing, 149–150
marker-assisted recurrent selection, 150–151
microsatellite markers, 134–135
molecular markers, 146t–148t
near-isogenic lines (NILs), 149
prospects, 53
restriction fragment length polymorphism (RFLP) markers, 134–135
safety duplication, 45
single nucleotide polymorphism (SNP) genotyping platforms, 135–136
in situ conservation, 42–44
species distribution, 39–40
speed breeding, 154–155
trait mapping, 143–145
transcript maps, 138
transcriptome profiles, 135
transcriptomics resources, 136–137
Genetic transformation, 13
Agrobacterium tumefaciens, 207–208
chickpea, 209t–211t
productivity constraints, 206
gene gun–mediated genetic transformation, 208–218
genetic base, 205–206
plant breeding, 206–218
root transformation, 219–220
selectable/scorable markers, 218–219
transgenic chickpea, 212t–217t
Genetic variability, 71–72
Genic markers
expressed sequence tag polymorphisms, 173
intron targeted primers (ITPs), 173

Genic molecular markers (GMMs), 138
Genome resequencing, 143
Genome sequencing, 143
Genome-wide association mapping (GWAS), 138, 168
Genome-wide association study (GWAS), 143–144
Genomic-estimated breeding values (GEBVs), 153
Genomics-assisted breeding, 145
Genomic selection (GS), 153–154
Genotyping by sequencing (GBS), 136, 175–177
Genus *Cicer*, 97–98
Germplasm characterization, 2–6, 45–53, 46f
Germplasm conservation, 2, 40–45
Global gene banks list, 133t
Global pulse economy, 228, 228f

H

Heat tolerance, 9–10
High-yielding genotypes., 63
Hybridization, 63, 70–71
- chromosome doubling prior to, 74
- somatic hybridization, 74–75

I

ICRISAT. *See* International Crops Research Institute for the Semi-Arid Tropics (ICRISAT)
Import, 239
Insertion-deletion markers, 173–174
In situ conservation, 42–44
International Chickpea Genome Sequencing Consortium (ICGSC), 143
International Crops Research Institute for the Semi-Arid Tropics (ICRISAT), 41–42, 43t, 45–46, 133–134
Interspecific hybridization, 74–76
Intervarietal/intraspecific hybridization, 63
Isozymes, 134

K

KASP-Hedgehog, 192

L

Linkage disequilibrium (LD), 153–154
Lysine plus threonine (LT) medium, 218–219

M

Marker-assisted backcrossing (MABC), 78, 132, 149–150
Marker assisted breeding (MAB), 177–178, 190–191
- blight resistance, 190–191
- double podding character, 191
- wilt resistance, 190–191

Marker-assisted recurrent selection (MARS), 132, 150–151
Marker-assisted selection (MAS), 5–6, 70, 177–178
Marker-trait associations, 12–13
Marketing channel, 247–248
Market segmentation, 245–247
Mass-pedigree, 95–96
4-Methylumbelliferyl-beta-D-glucuronide (MUG), 219
Microsatellite markers, 134–135
Microsatellite primed polymerase chain reaction (MP-PCR), 169
Minimum support prices (MSPs), 244–245
Molecular markers, 12–13, 146t–148t
- abiotic stress tolerance, 188–190
- agronomic traits, 191
- ascochyta blight resistance (ABR), 167
- bacterial artificial chromosome (BAC), 169
- *Cicer* arietinum sequence-tagged microsatellite site (CaSTMS) markers, 169
- disease resistance, genetic mapping of, 181–188
  - ABR, 183–185
  - agro-morphological traits, 185
  - monogenic/digenic traits, 185–186
  - polygenic traits, 186–188
  - QTLs, 183–185
  - wilt resistance, 181–183, 182t
- DNA-based markers, 169–172
  - expressed sequence tags (ESTs), 171–172
  - genic markers. *See* Genic markers
  - nongenic markers. *See* Nongenic markers
  - simple sequence repeat markers, 169–171
  - transcription factor gene–derived microsatellite markers, 172
- future perspectives, 192–194
- genome-wide association studies (GWAS), 168
- genotyping-by-sequencing (GBS), 175–177
  - chickpea SSRs transferability, 176–177
- KASP-Hedgehog, 192
- linkage maps, 177–180
- marker assisted breeding (MAB), 177–178, 190–191
- marker-assisted selection (MAS), 177–178
- microsatellite primed polymerase chain reaction (MP-PCR), 169

prediction genomic selection, 191
random amplified polymorphic DNA (RAPD), 167–168
restriction fragment length polymorphism (RFLP), 167–168
second-generation linkage map, 178–179
simple sequence repeats (SSRs), 167–168
single nucleotide polymorphisms (SNPs), 167–168
whole-genome sequencing (WGS), 169–170
Multiparent advanced generation intercrossing (MAGIC) lines, 69, 151
Multiple character breeding, 11–12
Mutagenesis, 71–72

N

Naphthalene acetic acid (NAA), 77
Narrow genetic bases, 67–68
Near-isogenic lines (NILs), 149
New germplasm/varieties, 75–77
Next-generation sequencing (NGS) technologies, 132
Nongenic markers
diversity arrays technology (DArT), 173
insertion-deletion markers, 173–174
single nucleotide polymorphisms, 174–175

P

Pedigree breeding, 95–96
Per capita availability, 232, 232t
Physical maps, 138, 139t–142t
Plant breeding, 206–218
Plant Genetic Resources (PGR) systematization, 45
Plant transformation, 74–75
Pod borer, 7–8, 66
Polygenic traits, agro-morphological traits, 186–188
Post-fertilization crossing barrier, 74–75
Powdery mildew, 66
Pre-fertilization crossing barrier, 74–75
Primary gene pool, 27–28
Pseudononis, 23
Puffed chickpea market, 247
Pulses
availability status, 233–237, 236t, 238t
chickpea flour market, 247
chickpea, global scenario, 229
chickpea, India, 232, 233t
Cicer arietinum, 226
contribution, 233, 235t, 237
cultivation cost, 244, 244t
direct food use market, 246
export, 239
food security, 225
global pulse economy, 228, 228f
government intervention, 248–249
import, 239
import duty, 243, 243t
India, 229–231, 230t–231t
marketing channel, 247–248
market segmentation, 245–247
minimum support prices (MSPs), 244–245
per capita availability, 232, 232t
prices, 245
productivity, 232–233, 234t
puffed chickpea market, 247
roasted chickpea market, 247
split (dal) chickpea market, 246–247
trade destination, 240, 242t

Q

Quantitative trait loci (QTLs), 78–80, 132, 138, 144, 151–152

R

Random amplified polymorphic DNA (RAPD), 167–168
Regression analysis, 107
Restriction fragment length polymorphism (RFLP), 134–135, 167–168
Roasted chickpea market, 247
Root traits length density (RLD), 48
Root transformation, 219–220

S

Safety duplication, 45
Secondary gene pool, 28–31
Seedlings, hybrid embryos, 115–116
Selectable/scorable markers, 218–219
Serial analysis of gene expression (SAGE), 137
Shuttle breeding, 95–96
Simple sequence repeat (SSR) markers, 5–6, 167–168
Single nucleotide polymorphism (SNP), 135–136, 167–168, 174–175
Single seed descent (SSD) method, 68–69
Small-seeded cultivars, 22
Somatic embryogenesis, 111–113, 114f
type 1 (horn-shaped embryo), 113
type 2 (monocotyledonary embryo), 114
type 3 (normal embryo), 114
type 4 (tricotyledonary embryo), 114
type 5 (multicotyledonary embryo), 115
type 6 (fasciated embryo), 115

Somatic hybridization, 74–75
Species distribution, 39–40
Speed breeding, 154–155
Split (dal) chickpea market, 246–247
Sprouted seeds, 21
Super serial analysis of gene expression (Super SAGE), 137
Suppression subtractive hybridization (SSH), 137

T

Tentative orthologous genes (TOGs), 138
Tertiary gene pool, 31
Trade destination, 240, 242t
Trait mapping, 143–145
Transcript maps, 138
Transcriptome profiles, 135
Transcriptomics resources, 136–137
Transgenic chickpea, 212t–217t
Transgressive segregants, 76

W

Western Regional Plant Introduction Station (WRPIS), Pullman, 51–52
Whole-genome sequencing (WGS), 169–170
Wide hybridization
  biotic stresses, transfer of resistance to, 77–79
  new germplasm/varieties, 75–77
Wild *Cicer* species, 72–73
Wilt resistance, 190–191

FSC
www.fsc.org
MIX
Papier | Fördert
gute Waldnutzung
FSC® C083411

Kolibri 360 GmbH
Ferdinand-Jühlke-Straße 7
99095 Erfurt, Deutschland
produktsicherheit@kolibri360.de